# Using Mobile Technologies in the Teaching and Learning of Mathematics

# MATHEMATICS EDUCATION IN THE DIGITAL ERA
Volume 12



The Mathematics Education in the Digital Era (MEDE) series explores ways in which digital technologies support mathematics teaching and the learning of Net Gen'ers, paying attention also to educational debates. Each volume will address one specific issue in mathematics education (e.g., visual mathematics and cyber-learning; inclusive and community based e-learning; teaching in the digital era), in an attempt to explore fundamental assumptions about teaching and learning mathematics in the presence of digital technologies. This series aims to attract diverse readers including: researchers in mathematics education, mathematicians, cognitive scientists and computer scientists, graduate students in education, policy-makers, educational software developers, administrators and teachers-practitioners. Among other things, the high quality scientific work published in this series will address questions related to the suitability of pedagogies and digital technologies for new generations of mathematics students. The series will also provide readers with deeper insight into how innovative teaching and assessment practices emerge, make their way into the classroom, and shape the learning of young students who have grown up with technology. The series will also look at how to bridge theory and practice to enhance the different learning styles of today's students and turn their motivation and natural interest in technology into an additional support for meaningful mathematics learning. The series provides the opportunity for the dissemination of findings that address the effects of digital technologies on learning outcomes and their integration into effective teaching practices; the potential of mathematics educational software for the transformation of instruction and curricula; and the power of the e-learning of mathematics, as inclusive and community-based, yet personalized and hands-on.

More information about this series at http://www.springer.com/series/10170

Nigel Calder · Kevin Larkin
Nathalie Sinclair
Editors

# Using Mobile Technologies in the Teaching and Learning of Mathematics

Springer

*Editors*
Nigel Calder
MSTE
University of Waikato
Tauranga, Bay of Plenty
New Zealand

Nathalie Sinclair
Faculty of Education
Simon Fraser University
Burnaby, BC
Canada

Kevin Larkin
School of Education and Professional Studies
Griffith University
Gold Coast, QLD
Australia

ISSN 2211-8136 ISSN 2211-8144 (electronic)
Mathematics Education in the Digital Era
ISBN 978-3-030-07953-6 ISBN 978-3-319-90179-4 (eBook)
https://doi.org/10.1007/978-3-319-90179-4



Printed on acid-free paper

This Springer imprint is published by the registered company Springer International Publishing AG part of Springer Nature
The registered company address is: Gewerbestrasse 11, 6330 Cham, Switzerland

# Contents

# About the Editors

**Dr. Nigel Calder** is an Associate Professor at the University of Waikato, Tauranga. He has taught in both primary and secondary schools. His research interests are predominantly in the use of digital technologies in mathematics education. His present and recent research projects are the influence of apps on primary school mathematics teaching and learning; the influence of mobile technologies on the engagement of 16–18 year-old reluctant learners; connecting science and mathematics instruction through exploratory talk and inquiry; and how mathematical thinking emerges through student-centered inquiry learning in secondary schools. He wrote the book: *Processing Mathematics through Digital Technologies: The Primary Years* and has published numerous resources (including Figure It Out books), chapters, and journal articles. Email: nigel.calder@waikato.ac.nz

**Dr. Kevin Larkin** is a Senior Lecturer (Mathematics Education) at Griffith University. He is a member of a number of research teams investigating: STEM education in early years education; mathematics education in primary and middle school contexts; and preservice teacher mathematics education. He has published widely in national and international publications in the areas of mathematics education, digital technologies, STEM, higher education, and activity theory. He is the current VP Communication for the Mathematics Education Research Group of Australasia (MERGA). Email: k.larkin@griffith.edu.au

**Nathalie Sinclair** is a Professor and the Canada Research Chair in Tangible Mathematics Learning at Simon Fraser University. She is the author of several books, including co-author of *Mathematics and the Body: Material Entanglements in the Classroom*. She has also led the design of educational technologies, including the touchscreen app *TouchCounts*. She is the founding editor of the journal *Digital Experiences in Mathematics Education*. Email: nathsinc@sfu.ca

# Mobile Technologies: How Might Using Mobile Technologies Reshape the Learning and Teaching of Mathematics?

**Nigel Calder, Kevin Larkin and Nathalie Sinclair**

As our attention moves to the opportunities and constraints that mobile technologies (MT) might afford, app developers, teachers and researchers have become more adept at identifying and enacting opportunities for enhancing mathematical thinking. These opportunities emerge through the various environments, both hardware (i.e., tablets) and software (i.e., applications), and the mathematical activity that these facilitate. The features of MT, for instance the ability to use in-built video and audio tools, allows users to capture authentic data in their everyday world and use the data for modelling, or statistical inference. Processing this data in situ changes the nature of the learning experience. Likewise, the potential for visual, interactive engagement with some learning experiences, coupled with the haptic and oral/aural affordances of the technology, change the nature of the mathematical activity. By inference, this changes the nature of the mathematical thinking. Engaging with number sequences by creating sets of objects that represent numbers as you touch the screen, using an oral count, using concrete materials, or learning a sequence by rote, are all different representations of number that might evoke a variety of understandings and ways of thinking mathematically. Being able to elegantly connect these various representations, and move between them, appears to offer opportunities for deeper conceptual understanding of mathematics concepts.

With MT being a relatively recent addition to the scope of digital technologies that might facilitate learning in mathematics, there is a need for research on the use

N. Calder (✉)
University of Waikato, Tauranga, New Zealand
e-mail: nigel.calder@waikato.ac.nz

K. Larkin
Griffith University, Gold Coast, Australia
e-mail: k.larkin@griffith.edu.au

N. Sinclair
Simon Fraser University, Burnaby, Canada
e-mail: nathsinc@sfu.ca

 
N. Calder et al. (eds.), *Using Mobile Technologies in the Teaching and Learning of Mathematics*, Mathematics Education in the Digital Era 12,
https://doi.org/10.1007/978-3-319-90179-4_1

of MT, and for this research to be presented in a coherent, multi-faceted manner. Such research is of particular importance as schools are investing heavily in mobile devices, often without a concomitant investment in developing practices regarding how such devices may be used to develop conceptual rather than procedural or declarative knowledge (e.g., Calder, 2011), nor of how such devices connect with other resources in the classroom. The effectiveness of their engagement in shifting conceptual understanding is also contingent on associated professional learning for teachers (O'Malley et al., 2013). Theoretical frameworks for teaching and learning with MT might also be influential in our understanding of the ways that using MT in the learning of mathematics might be examined.

Some researchers have noted a lack of theoretical rigour regarding the use of MT (e.g., Larkin, 2015), identifying issues in relation to the lack of mathematical quality of many mathematics apps. He also reported the lack of time and expertise for teachers to accurately evaluate them or their use. Nevertheless, MT offer vast potential to enhance mathematical learning. This book builds on international research (e.g., Attard, 2015; Calder & Campbell, 2016; Moyer-Packenham et al., 2015; Sinclair & Heyd-Metzuyanim, 2014) into the use of MT in mathematics education. It includes an examination of the ways MT might influence student engagement, cognition, collaboration and attitudes, through reshaping the learning experiences across a diverse range of year levels and contexts.

Central to learning mathematics through using MT is the nature of the tools and the apps utilised, the learning intentions of the teacher, and the type of activity that the students are engaged with. While there is frequently a focus in schools, and in the media, on *consumable* apps that is, those where students follow a set task at a specified level; more recently there has been a focus on apps that: enable students to create screencasts of their mathematical thinking; can be used for coding, including the programming of small robotic devices; enable students to create visual, dynamic representations of mathematical situations.

In this relatively new field of engaging with mathematics learning through using MT, this book reflects the growing understanding of how the learning experience might be reshaped to harness the opportunities that MT afford. It also incorporates an examination of using MT for developing mathematical thinking, enhancing teacher pedagogy, and understanding the embodied cognition inherent when using mobile, touch-screen devices. In addition, the broader assemblages incorporating underlying discourses and political elements are hugely influential in using MT effectively. The book proposes emerging frameworks, or new uses for existing frameworks that encourage educators to better interrogate student engagement and learning aspects as well as to evaluate the vast range of apps that continuously appear. As the field is always in flux, with researchers and teachers often scrambling to keep up with recent innovative developments, we are reminded to look beyond the specific to the general (Mason, 2005). Thus we look to find common themes and trends that enable us to gain insights into and evaluate MT, and the associated learning/teaching practices, from vantage points not dependent on understanding specific examples or apps. Not withstanding the need for a global approach, the examination of individual apps and experiences are crucial as they mark new initiatives and point toward potential innovation. And what of the methodologies that we use to theorise the terrain? There is an advantage in looking through a range of lenses, as what one lens doesn't highlight, another may. However, with the continual developments in technology, we need to also consider how MT might open up innovative approaches to methodology and research design. As a community, we need to continue to explore the edges, while incorporating the generic ways these innovations inform practice, reshape the learning experience and might enhance students' mathematical thinking.

This book draws from a diverse range of international studies, where MT have influenced the ways that learning might occur across a range of educational contexts. The book is divided into four sections: *Looking across the terrain; Traversing the learning and teaching landscape; Navigating content: focussing on particular concepts*; and *Exploring new forms of communication to make mathematical learning visible*. While the purpose of these groupings is to draw the reader's attention to particular themes within the chapters, there is nevertheless considerable overlap between the sections, and most chapters could have easily been situated in more than one section.

The first section, *Looking across the terrain* considers some generic aspects that straddle the diversity in the field. Larkin and Milford use cluster analysis to group apps based on particular processes, features and concepts. Undertaking this process enables the apps to be grouped independently from developers' marketing and highlights particular aspects that mathematics educators consider as important. Through this educators are supported in their selection of apps for the user's intended purpose, and hence enhance app use in mathematics classrooms. Calder and Murphy analyse aspects of a 2-year study involving primary children learning mathematics through the use of mobile devices and apps. With the teachers as co-researchers, they examined teacher practice and the inter-connectivity between teacher pedagogy and the affordances of the apps. The teachers used a diverse range of apps, including ones for screencasting and coding, with the various learning experiences and opportunities for influencing the learning outlined and considered. An interesting theme to emerge across the use of various MT was socio-material assemblages. The final chapter in this section also investigates the interplay between a range of contexts and the types of apps used in these contexts. Attard considers the notion of student engagement when using apps for mathematical learning. Her framework for interrogating aspects of engagement incorporates cognitive elements. She draws together common threads about engagement, while synthesising insights into a collection of different studies related to engagement in various contexts.

The second section, *Traversing the teaching and learning landscape,* includes chapters that link teaching and learning related to various learning processes that utilise particular processes or affordances of MT. Kyriakides and Meletiou-Mavrotheris discuss a multifaceted programme designed to provide a group of in-service teachers with the knowledge, skills, confidence, and practical experience required to effectively use tablet devices for enhancing mathematics teaching and learning. The teachers integrated the app *A.L.E.X* into their lesson plans and thus reshaped the students' learning experience. Sollervall, de la Iglesia, and Zbick explore how mathematics classroom teachers can implement an innovative mobile learning activity. They report on an ongoing, 5-year study into using a GPS app for geometry, with the focus of the activity involving GPS and spatial orientation tasks that are executed in outdoor settings. Sedaghatjou and Rodney's chapter considers how a particular multitouch app called *TouchCounts*, along with children's collaborative engagements, can enhance mathematical learning of number. They utilise *StudioCode* software to better understand children's collaborative, gestural practices within the *TouchCounts* environment. In the final chapter of this section, Bokhove, Clark-Wilson, and Pittalis consider two cases of how MT provided opportunities for "mathematics outside the classroom". The examples describe how using mobile phones with augmented reality allowed students to bridge between formal and informal mathematics learning. Their examples, a dynamic Ferris wheel and a static cathedral are used to demonstrate how educators can use *geo-location* and *augmented reality* to enhance the learning of mathematics through MT.

In the third section of the book, *Navigating content: focussing on particular concepts*, the authors primarily attend to specific mathematical concepts or processes. Pelton, Milford, and Francis Pelton use an app to develop children's understanding of time. Their chapter details the integration of a researcher-designed iPad app into a series of collaboratively created lessons to facilitate the learning of clock-reading and time concepts. The authors used a lesson study approach to design and refine the intervention that included teacher-led activities and structured use of the iPad app. Lommatsch, Tucker, Moyer-Packenham, and Symanzik examine what patterns were revealed when heatmaps were used with hierarchical clustering to examine pre-schoolers' performance with two touchscreen mathematics apps in two different learning sequences: counting and seriation. Their analysis highlighted changes in children's performance, speed, and developmental progressions after using the two apps. The use of hierarchical clustering analysis facilitated the analysis of individual and whole group data leading to the identification of young children's developmental progressions. Rosen, Palatnik, and Abrahamson explore an embodied-design for engaging particular mathematical concepts with an action level where the virtual objects were either generic (e.g., a circle), or situated, (e.g., a hot-air balloon). They evaluate an instructional methodology whereby students first learn to physically move objects on the screen before eventually generalising these movements as formal mathematical rules. Chorney and Sinclair describe a research project with first-grade children using a multi-touch, dynamic geometry app called *WebSketchpad* to study how the concept of symmetry arises. They analyse the data through the lens of inclusive materialism, considering the intra-actions involved in the child-device-geometry assemblages, and how new mathematical ideas might emerge from these assemblages. Their particular focus is how the multi-touch environment can provide the basis for emerging geometrical ideas. Ferrara and Savioli discuss a classroom-based intervention with a group of first-grade children using the multi-touch app *TouchCounts* to develop children's number sense. They investigate how understanding might emerge out of the relational entanglement of numbers, iPads, and learners, engendering new kinds of mathematical experiences with number and providing the basis for emerging relational meanings of number. In the final chapter of this section, Soldano and Arzarello examine an approach to geometry in a secondary-school context. Their chapter illustrates a way of using MT to support the transition from an empirical to a theoretical approach to geometry. Drawing on Zbiek's et al., (2007) notions of pedagogical, mathematical and cognitive fidelities, they implement group game-activities whereby students investigate the geometric property upon which the game is designed.

In the final section of the book, *Exploring new forms of communication to make mathematical learning visible,* the notion of screencasting is the focus. The use of screencasting opens up opportunities for mathematical thinking of learners to become more transparent as students and teachers might create individual explanations of their thinking using a blend of both digital tools and their associated social elements. Galligan and Hobohm examine a case study of the use of mobile devices and screencasting in university mathematics education teaching. They incorporate this with an evaluative tool for teachers and students to evaluate their own and others' screencasts, with the intention of developing pre-service teachers' understanding of mathematics and ways to teach it. The chapter concludes with recommendations for using screencasting to assist with developing mathematical understanding and pedagogical content knowledge. Prescott and Maher explore the ways primary-school students worked collaboratively to solve a problem, explaining their mathematical thinking. The students used screencasting apps such as *Explain Everything* and *Educreations* to produce create-alouds, which helped them to collaboratively understand and explain mathematical concepts. The apps also assisted teachers in providing formative assessment and feedback to the students. In the final chapter of this section, Ingram, Pratt, and Williamson-Leadley discuss how a Show and Tell app can make the students' thinking more observable in problem solving. They consider how using Show and Tell apps for problem solving can lead to improvements in the level and quality of student engagement. Students were encouraged to socially negotiate their understandings, making student thinking more visible during this process. The apps can also scaffold students in reflecting upon the processes they used for problem solving.

The learning opportunities provided, and the evolution of the ways of promoting engaging mathematics learning and thinking through MT, exist in a fast moving, dynamic space; one where the comparative costs for mobile devices and connectivity are dropping markedly. There are also emergent MT that might quickly come to dominate the field: virtual reality is already developing rapidly, as is artificial intelligence and robotics. Some trusts and educational systems are distributing, to all schools in their community, 3-D printers that can print using materials as diverse as wood and titanium. Due to the reduction in the costs of MT, many of the previous equity and accessibility issues are alleviated. The potential to envisage space, location, shape, number, movement and rates of change has already been transformed. Likewise, ways of analysing data to model real-life situations in situ have changed. Yet, despite the rapid change we have witnessed in recent times, we do not really know where the technology, or its potential as a digital pedagogical medium, is headed. What might the landscape look like in 5 years, let alone in 20 years?

Underpinning each of the chapters in this book is the understanding that mathematical thinking must be given primacy—in the end it is our guiding premise and intention. We need to ask ourselves whether what we do with MT enhances mathematical thinking and understanding and to reflect on how the MT might be changing what counts as mathematical activity. The chapters in the book contribute to mathematics teaching and learning by providing readers with opportunities to reflect on their practice. It is not possible, nor wise, to ignore the role of MT in enhancing teaching and learning in mathematics. Therefore, all educators need to be alert to the potential that MT provide to enhance student learning.

We thank all the authors for sharing their considerable experience and expertise, their engagement with the process, and the positive approach that they have taken to ensure this book was produced in a timely manner. We also thank the reviewers who have worked with the authors to strengthen the chapters in this book. Finally, we thank the editorial and publication team at Springer for their support in publishing this work.

## References

Attard, C. (2015). Introducing iPads into primary mathematics classrooms: Teachers' experiences and pedagogies. In M. Meletiou-Mavrotheris, K. Mavrou, & E. Paparistodemou (Eds.), *Integrating touch enabled and mobile devices into contemporary mathematics education* (pp. 197–217). Hershey, PA: IGI Global.

Calder, N. S. (2011). *Processing mathematics through digital technologies: The primary years*. Rotterdam, The Netherlands: Sense.

Calder, N. S. & Campbell, A. (2016). Using mathematical apps with reluctant learners. *Digital Experiences in Mathematics Education*. https://doi.org/10.1007/s40751-016-0011-y.

Larkin, K. (2015). "An App! An App! My Kingdom for An App": An 18-month quest to determine whether apps support mathematical knowledge building. In T. Lowrie & R. Jorgensen (Eds.), *Digital games and mathematics learning: potential, promises and pitfalls* (pp. 251–276). Netherlands: Springer.

Mason, J. (2005). Mediating mathematical thinking with e-screens. In S. Johnston-Wilder & D. Pimm (Eds.), *Teaching secondary mathematics with ICT* (pp. 81–100). Berkshire, UK: Open University Press.

Moyer-Packenham, P. S., Shumway, J. F., Bullock, E., Tucker, S. I., Anderson-Pence, K. L., Westenskow, et al. (2015). Young children's learning performance and efficiency when using virtual manipulative mathematics iPad apps. *Journal of Computers in Mathematics and Science Teaching, 34*(1), 41–69.

O'Malley, P., Jenkins, S., Wesley, B., Donehower, C., Rabuck, D., & Lewis, M. E. B. (2013). Effectiveness of using iPads to build math fluency. In *Presented at the council for effectiveness of using iPads to build math fluency, San Antonio, Texas, USA*. Retrieved from http://eric.ed.gov/?id=ED541158.

Sinclair, N., & Heyd-Metzuyanim, E. (2014). Learning number with *TouchCounts:* The role of emotions and the body in mathematical communication. *Technology, Knowledge and Learning, 19*(1), 81–99.

Zbiek, R. M., Heid, M. K., Blume, G. W., & Dick, T. P. (2007). Research on technology in mathematics education: A perspective of constructs. In F. K. Lester (Ed.), *Second handbook of research on mathematics teaching and learning* (pp. 1169–1207). Charlotte, NC: Information Age.

**Dr. Nigel Calder** is an Associate Professor at the University of Waikato, Tauranga. He has taught in both primary and secondary schools. His research interests are predominantly in the use of digital technologies in mathematics education. His present and recent research projects are: The influence of apps on primary-school mathematics teaching and learning; The influence of mobile technologies on the engagement of 16–18 year-old reluctant learners; Connecting science and mathematics instruction through exploratory talk and inquiry; and How mathematical thinking emerges through student-centred inquiry learning in secondary schools. He wrote the book: *Processing Mathematics through Digital Technologies: The primary years* and has published numerous resources (including Figure It Out books), book chapters and journal articles.

**Dr. Kevin Larkin** is a Senior Lecturer (Mathematics Education) at Griffith University. He is a member of a number of research teams investigating: STEM education in early years education; mathematics education in primary and middle school contexts; and pre-service teacher mathematics education. He has published widely in national and international publications in the areas of mathematics education, digital technologies, STEM, higher education, and Activity Theory. He is the current VP Communication for the Mathematics Education Research Group of Australasia (MERGA).

**Nathalie Sinclair** is a Professor and the Canada Research Chair in Tangible Mathematics Learning at Simon Fraser University. She is the author of several books, including co-author of *Mathematics and the Body: Material Entanglements in the Classroom*. She has also led the design of educational technologies, including the touchscreen app *TouchCounts*. She is the founding editor of the journal *Digital Experiences in Mathematics Education*.

# Part I
# Looking Across the Terrain

# Mathematics Apps—Stormy with the Weather Clearing: Using Cluster Analysis to Enhance App Use in Mathematics Classrooms

**Kevin Larkin and Todd Milford**

**Abstract** Mathematical apps are now used in many school settings. To support teachers in making appropriate pedagogical decisions regarding their increased use, empirical, quantitative analyses of apps are required. This chapter initially explores how cluster analysis can be used to identify elements within individual apps so that similar apps may be grouped together. This will assist teachers to make decisions regarding which apps might be most appropriate, either singularly or in groups, for various elements of their practice. Based upon selection criteria and ranking via four criterion-based scales, the cluster structure of 57 apps, primarily supporting number and algebraic thinking in elementary mathematics classrooms, is reported. The chapter then explores the homogeneity and heterogeneity of these clusters of apps and indicates when and how these apps may be used to enhance student mathematical learning. The chapter therefore makes both methodological and pedagogical contributions to the broader discussion of the use of apps in primary mathematics classrooms.

## Introduction

This research is an extension of a broader research project that has been on-going by the lead author since 2013. In brief, the initial part of the project (Phase 1), investigated the usefulness of modified versions of three measures, already used in published research, in evaluating the pedagogical appropriateness of approximately 200 apps categorised as educational at the iTunes store. The three modified

K. Larkin (✉)
Griffith University, Gold Coast, Australia
e-mail: k.larkin@griffith.edu.au

T. Milford
University of Victoria, Victoria, Canada
e-mail: tmilford@uvic.ca



N. Calder et al. (eds.), *Using Mobile Technologies in the Teaching and Learning of Mathematics*, Mathematics Education in the Digital Era 12,
https://doi.org/10.1007/978-3-319-90179-4_2

measures used in Phase 1 were the Haugland Scale (Haugland, 1999), Productive Pedagogies (Education Queensland, 2004), and Gee's Principles (2003). The use of these three measures is detailed in full later in this chapter. Based on the successful use of cluster analysis in evaluating 53 Geometry Apps (Larkin & Milford, 2018) we felt it necessary to revisit the earlier evaluation of Phase 1 apps (See Larkin, 2015) using cluster analysis to further understand why these apps were identified as being of high quality or otherwise.

Our research questions for this part of the research project are:

1. Whether and how the overall cumulative data on each app, gathered and discussed in Phase One, can be made more useful for teachers and researchers by additional analysis at a more granular level via cluster analysis; and
2. Whether and how specific number, algebra and a small number of non-number/algebra apps might be made more useful for specific teaching and learning activities if used in concert with other apps.

We initially argued in Larkin and Milford (2018) that cluster analysis provided a greater depth of valid and reliable information regarding the use of Geometry apps than had been provided in the earlier analysis by Larkin (2016). However, we seek here to increase the efficacy of cluster analysis for evaluating apps. The intent of this chapter is to, via a review of 57 apps using cluster analysis, assist teachers in choosing from these apps the most appropriate ones for their teaching whilst at the same time promote a methodology that more rigourously evaluates the quality of apps used individually and also in concert with other apps. Consequently, in this chapter, we offer an enhanced methodological approach to app research and uncover further pedagogical insights regarding app use.

## Literature Review

It is encouraging that since 2013 there has been an increase in the number of researchers investigating the use of mathematics apps in elementary and primary school classrooms (See Calder, 2015; Sinclair & Pimm, 2014). This is an important addition to the body of mathematics knowledge on the why and how of app use as apps are often used in schools without a strong, conceptual, pedagogical, or methodological underpinning. The use or misuse of novel technologies in mathematics is, of course, not a new educational experience, as previous waves of technology—e.g. calculators, computers, and virtual applets—have each impacted upon schools. What is perhaps different about the use of tablets is that, due to their rapid uptake in use in non-school contexts, there is additional pressure on schools to incorporate their use in classrooms.

## *App Affordances*

One strand of current research investigates the use of single or small groups of apps. There are currently very few such studies, and their focus is not specifically on supporting teachers choose which apps to use in their classrooms. For example, Sinclair, Chorney and Rodney (2016) investigated the affordances (academic, social and affective) of [*TouchCounts*], an app specifically designed and created for counting and doing arithmetic for early years students (3–8 years old). In this research, the authors identified rhythm as the primary unit of analysis and uncovered that the design of the app, which incorporated rhythm, worked as a motivational tool in terms of mathematics engagement as well as fostering the development of mathematical understanding in relation to early arithmetic. This work develops the notion of "finger gnosis" where direct and tactile engagement with [*TouchCounts*] fostered understanding of cardinality (Sinclair & Pimm, 2014).

Holgersson et al. (2016) used the app [*Fingu*], a multitouch virtual manipulative for understanding and mastering part-whole relationships for numbers 1–10 and established that the app provided valuable opportunities for early number development. In addition, they note that app design is a dynamic process and that [*Fingu*] continues to improve as an educational tool as newer versions are released. Lange and Meaney (2013), use Bishop's six mathematical activities and a Bernsteinian framework, to evaluate whether mathematical apps can support learning in preschool students. They suggest that mathematical apps provide opportunities for young learners to make their mathematical thinking more visible. Moyer-Packenham et al. (2016) report on the use of video to record 3–8 year olds interacting with 18 mathematics apps. Overall they found that although the apps aided student development, the level of development was varied, highlighting the difficulty in making broad educational claims regarding the use of mathematics apps. More recently, Lommatsch, Tucker, Moyer-Packenham, and Symanzik (2018 this volume) used cluster analysis to examine changes in the development progression of counting and seriation when supported by the use of pre-selected apps.

The work conducted by the range of authors above is promising and of significant value to teachers; however, as yet it is limited in scope to evaluating the affordances of pre-selected apps. As such, this research is limited in usefulness for teachers in determining the value of other apps.

## *Generic Reviews of Apps*

A second strand of recent research has involved reviews of apps, mainly iPad apps, due to the prevalence of iPads as the tablet of choice in elementary or primary schools. For example, Highfield and Goodwin (2013) evaluated 360, iTunes store apps in relation to age appropriateness, curriculum content and an initial classification of constructive-manipulable and manipulable-instructive. Powell (2014), in providing advice for teachers to find apps for their students, suggests that they begin with "an app search using the standard iTunes categories: 'Best New Apps,' 'Top Free Apps,' or 'Top Paid Apps'" (Powell, 2014, p. 21). Given the approximate 250 000 educational apps available (PocketGamer.Biz, Sept, 2016), the approach suggested by Powell is likely to be very time consuming and may not result in the discovery of useful mathematics apps.

Whilst the examples noted above often provide a useful starting point, reviews such as these are often generic and thus only provide teachers with a broad overview of the apps. In many cases, the broad overview may appear to meet some basic teacher requirements—e.g. drill and practice apps, apps that may keep early finishers engaged etc. As mathematics educators, what we suggest is also required is specific information regarding the types of mathematical knowledge inherent in the apps, the fidelity of the mathematics contained in the apps (Dick, 2008), or how the apps might be used productively in classrooms to support deeper mathematical learning. More recently, however, perhaps as a response to the growing demand for robust research into the quality of apps, a number of research articles have been published. For example, Namukasa, Gadanidis, Sarina, Scucuglia, and Aryee (2016) designed an instrument to evaluate apps according to their curriculum content, the range of affordances available in the app as a learning tool, user interactivity, and the quality of the overall design of the app. Their instrument offers a mechanism for teachers to evaluate apps use in upper primary and junior secondary mathematics, an area where there is a dearth of quality apps (Larkin, 2013, 2015) and is the type of broad based, robust, peer-reviewed research that is further required to enhance mathematics education using mathematics apps. While we support the work of Namukasa et al. (2016), our goal in this chapter is to add to extant research by evaluating a large number of apps using cluster analysis, and then examining of the types of mathematics promoted within each cluster of apps. We suggest that this approach provides more specific information regarding how teachers can coordinate the use of various elements of different apps to support mathematical learning beyond what can be achieved using individual apps.

## Study Design—Preparing the Data and Cluster Analysis

To answer the research questions we initially used cluster analysis, a collection of multivariate techniques that group individuals or objects (in our case apps) into clusters so that objects in the same cluster are more similar to one another than they are to objects in other clusters (Hair, Black, Babin, Anderson, & Tatham, 2006). In this way, cluster analysis maximizes the homogeneity within clusters, while at the same time maximizing the heterogeneity between clusters. In essence the approach attempts to keep more 'like things' together while simultaneously keeping 'unlike things' separate. In terms of apps and the clusters they are grouped into, this is useful information for teachers in that apps located within a cluster likely provide similar teaching and learning opportunities. Subsequently, teachers can then make more informed choices regarding selections of different apps to support different types of mathematical knowledge and skills.

## *Target Population and Criteria for Inclusion*

The initial population in this study were elementary mathematics apps labelled as educational for school children aged (5–11) at the iTunes App store. An initial search in 2013 on the term "mathematics education" returned 3740 apps thus the population was reduced with a targeted search using the following terms: elementary, primary, junior or infant mathematics. This still generated over 200 apps. A second level of quality control was then used (Larkin, 2013, 2015) by evaluating the apps using The Haugland Software Developmental Scale (Haugland, 1999). The Haugland Scale is a criterion based tool used to evaluate the appropriateness of web based applications and software for use by children. The Scale includes ten items including is the child in control of the learning, does the software cater for expanding complexity, is the software ethically sound and does the design of the app support independence and real world experiences. The scale was further modified for this research by clustering the ten items into three sub-dimensions (Child-Centred, Technical Design and Learning Design), and relating each dimension to an aspect of mathematics education. Each sub-dimension contributed to the overall score with child-centred scoring (0–4), technical design (0–3) and learning design (0–3). At the end of this evaluation procedure, 57 apps were determined as age appropriate and form the data set for this chapter. See http://tinyurl.com/ACARA-Apps for a full list of the reviewed apps. This link provides further details about each of the apps including price, Curriculum strand and sub-strand, Year Level appropriateness, type of knowledge developed (conceptual, procedural or declarative) as well as a lengthy review of the app in terms of its strengths and weaknesses.

## *Materials and Procedures*

The 57 apps were then evaluated by the lead author using two further measures Productive Pedagogies (Education Queensland, 2004) and Gee's Principles (2003). It is acknowledged here that this initial evaluation was based on a subjective evaluation; however, the lead author has over 30 years experience as a primary educator and 5 years teaching primary mathematics education at university. He has also written extensively about the review process in both professional and academic publications. In addition, in this chapter, international colleagues supported the further evaluation of the apps. These two measures were further modified to be quantitative and better targeted to evaluate mathematics apps. A full account of how Productive Pedagogies and Gee's Principles were used to score the initial apps is

**Table 1** Productive pedagogies (Education Queensland, 2004)

| Dimensions | Sub dimensions |
|---|---|
| Intellectual quality<br>*Total possible sub-dimension score = 30* | Higher order thinking/5<br>Deep knowledge/5<br>Deep understanding/5<br>Substantive conversation/5<br>Knowledge as problematic/5<br>Metalanguage/5 |
| Supportive environment<br>*Total possible sub-dimension score = 25* | Student direction/5<br>social support/5<br>Academic engagement/5<br>Performance criteria/5<br>Self regulation/5 |
| Connectedness<br>*Total possible sub-dimension score = 20* | Knowledge integration/5<br>Background knowledge/5<br>Connectedness to the world/5<br>Problem based/5 |
| *Total overall possible score = 75* | |

provided in Larkin (2015); what follows is a summary of the key aspects so that readers of this chapter understand how the apps were scored (see Tables 1 and 2).

The Productive Pedagogies are grouped under four dimensions: intellectual quality, supportive classroom environment, connectedness, and recognition of difference (Table 1). As very few apps attempted to cater for recognition of

**Table 2** Modified learning principles with definitions. (Adapted from Gee, 2003)

| Learning principle | Modified definitions |
|---|---|
| Active learning | All aspects of the app environment are set up to encourage active and critical, not passive, learning |
| Semiotic | Learning about and coming to appreciate interrelations with and across multiple sign systems as a complex system is core to tech learning experience |
| Achievement | For all learners there are intrinsic rewards from the beginning, customised to each learner's level and signalling the learner's ongoing achievements |
| Regime of competence | The learner operates within, but at the outer edge, of his/her level of competence so that there is both safety and challenge |
| Probing | Learning is a cycle of probing the world; reflecting in and on this action and, on this basis, forming a hypothesis for future testing |
| Multiple routes | There are many ways to complete the app, each of which caters for the strengths and interests of the learner |
| Situated learning | The meaning of signed are situated in embodied experiences and generated meanings are discovered bottom up |
| Practice | Learners get lots and lots of practice in a context where the practice is not boring and they therefore spend lots of time on the task |
| Discovery | Overt telling is kept to a minimum, allowing ample opportunity for the learner to experiment and make discoveries |
| Transfer | Learners are given ample opportunity to practice and transfer what they have learned to problems requiring adaptations and transformation |

difference, this dimension was discarded leaving three dimensions and fifteen Productive Pedagogies. The second measure used was a modified version of Gee's (2003) Principles. Based on the experience of the earlier evaluations, it became clear that many of the original 36 principles were not applicable for evaluating apps and that the entire 36 criteria would be too cumbersome (Jorgensen & Lowrie, 2012). For these reasons the number of principles was reduced to 10 (Table 2).

Each of the 57 apps in this study was evaluated by the first author on each of the sub-dimensions of the Productive Pedagogies and modified learning principles of the Gee. Each sub-dimension or learning principle was scored on a scale from low (1) to high (5), the range and score was variable. For example, there were 6 sub-dimensions for Intellectual Quality (IQ) resulting in a maximum possible score of 30. Following this break-down, the maximum possible score for Supportive Environment (SE) was 25, for Connectedness (C) 20 and, for the modified Gee's Principles (GP), 50 as there were 10 modified learning principles. A low score (1) on, for example the criteria of Transfer, indicates that any activity in the app is only relevant within the app (feeding an avatar to earn points) whereas a high score (5) indicates that the app fosters learning more broadly applicable (visualising rotations and reflections). A truncated version of descriptive statistics for the apps used in this study, in rank order, and based upon scores on the productive pedagogies and Gee's Principles, is presented in Table 3.

The internal reliability across the Productive Pedagogies in this study was calculated at $\alpha = 0.897$ and the reliability of the GP was calculated at $\alpha = 0.861$. To add robustness to the evaluation of the app and to offer further evidence of the psychometric quality of the scales used here, inter-rater reliability was calculated to determine whether these scales were consistent across more than one rater. Fifteen apps were randomly selected from the 57 and sent, along with accompanying documentation on each of the scales, to three graduate students who work with the second author at the University of Victoria. Fifteen apps were selected for the graduate students to confirm internal reliabilities as 25% of the total number of apps exceeds the informal 20% that is suggested when reliability is estimated from a sample. Other graduate students were given the same apps to evaluate. Their responses were then compared to the responses completed by the first author for

**Table 3** Top and bottom app scores for productive pedagogies and modified Gee Principles

| Selected apps | Productive pedagogies | | | Gee learning principles |
|---|---|---|---|---|
| | IQ | SE | C | |
| Mathemagica—Kids math | 28 | 23 | 20 | 36 |
| Area of rectangles | 28 | 22 | 16 | 37 |
| Early numbers: maths wizard counting | 24 | 22 | 14 | 26 |
| … | – | – | – | – |
| Telling time free | 11 | 10 | 7 | 18 |
| Math party | 11 | 11 | 6 | 16 |

inter-rater reliability. The alpha for each of IQ, SE, C and GP was 0.714, 0.766, 0.758, and 0.790 respectively (generally a value > 0.7 is considered acceptable for the inclusion of scales with non-critical consequences). It is not surprising that these values for alpha were lower than the authors, as they were not teachers nor experienced mathematics educators.

Building upon the internal consistency and inter-rater reliability calculated above, the rating of each of the 57 apps, as generated by the four scales (i.e., IQ, SE, C and GP), was used as data for a subsequent cluster analysis using SPSS v.22. We initially measured similarities as the squared Euclidian distances between each pair of apps on each of the four scale characteristics. In this way, smaller distances were viewed as indicating greater similarity. Once the similarity measures were calculated, a hierarchical procedure via the centroid cluster—which joins the apps in a weighted combination of the central points of the two individual clusters, where the weights are proportional to the sizes of the clusters—was applied to the clusters. Lastly, the number of clusters was determined, based upon the output, with the objective of generating the simplest structure possible while still representing homogeneous groupings. The number of clusters was determined by both the output and also a decision by the researchers to identify the simplest structure possible while still representing homogeneous groupings.

## Findings

Initial descriptives for the scales used to run the cluster analysis are presented in Table 4. All variables were presented in their original scale here (i.e., 30, 25, 20 and 50 respectively). There is no specific sample size required for cluster analysis; however, the data was screened for outliers and none were uncovered.

A correlational analysis (Table 5) was subsequently run on the four scales to determine if their inclusion in the cluster analysis would be appropriate, or if any overlap (i.e., multicolinearity—where two or more of the scales are highly correlated) might account for double counting (Hair et al., 2006). For example, the scores for Connectedness and Intellectual Quality are correlated at 0.812 and share over 64% of their variance.

Based upon this table, it was determined that the scales were all moderately to highly correlated (i.e., between 0.443 and 0.812) and thus multicolinearity was an

**Table 4** Scale descriptives

| Variable | N | Mean | Median | SD |
|---|---|---|---|---|
| Intellectual quality (30) | 57 | 17.2 | 17 | 4.23 |
| Supportive environment (25) | 57 | 15.6 | 15 | 4.10 |
| Connectedness (20) | 57 | 11.4 | 12 | 3.20 |
| Gee's principles (50) | 57 | 24.1 | 22 | 7.62 |

**Table 5** Correlations of the 4 scales

| | Intellectual quality | Supportive environment | Connectedness | Gee |
|---|---|---|---|---|
| Intellectual quality | 1.00 | | | |
| Supportive environment | 0.789** | 1.00 | | |
| Connectedness | 0.812** | 0.671** | 1.00 | |
| *Gee* | 0.597** | 0.433** | 0.553** | 1.00 |

**Correlation is significant at the 0.01 level (2-tailed)

issue with this data set. Hair et al. (2006) suggest that using Mahalanobis distance ($D^2$)—which bases clusters upon the central distance between clusters—can account for correlation among variables as it weights each variable equally. To account for this issue in SPSS, each scale score was standardized (i.e., mean deviated and divided by the standard deviation)—as distance scores are quite sensitive to differing magnitudes among the variables—and the centroid cluster option, which proportionally weights the apps, was applied to the clusters. The results of this second analysis are provided below.

Because we used an agglomerative method to determine clusters (i.e., each app started out as its own cluster), the dendrogram detailed in Fig. 1 should be read from left to right. Starting on the left with each of the 57 apps as its own cluster, using the centroid method of similarity, apps are combined one step at a time, based upon which two are the most similar, and are formed into a new cluster. The horizontal lines are indicative of homogeneity. The longer the horizontal line the more dissimilar the clusters are that are merged. For example, Fig. 1 indicates that [Case 38] is very homogenous to [Case 37]; in contrast the length of the connecting horizontal line indicates that [Case 1 and 2] are more heterogeneous to each other and also as a pair to [Case 6]. Based upon this distance measure, the vertical line was placed on the dendrogram to highlight the three-cluster solution.

From the dendrogram there are a number of apps (i.e., 1, 2, 6, 27, 36, 43, and 54) that are not captured in the three-cluster solution presented here. In order to convey any meaningful information, clusters need to contain at a minimum three objects and based upon this solution, none of these apps combined into a cluster. This is possibly due to the sample size as these apps may capture additional attributes not detailed in three clusters or these apps may be outliers. One additional display that helps to demonstrate why the three-cluster solution was selected is provided in Table 6. What is evident is that [*Mathemagica*] and [*Area of Rectangles*] always combine and are always separate from all other apps (except [*Math Galaxy Fractions*] until Cluster 6) regardless of where the cluster solution is placed. Likewise [*Hands on Maths*] remains separate to all other apps after Cluster 2. This indicates that whether a three, four, six or six cluster solution was tried, these seven apps consistently demonstrated heterogeneity from the other 50 apps evaluated via cluster analysis. This is significant as it means that (a) these seven apps are

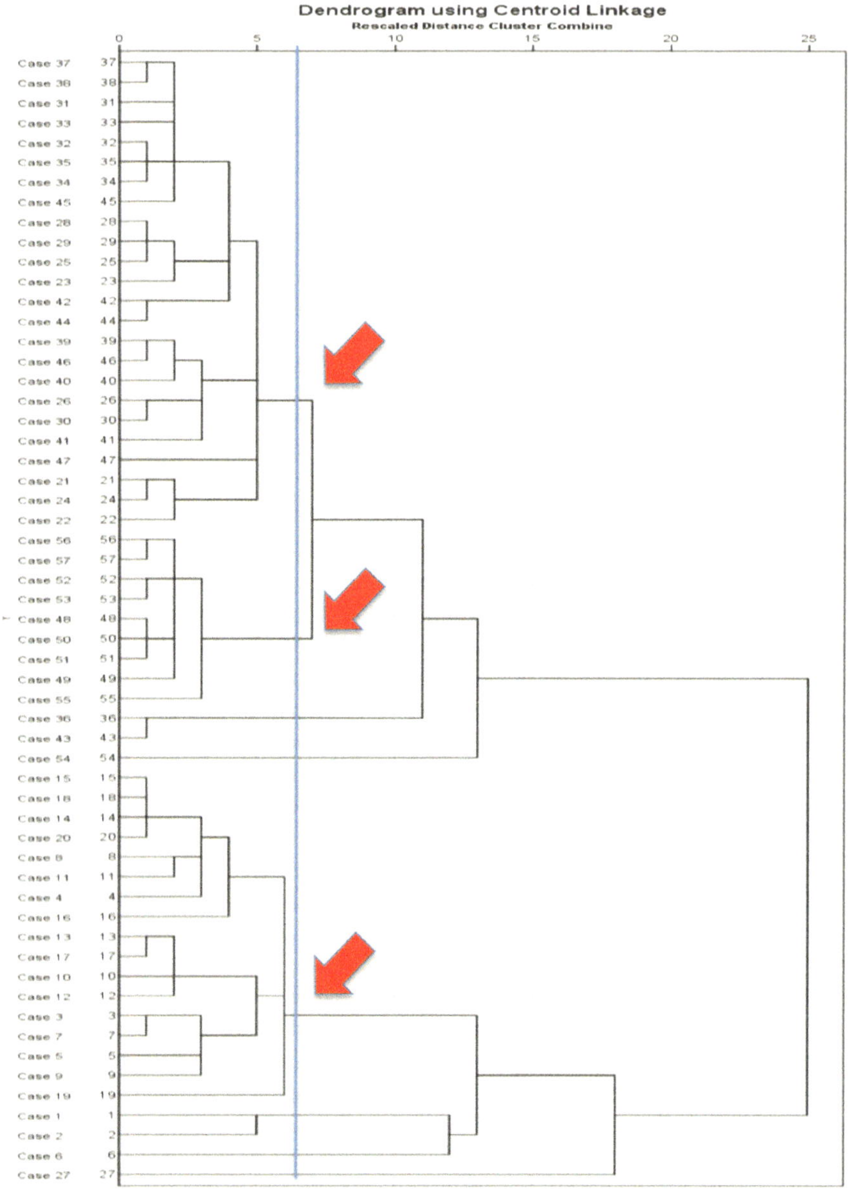


**Fig. 1** Dendrogram—with vertical line and arrows indicating point of cluster formation

un-clustered and (b) that there must be distinguishing aspects within each app which can be uncovered to account for their heterogeneity.

A final criterion for the selection of number of clusters is based upon the location where the distance coefficient makes the biggest jump (i.e., a simple percentage

**Table 6** Various possible cluster solutions of the 57 apps

| App | 8 cluster | 7 cluster | 6 cluster | 5 cluster | 4 cluster |
|---|---|---|---|---|---|
| Mathemagica—Kids math | 1 | 1 | 1 | 1 | 1 |
| Area of rectangles | 1 | 1 | 1 | 1 | 1 |
| Math galaxy fractions fun | 3 | 3 | 3 | 1 | 1 |
| Hands-on maths | 5 | 5 | 5 | 4 | 4 |
| Probability tools | 6 | 6 | 4 | 3 | 3 |
| Geometry 4 Kids | 6 | 6 | 4 | 3 | 3 |
| Tens frame | 8 | 7 | 6 | 5 | 3 |

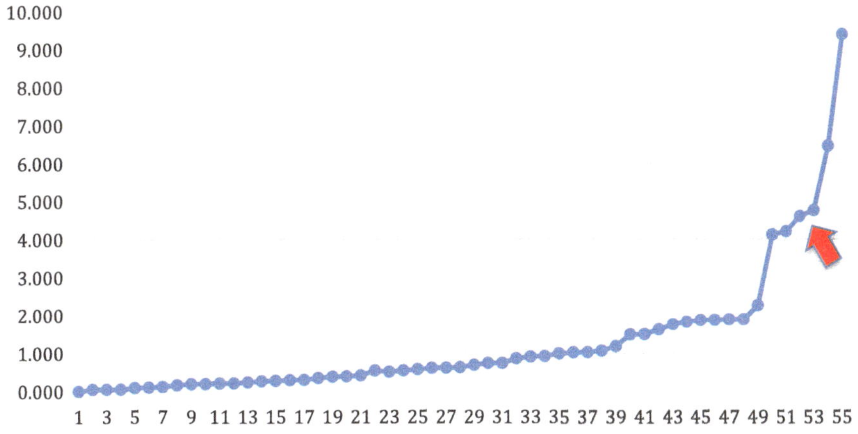


**Fig. 2** The scree plot

change in heterogeneity of the clusters). This is provided in the Scree Plot detailed in Fig. 2. In this case, the largest jump in the distance coefficient was from stage 54 to stage 55 and the number of clusters generated is based on the following algorithm [*the number of cases* subtract the largest *distance coefficient* equals the *number of clusters*]. In our example, this equates to 57 – 54 = 3. Thus, based upon both a graphical depiction, and a percentage change in heterogeneity, a three-cluster solution was accepted as best representing the homogeneity and heterogeneity of the apps.

## Discussion

As indicated earlier, a limitation of our previous research (Larkin, 2015, 2016; Larkin & Milford, 2018) was that only the lead author performed the analysis and synthesis of the clusters formed. Therefore, in order to enhance the validity of this research, two international mathematics educators, one from Canada and one from New Zealand, as well as the lead author, independently examined the formation of the clusters and independently identified themes that were apparent from the clusters that were formed.

## *Types of Mathematics Knowledge*

In analysing the formation of the clusters, in terms of their homogeneity and heterogeneity, and reflecting upon the contributions of our international colleagues, one logical way to explain the formation of the clusters is in relation to the types of mathematical knowledge that they develop. Here we draw on the work of Miller and Hudson (2007) and others who proposed three types of knowledge—conceptual, procedural and declarative. A full description of how the apps were evaluated for high versus low conceptual and procedural knowledge can be found in Larkin (2015, 2016). Here we are evaluating whether cluster analysis adds further information as to how the apps, regardless of their high/low quality, might be used either individually, or in concert with other apps, for specific pedagogical purpose. According to Goldman and Hasselbring (1997) conceptual knowledge refers to a "connected web of information in which the linking relationships are as important as the pieces of discrete information that are linked" (p. 4). A student's conceptual knowledge is increased, for example, when they recognise relationship between multiplication and division or common and decimal fractions as opposed to when these concepts are only understood in isolation.

Procedural knowledge is the ability to follow a set of sequential steps to solve a mathematical task (Goldman & Hasselbring, 1997; Miller & Hudson, 2007) and is primarily used to solve computational tasks—e.g. finding areas or calculating change. Declarative knowledge is knowledge that students are able to efficiently recall from memory without hesitation—e.g. subitising small amounts or fluent processing of number facts. Of some concern to Miller and Hudson (2007), and also the Australian Curriculum, Assessment and Reporting Authority ACARA (2016), is the observation that mathematics educators have traditionally placed a heavy emphasis on the development of declarative and procedural knowledge and this emphasis is reflected in the large percentage of mathematics apps that develop these latter two forms of knowledge (Larkin, 2013, 2015). Our argument here is that mathematics apps require a balance between the three knowledge areas; either within one app [e.g. *Mathemagica; Area of Rectangles*] or in groups of apps on a specific topic, e.g. Fractions [*Hands On Number Sense*—Conceptual; *Fraction Time*—Procedural; and *Subtracting Like Fractions*—Declarative]. Unfortunately, in our view, this balance is not evident in the range of apps that are available to school mathematics educators and mirrors the findings of Namukasa et al. (2016) who reported that only four of the 80 apps they reviewed "focused on building understanding of concepts" (p. 290).

We now turn our attention to the constitution of each of the clusters. The labels of the clusters were generated after determining the types of mathematics content and pedagogy they contained. In discussing these apps, an important observation is that apps did not cluster according to content; therefore, there are apps spread across the three clusters developing content from a wide range of curriculum sub-strands e.g. Fractions, Place Value, Patterning, Chance or Statistics. This is valuable knowledge as teachers may suspect that quality apps are more likely to develop particular content and poor apps other content, when the reverse is true; there are both quality apps and poor apps developing the same content areas, e.g. [*Early Numbers: Maths Wizard Counting*] and [*Letz Learn Counting*] are respectively very high and very poor in the early counting domain.

## *Cluster 2—Conceptual and Procedural Knowledge*

By and large the 17 apps within this cluster develop both conceptual and procedural knowledge across a range of content areas including early number, computations, algebra, statistics and place value. The exceptions to this general rule are the apps [*Marble Math Junior*—Procedural only] and [*Math Model*—Conceptual only]. In both cases, an examination of where these particular apps are positioned in the dendrogram indicates that they are only loosely connected to this Cluster; in other words, a different cluster formation, or a review of more apps, might see these apps clustered with other apps that are solely procedural or solely conceptual.

What was consistently the case for the remaining 15 apps in this cluster is that they all had multiple components such that the students could be developing conceptual knowledge using one component of the app and procedural knowledge when using a different component of the same app. Variance within the cluster occurred as some apps were stronger at one knowledge element than others (while still developing both). For example, [*Place Value Chart*] was very strong in developing conceptual knowledge, as students were free to explore the app and modify a range of settings. At the other end of the spectrum within this cluster [*Friends of Ten*] was highly scaffolded and thus developed procedural knowledge around place value but did not afford the level of self-direction likely needed for deep conceptual knowledge. Thus teachers can use both apps concurrently depending on the pedagogical intent of the learning.

From a design perspective, these apps all used representations or icons that can be manipulated/moved, but in simple ways. Variance within this cluster is largely accounted for in terms of the Gee Principles with some of the more supportive apps in terms of student environment in Productive Pedagogies attaining a high score in this sub-dimension at the expense of opportunities for Probing, Discovery and Multiple Routes in Gee; all considered important elements in conceptual knowledge development. Overall, this cluster is very useful for teachers in the early conceptual development stage of a range of content areas and then; using different elements of the same app, they can later develop procedural knowledge around the initial concept.

## *Cluster 4—Procedural and Declarative or Solely Procedural Knowledge*

Cluster 4 was the largest cluster and contained 24 apps. Given its size, it is perhaps unsurprising that it was the most disparate in terms of the types of knowledge developed with apps either solely procedural; largely procedural with some declarative knowledge aspects; or in two cases conceptual [*Fun Count App*] and [*Patterns, Colors and Shapes*] but at a very low level which excluded these two apps from Cluster 2 where both conceptual and procedural knowledge were developed. In examining the dendrogram for the other 22 apps, it is evident that at the 6 cluster stage they are linked with the final Cluster 7 (see below). This might indicate a close relationship with Cluster 7 in terms of the skills and processes developed as they pertain more towards declarative knowledge rather than conceptual understanding of the mathematics as noted in Cluster 2. Many of the apps that scored more poorly in this cluster offered a behaviourist approach that uses rewards for correct answers and lack of progress for incorrect ones. Hence the focus, in these lower scoring apps, is on extrinsic motivators that work more effectively in declarative than procedural development modes. From a design perspective, there is a range in the quality of the visual representations in each app; in some the visual representations are used creatively and promote student thinking; however, in most, the representations are used solely to represent procedures or processes rather than supporting students to make conjectures or establish their own patterns of thought evident in the more conceptually oriented apps in Cluster 2. Most of the apps in Cluster 4 are, therefore, of limited use as they are largely procedural apps that use extrinsic motivators, and the visual representations do not necessarily enhance learning. However, they may have specific use in targeted scenarios such as developing fluency or reinforcing area formulas once conceptual understanding has been developed.

## *Cluster 7—Declarative Knowledge or Declarative and Minimal Procedural Knowledge*

This cluster consisted of nine apps that generally scored poorly in both Productive Pedagogies and to a slightly lesser extent Gee Principles. Perhaps as a consequence of predominantly dealing with the content areas of time and mass, two areas where conceptual development is difficult due to the abstract nature of both concepts in terms of mass being independent of size and time being a non-visible attribute; these apps were all highly directive in nature focussing mainly on reading clock faces or conversions of mass units. The predominant design feature of the apps was accuracy and speed of feedback without any opportunity for any exploration and investigation (important for conceptual development) or process/skills work (necessary for procedural development). In terms of the Productive Pedagogies they were weak overall in the sub-dimension of Connectedness indicating that they were primarily focused on one content area (in each app) or on one knowledge aspect (mainly declarative knowledge—e.g. point of time activities). Likewise, in the Gee Principles, they scored poorly in the probing and transfer principles reinforcing their limited attempt at connecting to either the real world experience of the students or to a range of related mathematics content areas. These factors constrain the usefulness of these clusters of apps. Of the time apps, [*Tillie's Time Shop*] showed most promise, scoring in the mid-range on the Gee scores and best of the rest in the Productive Pedagogies.

## *Miscellaneous—i.e. Non-clustered Apps*

Of most interest to us, and an occurrence not evident in our earlier use of cluster analysis, are the seven apps: *Mathemagica—Kids Math* [Case 1]; *Area of Rectangles*—now labelled in the iTunes store as *Area of Figures* [Case 2]; *Math Galaxy Fun* [Case 6]; *Hands on Maths* [Case 27]; *Probability Tools* [Case 36]; *Geometry for Kids* [Case 43]; and *Tens Frame* [Case 54] which either only formed a cluster with one other app [Case 1 and 2] and [Case 36 and 43] or never formed a cluster [Cases 6, 27 and 54]. This is an indication that cluster analysis provides a lens for further analysis of apps that were rated in earlier research as similar to other apps in a range of clusters. In other words these miscellaneous apps were across the spectrum of conceptual, procedural and declarative knowledge yet different to the apps within the three clusters. What this indicates is that cluster analysis provides more detailed information than supplied in the earlier research that largely rated apps in terms of raw scores and the content that individual apps developed. Given that [Cases 1 and 2] and [Cases 36 and 43] clustered consistently together, each will largely be discussed as one entity and [Cases 6, 27 and 54] which never clustered, will be considered individually in the following discussion.

[Case 1 and 2] are the standout apps scoring highly in terms of raw scores in the Productive Pedagogies and Gee, respectively. What distinguishes these two apps from all other apps is that they (a) provide opportunities within the one app for conceptual, procedural and declarative knowledge development; (b) they do so at a high level across each of the three types of knowledge; and (c) they support student learning with high quality external representations. This understanding would not be possible by just looking at the individual raw scores for these two apps. This latter point is very critical in the success of [Case 2] which teaches about four shapes. Within each shape component there are four learning episodes: Manipulative and Challenge which develop conceptual knowledge; Lesson which develops procedural knowledge; and Questions which develops declarative knowledge. If teachers have a limited budget, these two apps provide a scaffolded learning sequence across the topics of number and area respectively and are a sure thing to be added "to the cart".

In contrast, [Case 36 and 43] are different from the other apps in that they develop only conceptual knowledge [Case 36] or primarily conceptual knowledge and some declarative knowledge [Case 43]. This indicates that both apps are useful when students are developing conceptual knowledge but that teachers will need to value add skills/processes and fluency in the teaching sequence. A second point of interest is that out of the entire 57 apps, these two apps had the lowest scores (respectively 9 and 8 out of 25) for Supportive Environment on the Productive Pedagogies, indicating that the students are left pretty much to fend for themselves in developing pedagogical knowledge, due in part to the lack of quality external representations or scaffolding in developing conceptual knowledge. These apps therefore need to be used judiciously by teachers.

The three remaining cases, [Cases 6, 27 and 54] span the gamut of quality but remain discrete in terms of the cluster analysis. By rights, according to the overall scoring pattern, [Case 6] should be considered Conceptual and Procedural, [Case 27] Procedural and possibly Weak Declarative and [Case 54] Declarative and possibly Weak Procedural. However, [Case 6] is unique in that it only develops Procedural knowledge—via a comprehensive range of tutorials on a broad range of fractions content—but does so in an exceptionally robust way with high quality external representations and appropriate language. The app, however, assumes prior conceptual knowledge and therefore only develops skills and processes. Thus it is a useful app to consolidate and practice fraction procedures once conceptual development has occurred. By contrast, [Case 27] is only conceptual; as was the case with [*Math Model*], but is not as effective due to a lack of authentic external representations and limited variety in concept development—students merely drag a variety of different sized and coloured prototypical shapes into a Venn Diagram to determine the "rule" for their sorting. In addition, no information is provided to assist students develop either skill or fluency in the sorting. Finally, [Case 54] never clusters, as it does not appropriately develop any form of knowledge. It might be described somewhat similarly to [Case 27] in that clearly no procedural or declarative knowledge is developed. However, it is markedly less useful than [Case 27] in that it largely operates as a "place value sandpit" where students can freely play with the place value counters but not necessarily be developing any concept of base ten. This differs to the high quality [*Place Value Chart*], which also allows free play but, depending on the mode, either implicitly or explicitly develops the concept of "ten-ness". [Case 27] is therefore not recommended for classroom use as it requires explicit teacher input and offers nothing extra than what the aforementioned [*Place Value Chart*] delivers at a higher quality.

Perhaps, the take away for teachers from the discussion concerning the non-clustered apps is that the previous clusters offer a good place to go for teachers seeking support for student learning in the specific areas of conceptual, procedural, and declarative knowledge. However, if teachers are seeking apps that score well on measures such as the sub-domains of the Productive Pedagogies and modified learning principles of the Gee, yet capture multiple areas of conceptual, procedural, and declarative knowledge, this would be the cluster to further explore for their students.

## Limitations

Before concluding, it is appropriate to discuss two main limitations in relation to this relatively novel approach to evaluating apps. Firstly, in terms of cluster analysis, limitations include that different approaches can often give different clusters, and that the analysis is highly dependent upon the variables used to differentiate the cases (in this case the apps). This differing cluster structure, dependent on variables used to differentiate cases, is something that may influence the findings for others who would seek to replicate this study with a different selection of apps. However, the application of this methodology with a new and revised selection of apps may find similarity in findings lending support to our approach. Secondly, from a data collection perspective, the apps reviewed are from 2013—consequently quality new apps that we are aware of such as [*TouchCounts*] or [*MotionMath Zoom*] are not reviewed, as they were not available at the iTunes store when the initial review was conducted. From a methodological point of view it was not appropriate to hand pick more recent quality apps and include them as it defeats the purpose of the a priori clustering of all the apps. In addition, other quality apps have disappeared e.g. [*Mathemagica*] is no longer available at the Australian iTunes store (but is still available at other iTunes stores).

## Conclusion

In this chapter we have further developed our understanding of the use of a multivariate data analysis tool (i.e., cluster analysis), a relatively novel methodological approach in educational research, to form combinations of internally homogenous and yet externally heterogeneous groups. Thus the cluster analysis process resulted in the identification of apps that, although they scored similarly in the overall scores on the Productive Pedagogies and Gee's, were quite different in terms of their pedagogical utility. These cluster categories therefore offer a more nuanced opportunity for the teacher to align their choice of classroom apps to the type of pedagogical knowledge they intend to develop with their students. In addition, they are able to do so at the level of a set of apps rather than only on the individual basis made possible in the earlier research. This resolves a problem identified in the earlier research that found that most apps were designed as stand-alone apps targeting a particular type of knowledge or content area (e.g., adding common fractions). Thus the limitation of highly specific content, in apps that otherwise are strong in developing conceptual or procedural knowledge, can be offset by the use of a set of apps developing the same type of knowledge, in this case procedural, but based on different content, e.g. subtraction of two digit numbers. The overall outcome of the cluster analysis has confirmed that the majority of the 57 apps, identified as age appropriate by the Haugland Scale, could be broadly classified according to the type of mathematics knowledge they likely support—conceptual, procedural or declarative—or combinations of the three. Although it is difficult to firmly recommend one app over another—given that how they are used in clusters impact upon their quality—we recommend that teachers using apps for the first time consider *Mathemagica—Kids Math* or *Area of Rectangles*—now labelled in the iTunes store as *Area of Figures* [Case 2] to support their classroom practice.

A distinct finding in this research, in contrast to our earlier use of cluster analysis on Geometry apps where five clearly distinct clusters of apps were generated (Larkin & Milford, 2018), is that here there were three apps that did not cluster at all and a further four apps that clustered in pairs respectively regardless of whether a larger or small cluster formation point, as suggested in either the dendrogram (Fig. 1) or the Scree Plot (Fig. 2), was chosen. This has clear implications for teachers who (a) can confidently use the apps from individual clusters for specific types of maths teaching; and (b) if they are intend to use any of the very strong apps that either did not cluster, or only clustered with one other app, they will need to be much more precise in their decision-making regarding the intended learning outcomes that they want for their students. This is additional knowledge, developed via the use of cluster analysis, and not generated in the earlier analysis of these apps that only investigated them as stand-alone entities.

The amount of time that the authors have spent on initially finding, scoring and evaluating the apps, when combined with the time pressures on teachers and the continuing expansion of the number of apps targeting primary aged students (and their teachers and parents), supports our claim that robust research such as this is vital in assisting largely time poor classroom teachers to select appropriate mathematics apps. We also suggest that we have further developed the use of cluster analysis as an important research methodology for uncovering connections between the apps that were difficult to identify in the earlier research. These connections therefore provide a fresh perspective in evaluating how combinations of apps can be used for specific teaching purposes.

**Acknowledgements** This project was initially funded through a Griffith University, School of Education and Professional Studies, research grant. Aspects of this book chapter were presented at an earlier MERGA conference. The data set for the analysis is a component of the larger dataset used in (Larkin, 2015). However, this chapter is a substantive development of the earlier work. We gratefully acknowledge the contribution of Professor Nathalie Sinclair (Simon Fraser University, Canada) and Associate Professor Nigel Calder (Waikato University, New Zealand) for their insights regarding the clustering of the apps. We also acknowledge the contribution of post-graduate students from the University of Victoria, Canada—Aishah Bakhtiar (Ph.D. Candidate), Sarah Davis (Ph.D. Candidate) and Rebecca Edwards (MA Candidate) who evaluated a random sample of the 57 apps to provide inter-rater reliability on the initial scoring of the apps.

## References

Australian Curriculum, Assessment and Reporting Authority (ACARA). (2016). *Australian curriculum: Mathematics—Rationale*. Retrieved from http://www.australiancurriculum.edu.au/mathematics/rationale.

Calder, N. (2015). Apps: Appropriate, applicable and appealing? In: T. Lowrie, & R. Jorgensen (Eds.), *Digital games and mathematics learning: Potential, promises and pitfalls* (Vol. 4). The Netherlands: Springer.

Dick, T. P. (2008). Fidelity in technological tools for mathematics education. In G. W. Blume, & M. K. Reid (Eds.), *Research on technology and the teaching and learning of mathematics: Volume 2. Cases and perspectives* (pp. 333–339). Charlotte, NC: Information Age Publishing.

Education Queensland. (2004). *Productive pedagogies: Classroom reflection manual.* Retrieved from http://education.qld.gov.au/public_media/reports/curriculum-framework/productive-pedagogies/. html.

Gee, J. P. (2003). *What video games have to teach us about learning and literacy*. New York: Palgrave Macmillan.

Goldman, S. R., & Hasselbring, T. S., The Cognition and Technology Group at Vanderbilt. (1997). Achieving meaningful mathematics literacy for students with learning disabilities. *Journal of Learning Disabilities, 30*(2), 198–208.

Hair, J. F., Black, W. C., Babin, B. J., Anderson, R. E., & Tatham, R. L. (2006). *Multivariate data analysis* (6th ed.). Upper Saddle River, NJ: Pearson.

Haugland, S. (1999). Computers and young children: The newest software that meets the developmental needs of young children. *Early Childhood Education Journal, 26*(4), 245–254.

Highfield, K., & Goodwin, K. (2013). Apps for mathematics learning: A review of 'Educational' Apps from the iTunes App Store. In: V. Steinle, L. Ball, & C. Bardini (Eds.), *Mathematics education: Yesterday, Today and Tomorrow (Proceedings of the 36th Annual Conference of the Mathematics Education Research Group of Australasia)* (pp. 378–385). Melbourne, VIC: MERGA.

Holgersson, I., Barendregt, W., Emanuelsson, J., Ottosson, T., Rietz, E., & Lindström, B. (2016). Fingu-A game to support children's development of arithmetic competence: Theory, design and empirical research. In: P. S. Moyer-Packenham (Ed.), *International perspectives on teaching and learning mathematics with virtual manipulatives* (Vol. 7). Mathematics education in the digital era. https://doi.org/10.1007/978-3-319-32718-1_6.

Jorgensen, R., & Lowrie, T. (2012). Digital games for learning mathematics: Possibilities and limitations. In: J. Dindyal, L. P. Cheng, & S. F. Ng (Eds.), *Mathematics Education: Expanding Horizons (Proceedings of the 35th Annual Conference of the Mathematics Education Research Group of Australasia)*, (pp. 378–384). Singapore: MERGA.

Lange, T., & Meaney, T. (2013). iPads and mathematical play: A new kind of sandpit for young children? In: *Eight Congress of European Research in Mathematics Education*. Retrieved from http://cerme8.metu.edu.tr/wgpapers/WG13/WG13_Lange_Meaney%20.pdf.

Larkin, K. (2013). Mathematics Education. Is there an App for that? In: V. Steinle, L. Ball, & C. Bardini (Eds.), *Mathematics Education: Yesterday, Today and Tomorrow (Proceedings of the 36th Annual Conference of the Mathematics Education Research Group of Australasia)*, (pp 426–433). Melbourne, VIC: MERGA.

Larkin, K. (2015). "An App! An App! My Kingdom for An App": An 18-month quest to determine whether apps support mathematical knowledge building. In: T. Lowrie, & R. Jorgensen (Eds.), *Digital games and mathematics learning: Potential, promises and pitfalls* (Vol. 4, pp. 251–276). The Netherlands: Springer.

Larkin, K. (2016). Geometry and iPads in primary schools: Does their usefulness extend beyond tracing an oblong? In P. S. Moyer-Packenham (Ed.), *International perspectives on teaching and learning mathematics with virtual manipulatives* (pp. 247–274). Cham: Springer International Publishing.

Larkin, K., & Milford, T. M. (2018). Using cluster analysis to enhance student learning when using geometry mathematics apps. In L. Ball, P. Drijvers, S. Ladel, H. Siller, M. Tabach, & C. Vale (Eds.), *ICMI – 13 Monographs. Uses of technology in primary and secondary school mathematics education*. Dordrecht, The Netherlands: Springer.

Lommatsch, C. W., Tucker, S. I., Moyer-Packenham, P. S., & Symanzik, J. (2018 this volume). Heatmap and hierarchical clustering analysis to highlight changes in young children's developmental progressions using virtual manipulative mathematics apps. In: N. Calder, K. Larkin, & N. Sinclair (Eds.), *Using mobile technologies in the teaching and learning of mathematics*. Mathematics education in the digital era. Springer.

Miller, S., & Hudson, P. (2007). Using evidence-based practices to build mathematics competence related to conceptual, procedural, and declarative knowledge. *Learning Disabilities Research & Practice, 22*(1), 47–57.
Moyer-Packenham, P. S., Bullock, E. K., Shumway, J. F., Tucker, S. I., Watts, C. M., Westenskow, A., et al. (2016). The role of affordances in children's learning performance and efficiency when using virtual manipulative mathematics touch-screen apps. *Mathematics Education Research Journal, 28*(1), 79–105. https://doi.org/10.1007/s13394-015-0161-z.
Namukasa, I. K., Gadanidis, G., Sarina, V., Scucuglia, S., & Aryee, K. (2016). Selection of apps for teaching difficult mathematics topics: An instrument to evaluate touch-screen tablet and smartphone mathematics apps. In P. S. Moyer-Packenham (Ed.), *International perspectives on teaching and learning mathematics with virtual manipulatives* (pp. 275–300). Cham: Springer International Publishing.
PocketGamer.biz. (2016). App store metrics. Retrieved from http://www.pocketgamer.biz/metrics/app-store/.
Powell, S. (2014). Choosing iPad Apps with a purpose: Aligning skills and standards. *Teaching Exceptional Children, 47*(1), 20–26.
Sinclair, N., Chorney, S., & Rodney, S. (2016). Rhythm in number: Exploring the affective, social and mathematical dimensions of using TouchCounts. *Mathematics Education and Mobile Technologies—Special Issue of the Mathematics Education and Research Journal, 28*(1), 31–51. https://doi.org/10.1007/s13394-015-0154-y.
Sinclair, N., & Pimm, D. (2014). Number's subtle touch: Expanding finger gnosis in the era of multi-touch technologies. In: *Proceeding of PME 38 and PME-NA 36, 5* (pp. 209–216).

**Dr. Kevin Larkin** *EdD (Griffith), M Ed Studies (Griffith), M Ed Leadership (ACU), Grad Dip Arts (ACU), B Ed (ACU), DipTeach (Kelvin Grove CAE).* Kevin is a Senior Lecturer (Mathematics Education) at Griffith University. He is a member of a number of research teams investigating: STEM education in early years education; mathematics education in primary and middle school contexts; and pre-service teacher mathematics education. He has published widely in national and international publications in the areas of mathematics education, digital technologies, STEM, higher education, and Activity Theory. He is the current VP Communication for the Mathematics Education Research Group of Australasia (MERGA).

**Dr. Todd Milford** (BSc-1994, BEd-1997, Dip SpecEd-2000, MEd-2004, PhD-2009; University of Victoria) is an associate professor in science education and research methodologies at the University of Victoria. He was a lecturer in the Art, Law, and Education Group at Griffith, University in Brisbane Australia. He has mathematics, science and special education teaching experience at the elementary and secondary level (Victoria and Vancouver, BC) as well as in the on-line environment (SIDES, Saanich, BC). He has been teaching at the postsecondary level since 2005 primarily in the areas of science education, mathematics education, and classroom assessment. His research has been and continues to be varied; however, the constant theme is on using data and data analysis to help teachers and students in the classroom.

# How Might Apps Reshape the Mathematical Learning Experience?

**Nigel Calder and Carol Murphy**

**Abstract** This chapter reports on how the use of mathematics apps has the potential to reshape the *learning experience*, a particular aspect of learning with apps that emerged from a research project examining the ways mobile technologies are used in primary-school mathematics. The chapter will consider a number of key themes related to student learning that have emerged through the research. When using some of the apps in the study, students used different digital tools within the app to solve word problems, while the affordances, including multi-representation, dynamic and haptic, made the learning experience different from when using pencil-and-paper technology. Other themes that were identified included: collaboration, socio-material assemblages, and personalisation. All of these appeared influential in the development of mathematical thinking. While the affordances of the mobile technologies are important, the teacher's pedagogical approach and the dialogue that the apps evoked were central in the learning.

**Keywords** Affordances · Collaboration · Differentiation · Engagement
Mathematical thinking · Primary-school mathematics · Assemblages
TPACK · Video analysis

---

N. Calder (✉) · C. Murphy
University of Waikato, Hamilton, New Zealand
e-mail: nigel.calder@waikato.ac.nz

C. Murphy
e-mail: carol.murphy@utas.edu.au

 
N. Calder et al. (eds.), *Using Mobile Technologies in the Teaching and Learning of Mathematics*, Mathematics Education in the Digital Era 12,
https://doi.org/10.1007/978-3-319-90179-4_3

## Introduction

Mobile technologies are everywhere! The use of mobile technologies (MT), especially smart-phones, has grown markedly, as has the availability of WiFi, both in educative settings and in community settings. The accompanying growth in the number of apps has likewise been recognised (Larkin, 2015). There is a lot of excitement regarding their potential to transform the learning experience and enhance mathematics learning opportunities. Their ease of operation, allied with the students' interaction being focused primarily on touch and sight, can make their use intuitive for learners (Calder & Campbell, 2016). While concerns have been raised regarding the suitability of the pedagogical approaches utilized through and with apps (e.g., Philip & Garcia, 2014) other research has highlighted their effectiveness in various aspects of mathematics learning (Attard, 2015; Carr, 2012). Also, as MT have become a more enduring element of the evolving digital world, we need to consider their potential for learning. This chapter reports on an aspect of a research project examining the ways tablets, as examples of MT, are used in primary-school mathematics. The project considers the pedagogy that might best facilitate the learning with students (ages 5–11) when engaging in mathematics using MT. One aim of the research was to identify aspects of the learning process that influenced the mathematical learning, when students engaged with mathematics using apps. What were features of the learning that emerged through using MT, apps in particular? The chapter reports on the themes related to pedagogy that emerged from the research and how they might be seen to reshape learning experiences in primary mathematics.

In the chapter, we will concentrate on the emerging themes related to the ways that teachers are using MT in their classrooms and mathematics programmes. The themes and framework emerged from an iterative process of co-construction by the research team, including 12 teacher researchers. The themes are: affordances, collaboration, socio-material assemblages, and personalisation/differentiation. The ways that mathematical thinking is hinged to each of these themes will be inherent in the discussion of each, as will the interconnectedness and relationships between the themes. Each theme will be considered in a separate section, prefaced by some informative, and influential, theoretical and research perspectives. The chapter will conclude with how some themes might overlay and influence each other, as well as some perspectives that emerged from the project overall.

## Methodology

The research project used an interpretive methodology that relates to building knowledge and developing research capability through collaborative analysis and critical reflection of classroom practice and student learning. The research design was aligned with teacher and researcher co-inquiry whereby the university researchers and practicing teachers work as co-inquirers and co-learners (Hennessy, 2014). Allied to this is an emphasis on collaborative knowledge building. This research method is based on a transformational partnership arrangement that generates new professional knowledge for both academic researchers and teachers (Groundwater-Smith et al., 2013). There is joint scrutiny of the reflections and evaluations, and hence joint scrutiny of an educational practice. This scrutiny informs new forms of awareness for teachers and researchers (Hennessy, 2014). Three teachers, all experienced with using MT in their programmes, were involved in the first year of the study. One teacher taught a year-4 class in a school using a approach, while the other two teachers team-taught in a year-5 & 6 class, in a school with 1-1 iPad provision. The data were analysed using NVivo. The themes developed from the observed use of MT in classes, with data collected by video, teacher semi-structured interviews and student blogs, and from collaboration in research meetings with teachers viewing one or two extracts of video each time. The video extracts were of the students working in class during their mathematics lessons. Ethical approval was obtained and pseudonyms are used for participants.

In the second year of the project, nine other teachers joined the research team. These teachers were across a range of year levels (years 1–6) and experience with using apps in their mathematics programmes (from using apps for students to practice a particular skill, to using apps such as *Math Shake* with screen-casting ability, for students to explain their strategies and solutions). See Table 1 for demographic information of teachers discussed in this chapter. The themes from the first year were carried forward into year two. Refinement of the identified themes occurred through joint critical reflection between the teacher practitioners and academic researchers in research meetings. As the chapter focuses on the themes, the teacher data from both years were considered, while the student data is only from the first year of the project.

**Table 1** The teachers, their context, and experience with MT at the start of the project

| Teacher | School | Year level | Experience with MT in mathematics programme |
|---|---|---|---|
| Anna | 1-1 iPad | Y 2 | Uses apps to support skill development |
| Sarah | BYOD | Y 5 | Very experienced, but uses apps to support skill development |
| Brad | 1-1 iPad | Y 5 & 6 | Integrates MT into many aspects of programme including coding. Uses screen-casting effectively |
| Jane | BYOD | Y 4 | Integrates MT into many aspects of programme. Uses screen-casting effectively |
| Alan | 1-1 iPad | Y 6 | Limited, uses apps to support skill development |
| Trish | 1-1 iPad | Y 5 & 6 | Integrates MT into many aspects of programme. Uses screen-casting effectively |
| Joy | 1-1 iPad | Y 1 | Limited, uses apps to support skill development |

In this chapter we present extracts of data from the study that were considered in relation to the emerging themes of the project. The following sections consider some theoretical perspectives of each theme, along with data from the research project that illustrate how engaging with the mathematics through the apps, might reshape the mathematical learning.

## Affordances

In relation to Gibson's (1977) notion of affordance as the complementarity of the learner and the environment, the affordances of MT, including visual, haptic and dynamic, may be seen to fashion the learning experience in distinctive ways. This gives opportunity to reposition students' engagement with mathematics. Affordances can be thought of as the potential opportunities and constraints in the relationship between the digital object and the user (Calder, 2011).

An affordance frequently associated with digital environments is the notion of multiple representations. The ability to link and simultaneously interact with visual, symbolic, and numerical representations in a dynamic way has been acknowledged extensively in research (e.g., Calder, 2011). In a similar way, various studies involving dynamic geometry software, report that the dynamic, visual representations enhanced mathematical understanding (e.g., Falcade, Laborde, & Mariotti, 2007). This dynamic affordance, coupled with the instant feedback to input, opens opportunity for reshaping the learning experience.

Virtual manipulatives (VM) are frequently part of mathematical apps. They are described as interactive, web-based visual representations of dynamic objects (Moyer, Bolyard, & Spikell, 2002) that might afford opportunities for mathematical thinking. VM offer potential to extend the learning experiences with representations beyond those with pencil-and-paper medium (Arcavi & Hadas, 2000). In *Math Shake*, for example, word problems are generated at various levels, and it provides a range of digital pedagogical tools (e.g., empty number lines, counters, ten frames), that students can select to help with their solutions.

Moyer-Packenham and Westenskow (2013) identified the affordances of focused constraint, creative variation, simultaneous linking, efficient precision, and motivation when students used apps in their mathematical learning. While these affordances interact, and appear to be mutually influential of each other, three of them resonate with the other emerging themes. These are: focused restraint, where the app might focus students' attention on particular mathematical concepts or processes; creative variation, where the app might encourage creativity, hence evoking a range of student approaches and potential solutions; and simultaneous linking, where the app might link representations simultaneously and connect them to student activity (Moyer-Packenham & Westenskow, 2013).

The interface of an iPad offers a further affordance through touch. Student interaction is more direct and tactile than when working on desktops, further enhancing the relatively high agency of the medium. There is direct interaction with the phenomena, rather than being mediated through a mouse or keyboard, making the iPad more suitable for young children than desktop computers (Sinclair & Heyd-Metzuyanim, 2014). Some apps make use of this haptic affordance (e.g., with *Multiplier*, where within the task, the student drags out the visual area matrix associated with multiplication facts). This app also evokes multi-touch functionality, enabling students to make sense of individual effects of particular screen touches (Hegedus, 2013; Jackiw, 2013), and to create personal explanations of their thinking (e.g., making a screencast of their problem solving strategy). This is similar to the simultaneous linking and creative variation that Moyer-Packenham and Westenskow (2013) identified (Fig. 1).

Much of the discussion regarding the ways iPads and apps might transform the learning experience is centred on the notion of student engagement (e.g., Attard, 2015, also see chapter in this book). Meanwhile other researchers have reported improved high-level reasoning and problem solving linked to learners' investigations in digital environments (Sandholtz, Ringstaff, & Dwyer, 1997). Many apps provide affordances of interactivity and non-threatening instantaneous feedback that foster the learner's willingness to experiment and take cognitive risks with their learning (Calder & Campbell, 2016). These types of apps allow students to model in a dynamic, reflective way. Students in the study took risks while using *Multiplier* by trying different arrays. They would try a number of possibilities, some of which were unconventional and sometimes incorrect, before settling on their preferred option.

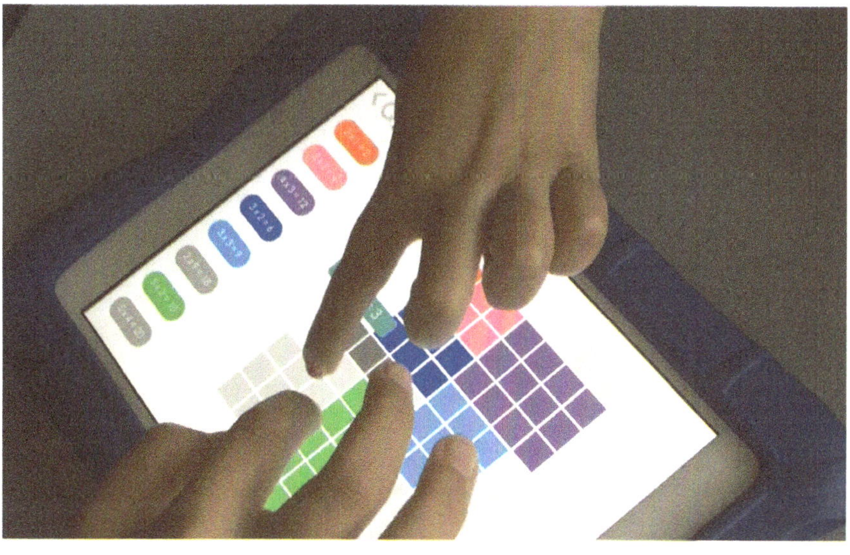

**Fig. 1** Image from video data illustrating the visual, haptic and interactive affordances of *Multiplier*

Apps such as *Explain Everything* and *Math Shake* allow students to record individual or group presentations of mathematical processes, strategies and solutions. The screen-casting feature of the app, and the simplicity with which it is enabled on an iPad, opens up other learning opportunities that would not be possible with pencil-and-paper technology. Such apps introduce a further multi-representational affordance whereby students can create an aural representation that other students can listen to.

One teacher, Anna, commented on the direct interface of the iPad screen, suggesting that the students were interacting more directly with the content of the mathematics—"Like a physical object that they're interacting with." This suggests the haptic affordance and focused restraint as the teacher perceived the app facilitating more direct interaction with mathematical content. Some students commented how the feedback and opportunity to record their solutions had helped their learning in mathematics:

> Jake: *We can write things down and answer questions to see if we are right or wrong.*

Sometimes this feedback was directly from the app, and at times it would be from other students, or the teacher, after they had viewed the screencasts in *Google classroom*. One student comment identified the range of digital tools, such as those that enabled screencasting, as being beneficial for learning.

> Josh: *The most helpful app for me is* Explain Everything *as it has lots of tools and options to help learning rather than doing it on paper with a pencil.*

Teacher comment likewise indicated that the features of the MT medium afforded particular teaching and learning opportunities:

> Sarah: *One to help me as a teacher... a teaching tool... to explain things or to use a learning object like an interactive number line or arrow cards or voice recordings as a teacher tool.*

> Brad: *They had to use an app called* Tickle *to program some robots to draw those same shapes on the map in real life – which was really cool because there's quite a little bit of shift in the mathematics thinking because you couldn't just use the internal angles, you had to convert from how much the robot has to turn using the internal angles as a reference point ... and the kids had to work out why that worked and what to do to get that to work.*

This also involved the students working together, trying ideas out in practice and negotiating possible solutions. Hence, the collaboration theme was identified.

## Collaboration

Simply put, collaborative learning may occur when two or more students are engaged in an activity and learning together (Dillenbourg, 1999). Such a perspective on learning in mathematics shifts from individual acquisition to participation in a social practice (e.g., Cobb & Bowers, 1999; Sfard, 1998). Educational engagement and collaboration associated with joint problem solving has been connected to academic success. For example, Mercer and Sams (2006) showed how collaboration with students engaged in an online task supported learning outcomes in mathematics. More recently Mercier and Higgins' (2013) study has shown how the collaborative use of digital technologies can support students in developing more flexible approaches to problem solving.

The ability of iPads to support collaboration would seem a key aspect of reshaping the learning of students in mathematics. The iPads potential for social computing has been acknowledged (della Cava, 2010), but this potential is still to be fully explored in the mathematics classroom. Zurita and Nussbaum (2004) noted how the flexibility of MT allows "students to engage in highly collaborative activities anywhere, at any time" (p. 293), and Fisher, Lucas, and Galstyan (2013) indicated how the portability and tactile interface of the iPad allows students to work both privately and publicly and to transition easily between the two.

In the same way that Fisher et al. (2013) noted the transition between private and public use, Looi et al. (2009) noted how the mobility of iPads allows students to not only make choices regarding where they are working, but also whom they may wish to work with. For example, there is the ability for a student to easily "...swivel and show..." another student what they are doing or share what they had recorded earlier (Looi et al., 2009), and hence share their thinking on the iPad with one or more of their fellow students. In this way the use of iPads encouraged incidental collaboration between the students in the classrooms we studied (see Fig. 2).

**Fig. 2** Students collaborating on volume models related to *Minecraft* activity

In the interviews, the teachers also reported that the flexible learning environment encouraged collaboration:

> Jane: *I was surprised at how much collaboration went on because they were allowed to sit anywhere they wanted ... they would just ask their neighbour something and then there was this little conversation.*

While this might occur in a learning situation without MT, having the MT as the medium for learning enhanced student collaboration in the classroom setting.

Zurita and Nussbaum (2004) described different collaboration areas including the use of MT devices to coordinate task activities, where the mobility of the device allowed students to move with the device and work with other students at different locations. The joint coordination of a task enables students to communicate and negotiate in order to support decision-making (Zurita & Nussbaum, 2004), and, as such, they are involved in "a coordinated joint commitment to a shared goal" (Mercer & Littleton, 2007, p. 23). Teachers used iPads directly for joint collaborative activities with their students, for example when students worked together to create a screen cast together.

Alan set a collaborative task in relation to students' calculation strategies and noted:

> *That's another thing we did – we sent them off in groups to work on a strategy – they each used a different strategy, video recorded their thinking, came back together, argued about which strategy was the best by watching the videos and then decided.*

Here, the screencasting feature of the app appeared to better facilitate the ease with which the students could video their explanations, review them as a group, and then debate the merits of each before collaborating to decide the content of their group's final screencast. Two groups of students working on a problem using *Minecraft*, on a single iPad, suggested two aspects of collaboration. The first related to the contestation of ideas and processes:

| | |
|---|---|
| Aaron | *Okay, 5 lots of 5 blocks* |
| Zac | *Yep, 5 blocks* |
| Don | *Shall we use a line?* (He indicated where the 5 blocks might go on the screen) |
| Zac | *No, not 5 blocks up!* |
| Aaron | *Yes, you need to use it there* |
| Don | *Yeah, there* |
| Zac | *Is it? No this one* (pointed to the screen) |
| Aaron | *You need the 5 blocks across and going up* (indicated on the screen) |
| Zac | *Oh yeah, yeah now I see.* |

Here, Zac's understanding of the solution and the process changes through the discussion related to the group's direct interaction with the iPad screen. It was the visual tension evoked from touching the screen, and the immediate impact from that action, that initiated the dialogue and also enabled, in conjunction with the dialogue, the transition in Zac's understanding. The second excerpt relates to the sharing of knowledge and ideas:

| | |
|---|---|
| Ali | *You can fly too, you realize* (demonstrates by touching the arrows on the screen). *Double click the jump button.* (Whetu double clicks the button) |
| Ali | *That isn't the jump button!* (Ali demonstrates the button and how to fly again. Whetu takes over) |
| Whetu | *I can fly, fly high in the sky!* |
| Ali | *So you can control your flight with that one* (Whetu takes over the arrow controls) |
| Whetu | *Going up! Weee! Now I want to go down now.* |
| Ali | *Use the other one then* (points to the screen). (Whetu changes buttons and brings the "flight" down). |

Ali peer shares her knowledge of the app (how to fly in that particular digital environment) and then peer teaches Whetu so that her understanding of the process and potential of the app is enhanced. With both excerpts, the particular visual, interactive affordances of the app, coupled with the instantaneous visual feedback to their input, influenced the dialogue and the interaction with the task. This enabled the collaboration and learning process to evolve in a manner that is distinctive from pencil-and-paper approaches.

Mercer and Littleton's (2007) definition of collaborative learning goes beyond the sharing of ideas and task coordination to "reciprocity, mutuality and the continual (re)negotiation of meaning" (p. 23). Collaborative learning in line with this definition was identified by one student, Tui, in commenting on how the MT facilitated collaboration through the utilization of individual understandings and expertise:

> Tui: *... we can work through it together because I might be smarter with the device and I can help you with the device but you can help me with my maths so when we ... we can go and work together and solve things.*

Hence, it was noted that apps enabled collaborative approaches to learning when a MT was being used in a jointly coordinated shared task, as well as incidentally when students were working individually and informal opportunities to share arose. In both instances, the use of apps initiated discussion, with the potential to renegotiate thinking, which in turn initiated further engagement with the MT (Fig. 3). In this way the learning through apps took place within interconnected groupings of digital elements and the social aspects that they evoked. We have identified this relationship as socio-material assemblages.

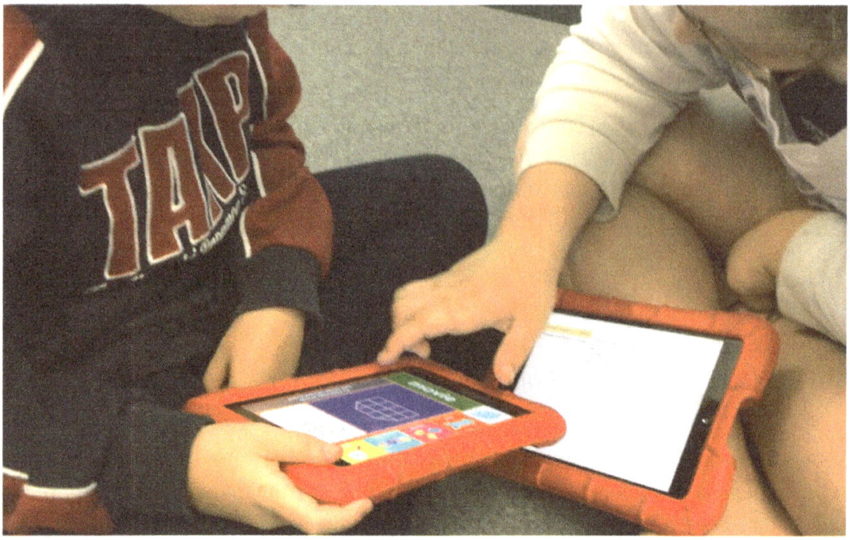

**Fig. 3** Incidental collaboration on an area task using *Brainpop*

## Socio-material Assemblages

It has been suggested that MT offer a socio-material bricolage for learning (Meyer, 2015). Drawing on Fenwick and Edwards' (2012) notion of socio-material approaches to learning, Meyer envisaged interconnected systems where resources interact with knowledge that is socially distributed. A range of people, communities and sites of practice might be influential in assisting student learning. Meyer (2015) used the term socio-material bricolage to describe the "ecological entanglement of material and social aspects of teaching and learning with technology" (p. 28).

The notion of bricolage suggests that there is a mutually influential collective of tools and users affecting the dialogue, learning experience, and mathematical thinking, in particular and personalised ways. For example, when students collaborate on a task, they incorporate input from the wider class, school and 'home' communities, while also drawing from the broader underlying discourses, such as political or socio-cultural elements that influence their pre-conceptions about the task and mathematical activity. De Freitas and Sinclair (2014) discussed 'thought' as being distributed across both social and physical environs and influencers. We consider that thought evolves in a complex material and social milieu. When screencasting their strategy and solution(s) the students might incorporate a range of digital, visual, and concrete material resources in mutually interdependent ways. All of this activity has associated social elements, both immediate interaction as well as the drawing forth of the underlying discourses. The resulting process is not just the accumulation of the various 'bits', but also a new mesh of the social and material elements.

Johri (2011) argues that in learning, the education participants often make do with the tools available to them, with what is at hand, rather than following planned approaches with tools not immediately available. He contends that a socio-material bricolage supports the interwoven social and material nature of human practices. Sandholtz et al. (1997) indicated that affordances of digital technologies, together with the associated dialogue and social interaction, may lead to students exploring powerful ideas in mathematics, learning to pose problems, and creating explanations of their own. Various aspects overlap and interlace. Students are seen to have a choice in how they imbricate their perceptions with the material, related to the features of the iPad and the app, within the context of a mathematical problem. In turn, the material has the potential to influence the imbrication. The data were relatively cohesive, in terms of being influential in the learning process, regarding the connection between the use of the apps, other technologies such as concrete materials, and the dialogue and social interaction that engagement with them evoked (Meyer, 2015). For instance, Trish commented:

> *They used the iPad to watch a video and they took a brainstorm on a piece of paper about what a triangle is and different types of triangles – what internal angles are and external angles and things like that and then we... the kids used that information to create some triangles.*

Here, we see the use of different technologies (including paper) yet it is the interconnected, mutually influential social elements, such as, brainstorming and using the information to create, which become part of the socio-material assemblage. For example, students were observed using the iPad to investigate a problem in context, then using counters and rods, all the time interacting with each other and the range of tools. They used an empty number line in the app and a white table for story boarding the screencast of their strategy and solution. This was then loaded into a *Google classroom* site that the teacher could access for review and feedback.

One teacher, Brad, saw this tapestry of material and social elements as an ecosystem:

> Brad discussing *Hopscotch*: *There's a really big app eco system – I don't think there's many other devices that you can program on the iPad and then program robots and record your voice and make videos and all that stuff – it's a very rich ecosystem.*

There were also instances where concrete materials were used in conjunction with apps. For example, Joy talks mainly of an assemblage of material elements, with the associated social aspects, including the relationships and interaction between students, teacher, school community and the broader societal discourses inherent in the activity described:

> *You might do something with those Cuisenaire rods... those plastic things... there's also an app that would do it as a lesson and then there's an app that actually has the rods in it so the kids can go away and practice moving them around the screen after they've done it with you physically... so there's a nice connection.*

The students recognized the same potential for using a mixture of technologies at the appropriate time for their learning:

**Fig. 4** Students solving a number problem in pairs using a mixture of writing and digital activity

> Tim: *Sometimes I make a plan (on paper) to work out my word problem, then I can put the pictures on and record my answer on the iPad.*

> Whitu: *So sometimes things are better to work out on paper, but other things are better on my iPad.*

These student blog data were examples of the students integrating different technologies to best investigate and solve a problem (Fig. 4).

In our observations we saw that students moved relatively seamlessly between pencil-and-paper and digital technology and utilized the type that they found best facilitated the learning process for them. This indicates that they chose the technology, and the way that they used it, to suit their individual or group requirements—a form of personalisation. The next section considers this theme.

## Personalisation/Differentiation

Current perspectives on, and manifestations of, personalisation vary markedly, often in conflicting ways. Some advocate that student choice is paramount, while in other perspectives personalisation is something the teacher directs, with no student input. An and Reigeluth (2011) note that personalised learning involves teachers paying close attention to individual student's knowledge and skills and using this knowledge to provide personalised experiences and support in learning. Contrastingly, Leadbetter (2005) contends that personalised learning is being focused on motivating students to become engaged in their own learning by allowing them to make personal choices about it. In this view the teacher's aim would be to create an environment where students are empowered to make decisions.

Waldrip et al. (2014) note that personalised learning relates to structured activities that students engage in with scaffolding from their teachers such as "… modeling, guidance in goal-setting and timely feedback" (p. 357). Using Tomlinson's (2009) model of student variance as interest, readiness, and learning profile, the use of apps such as *Explain Everything* can be examined in relation to socio-material assemblages. The entanglement of the social (students' interests, readiness and learning profile) and the material (hardware and software) suggests evidence of reconfiguration related to human and material agency. Interestingly, Tomlinson (2009) used the term differentiation to describe this sort of concept and concluded that teaching with an emphasis on student variance and choice should elicit conceptual understanding. The use of apps such as *Explain Everything* can be examined while also enhancing student efficacy and ownership of learning. Others contend that MT can provide new forms of personal ownership (e.g., Meyer, 2015) that in turn supports learners' personal understanding and conceptual frames (Melhuish & Falloon, 2010). Whatever the definition of personalised learning, a key tenet is that students choose the tools and the contexts for the task; for instance, students' personal use of images and recordings when using the *Explain Everything* app. iPads, as a type of mobile technology, have been identified as having the potential to enhance personalised learning due to two key characteristics, the mobility of the devices, and their ability to continually change contexts (Looi et al., 2013).

The feature of mobility suggests that iPads can be used anytime and anywhere. When used in education this serves to change the definition of what is considered a learning space. Learning is no longer in a particular place or time, but can be anytime and anywhere (Melhuish & Falloon, 2010). This mobility extends beyond the classroom as the iPads can be used seamlessly, between school, home and further afield such as on field trips (Calder & Campbell, 2016) (Fig. 5).

**Fig. 5** Students in the study used a variety of work spaces

Another way in which different contexts are created is through the ability of apps to shape experiences to meet specific needs. There are specific apps (e.g., *MathsBlaster*) designed to meet various learning stages and steps, meaning that the selection of apps can be personalised to meet a range of students' needs (Clark & Luckin, 2013). Furthermore, some apps provide specialized features that enable specific learning and instruction within the app (e.g., *Brainpop*). This means that not only can the choice of apps be personalised, but also what happens within them can be modified, enabling interaction with the apps to be personalised to a student's specific learning needs, perhaps through the type or level of question (Calder & Campbell, 2016; O'Malley et al., 2013). Teacher-initiated personalisation resonates with Cutler, Waine and Brehony's (2007) contention that personalisation is about the raising of achievement. In the classroom situation, the teacher's ability to choose apps, and levels and tasks within apps to suit specific learning needs, has facilitated differentiation of the learning for specific students (Clark & Luckin, 2013; O'Malley et al., 2013).

Changing contexts can also be associated with personalising the features of the iPad working environment, such as the font and colour in their presentations (Robinson & Sebba, 2010). However, within this customization, there are concerns with Looi et al., (2009) noting that the endless customization features of mobile technologies led a student in their study to spend an excessive amount of time on the aesthetic features, rather than focusing on the intended learning. A blog post from a student indicated the impact of customizing features:

> Ella: *You can make math more interesting by changing the colour, font and size, and you can use pictures from the internet.*

Another identified the features in *Explain Everything* as being both motivational and helpful for the learning:

> Kate: *I use* Explain Everything *with my Thinking Boards. I use the voice recorder, the drawing pen, different colours, I can pick the size of my pictures, duplicate things. I can move things to show my thinking.*

A sense of ownership and individuality can also be expressed in the images the students used as screen-savers on their iPads, which might lead to an emotional attachment. For instance, one boy commented when asked to leave his iPad in the classroom "Goodbye my darling," as he hugged it goodbye! Importantly, the apps can facilitate the differentiation of the learning associated with cognitive understanding, either for individuals or for groups, sometimes linked to accuracy and speed. Student blog data was illustrative of this:

> Ethan: *I sometimes do* Skoolbo *and I have* Maths Sums *where I get to choose what sum you wanna do and you get to choose the level, and you have to unlock the level, and you get 20 seconds to answer the question.*

> Julie: Math Shake *is a great learning tool because it can help you with your problem solving. So you can choose a level for you, so just say you were genius or easy or confident or even beginner, there are a lot of levels to choose from. And there are also some amazing*

*tools to help you solve your word problem for instance number lines, fractions, counters, and there is also different coloured pencils that you have to earn.*

Teresa: *In* Money Mind NZ, *I like to go shopping and I have to work out how much I have to pay (I like choosing my items to buy), and I like getting it right.*

Teacher comment also indicated that in teacher directed differentiation apps could be selected based on the basis of their suitability for particular levels of learning. The teacher could shape the learning experience based on their knowledge of the students, including their conceptual and technological understanding (Fig. 6).

Two teacher comments were particularly indicative of this, with the first related to teaching a group of high achieving mathematics students and the second teaching a group needing more support with their mathematics learning:

Brad: *An extension app that I love to use which is a web app is called* Lure of the Labyrinth *which has been really good at high end critical thinking and things like... the kids, we were converting like between base 10 numbers... base the total, like base 6, 7, 8 and 9 numbers and binary to solve puzzles.*

Trish: *They've made stop motion animations on polygons – this is the lower group – like what is a hexagon, what is an octagon, what is a triangle and they use little stop motion animation and match sticks to make those shapes and animate them then talk about them.*

While there is some fluidity in use and meaning associated with personalisation and differentiation of the learning experience, in this chapter we have considered

**Fig. 6** Solving a problem using a personal workspace and images

two versions in particular. One that involves teachers paying close attention to individual student's knowledge and skills and using this knowledge to provide personalised experiences and support in learning (An & Reigeluth, 2011). Meanwhile, the second advocates that the teacher focus on motivating students to become engaged in their own learning by allowing them to make personal choices about it (Leadbetter, 2005). There are areas of convergence and contrast in these two perspectives, with a key element of both being the teacher trying to optimize the students' engagement with, and understanding of, the mathematics. The intention is to differentiate the learning experience to best facilitate mathematical learning for the individual or group of students.

Having a classroom culture and mathematical activities that promote individual student choice is intended to engage and motivate the students through a sense of ownership of the learning so that they might be more receptive to the mathematics learning. These two perspectives are not distinct, however. They may operate in tandem, and there is a continuum of the possible inter-relatedness of both in the learning experience. In a similar way, the themes identified in this research can overlap and be mutually influential in the mathematics learning. The next section draws together the four themes to consider the ways that they might facilitate the mathematics learning.

## The Weaving of Themes

Although the four themes are different and influence the mathematical learning in varying ways, they are not discrete or necessarily independent. While the personalisation of the working environment seemed to motivate the students, the affordances of the apps coupled with the pedagogical approach and culture of the classroom, appeared to be influential in personalising the learning experience, and for differentiating the individual learning needs and preferences. Jane's comment is indicative of other teacher comments:

> *Students were asked to explore a mathematics strategy with a buddy, creating a video in* Explain Everything *to explain how their selected strategy worked, and what it was good for. Students were free to select any strategy they liked, and engage in their learning how they wanted. They were observed exploring various strategies, such as equal addition and reversibility, with their methods and recording occurring in a multitude of ways, such as through the use of whiteboards, calculators and discussions.*

This excerpt of data also suggests other themes and indicates their inter-connectedness. The students use the affordance of the MT, they collaborate, and there is reference to the use of a multiple of ways, including whiteboards, calculators and discussion—a socio-material assemblage.

Brad's comment below focused on developing the cognitive understanding of triangles and exploring the relationships in their properties. It identified the use of the affordances of the apps, collaboration (both between students and with the teacher), and socio-technical assemblages with an app, a concrete resource (*Sphero*) and social aspects being integrated. Evident is the student choice and potential for differentiated learning, while the key focus throughout is the students' conceptual understanding of triangles:

> *The app called* Tickle *was used whilst trying to program the* Spheros *(little robotic balls) to move in triangles for our project. This helped by showing us the way triangles were made, and improved our patience when the programming didn't work.* Tickle *is the actual app used for the programming.* Hopscotch, *is another programming app, but used to program a virtual character of your choice. It is the same, but it is different to use, different commands. This is helping by helping us discover the degrees and angles of the triangles.*

## Concluding Comments

The characteristics of learning mathematics through MT, including apps are important. There are some that better facilitate individual mathematical thinking and understanding, while others reshape the nature of the learning experience through the affordances, such as dynamic, visual and haptic experiences, not easily obtained with other pedagogical media. Others offer non-threatening, instantaneous feedback that enable better opportunity for investigative approaches and differentiation of the learning. However, the research project on which this chapter is based suggests that it is more than the qualities of the app that are influential in optimizing the learning opportunities. The themes and the associated data were relatively coherent that the expertise and experience of the teacher, manifested through their technological pedagogical and content knowledge, were vital elements of the learning process.

Other understandings that are beginning to emerge for the project are the importance of pedagogy over app quality. This is in relation to student engagement and learning. Another finding is the ways that apps and other technologies (e.g., equipment and concrete materials) can be integrated effectively, with the transition of students between them, seeming to help build relational understanding of mathematical concepts. A key finding identified by the extended research group relates to the ways that these groupings of technologies become part of socio-material assemblages through evoking social engagement and dialogue. While these findings need further research, and their connection to student mathematical understanding better understood and articulated, they nevertheless indicate the potential of MT to transform the mathematics experience. This, in turn, will enhance both the engagement and mathematical thinking of primary and secondary school students.

## References

Attard, C. (2015). Introducing iPads into primary mathematics classrooms: Teachers' experiences and pedagogies. In M. Meletiou-Mavrotheris, K. Mavrou, & E. Paparistodemou (Eds.), *Integrating touch enabled and mobile devices into contemporary mathematics education* (pp. 197–217). Hershey, PA: IGI Global.

An, Y.-J., & Reigeluth, C. (2011). Creating technology-enhanced, learner-centered classrooms: K-12 teachers' beliefs, perceptions, barriers, and support needs. *Journal of Digital Learning in Teacher Education, 28*(2), 54–62.

Arcavi, A., & Hadas, N. (2000). Computer mediated learning: An example of an approach. *International Journal for Computer for Mathematics Learning, 5,* 25–45.

Calder, N. S. (2011). *Processing mathematics through digital technologies: The primary years.* Rotterdam, The Netherlands: Sense.

Calder, N. S., & Campbell, A. (2016). Using mathematical apps with reluctant learners. *Digital Experiences in Mathematics Education.* https://doi.org/10.1007/s40751-016-0011-y.

Carr, J. (2012). Does math achievement "h'APP'en" when iPads and game-based learning are incorporated into fifth-grade mathematics instruction? *Journal of Information Technology Education, 11,* 269–286.

Clark, W., & Luckin, R. (2013). *What the research says: iPads in the classroom.* University of London, UK: Institute of Education.

Cobb, P., & Bowers, J. S. (1999). Cognitive and situated learning perspectives in theory and practice. *Educational Researcher, 28*(2), 4–15.

Cutler, T., Waine, B., & Brehony, K. (2007). A new epoch of individualization? Problems with the "personalization" of public sector services. *Public Administration, 85*(3), 847–855.

della Cava, M. R. (2010). Does the iPad have the magic to bring people together? *USA Today.* Retrieved from http://www.usatoday.com/life/lifestyle/2010-06-07-ipadculture07_CV_N.htm.

De Freitas, E. & Sinclair, N. (2014). *Mathematics and the body: Material entanglements in the classroom.* New York, NY: Cambridge University Press.

Dillenbourg, P. (1999) *Collaborative learning: Cognitive and computational approaches.* Advances in learning and instruction series. New York, NY: Elsevier Science, Inc.

Falcade, R., Laborde, C., & Mariotti, M. A. (2007). Approaching functions: Cabri tools as instruments of semiotic mediation. *Educational Studies in Mathematics, 66*(3), 317–333.

Fenwick, T., & Edwards, R. (Eds.). (2012). *Researching education through actor network theory.* Sussex: Wiley-Blackwell.

Fisher, B., Lucas, T., & Galstyan, A. (2013). The role of iPads in constructing collaborative learning spaces. *Technology, Knowledge, and Learning, 18*(3), 165–178.

Gibson, J. J. (1977). The theory of affordances. In R. Shaw & J. Bransford (Eds.), *Perceiving, Acting, and Knowing: Toward an Ecological Psychology* (pp. 67–82). Hillsdale, NJ: Lawrence Erlbaum.

Groundwater-Smith, S., Mitchell, J., Mockler, N., Ponte, P., & Ronnerman, K. (2013). *Facilitating practitioner research.* London, England: Routledge.

Hegedus, S. (2013). Young children investigating advanced mathematical concepts with haptic technologies: Future design perspectives. *The Mathematics Educator, 10*(1&2), 87–107.

Hennessy, S. (2014). *Bridging between research and practice: Supporting professional development through collaborative studies of classroom teaching with technology*. Rotterdam, The Netherlands: Sense.

Jackiw, N. (2013). Touch & multitouch in dynamic geometry: Sketchpad explorer and "digital" mathematics, In *Proceedings of ICTMT12* (pp. 149–155), Bari, Italy.

Johri, A. (2011). The socio-materiality of learning practices and implications for the field of learning technology. *Research in Learning Technology, 19*(3), 207–217.

Larkin, K. (2015). "An App! An App! My Kingdom for An App": An 18-month quest to determine whether apps support mathematical knowledge building. In T. Lowrie & R. Jorgensen (Eds.), *Digital Games and Mathematics Learning: Potential, Promises and Pitfalls*. (Vol. 4, pp. 251–276). The Netherlands: Springer.

Leadbetter, C. (2005). The shape of things to come: Personalised learning through collaboration. In *Department for education and skills*. Retrieved from http://www.standards.dfes.go.uk/innovation-uni.

Looi, C.-K., Wong, L.-H., So, H.-J., Seow, P., Toh, Y., Chen, W., ... Soloway, E. (2009). Anatomy of a mobilized lesson: Learning my way. *Computers & Education, 53*(4), 1120–1132. http://doi.org/10.1016/j.compedu.2009.05.021.

Looi, C.-K., Wong, L.-H., & Song, Y. (2013). Mobile computer-supported collaborative learning. In C. E. Hmelo-Silver, C. A. Chinn, & C. Chan (Eds.), *The international handbook of collaborative learning* (pp. 420–436). Florence, KY, USA: Routledge.

Melhuish, K., & Falloon, G. (2010). Looking to the future: M-learning with the iPad. *Computers in New Zealand Schools: Learning, Leading. Technology, 22*(3), 1–15.

Mercer, N., & Littleton, K. (2007). *Dialogue and the development of children's thinking: A sociocultural approach*. London, UK: Routledge.

Mercer, N., & Sams, C. (2006). Teaching children how to use language to solve mathematics problems. *Language and Education, 20*(6), 507–528.

Mercier, E., & Higgins, S. (2013). Collaborative learning with multi-touch technology: Developing adaptive expertise. *Learning and Instruction, 25,* 13–23.

Meyer, B. (2015). iPads in inclusive classrooms: Ecologies of learning. In P. Isaias, J. M. Spector, & D. Ifenthaler (Eds.), *E-learning systems, environments and approaches theory and implementation*. Dordrecht: Springer International Publishing.

Moyer, P. S., Bolyard, J. J., & Spikell, M. A. (2002). What are virtual manipulatives? *Teaching Children Mathematics, 8*(6), 372–377.

Moyer-Packenham, P. S., & Westenskow, A. (2013). Effects of virtual manipulatives on student achievement and mathematics learning. *International Journal of Virtual and Personal Learning Environments, 4*(3), 35–50.

O'Malley, P., Jenkins, S., Wesley, B., Donehower, C., Rabuck, D., & Lewis, M. E. B. (2013). Effectiveness of using iPads to build math fluency. In *Presented at the Council for Effectiveness of using iPads to build math fluency, San Antonio. Texas, USA*. Retrieved from http://eric.ed.gov/?id=ED541158.

Philip, T. M., & Garcia, A. (2014). Schooling mobile phones: Assumptions about proximal benefits, the challenges of shifting meanings, and the politics of teaching. *Educational Policy, 29*(4), 676–707.

Robinson, C., & Sebba, J. (2010). Personalising learning through the use of technology. *Computers & Education, 54*(3), 767–775.

Sandholtz, J. H., Ringstaff, C., & Dwyer, D. C. (1997). *Teaching with technology: Creating a student centred classroom*. New York: Teachers' College Press.

Sfard, A. (1998). On two metaphors for learning and the dangers of choosing just one. *Educational Researcher, 27*(2), 4–13.

Sinclair, N., & Heyd-Metzuyanim, E. (2014). Learning number with *TouchCounts:* The role of emotions and the body in mathematical communication. *Technology, Knowledge and Learning, 19*(1), 81–99.
Tomlinson, C. (2009). The Goals of Differentiation. In M. Scherer (Ed.), *Supporting the whole child: Reflections on best practices in learning, teaching, and leadership* (pp. 3–11). Alexandria, Virginia, USA: ASCD.
Waldrip, B., Cox, P., Deed, C., Dorman, J., Edwards, D., Farrelly, C., … Yager, Z. (2014). Student perceptions of personalised learning: development and validation of a questionnaire with regional secondary students. *Learning Environments Research, 17*(3), 355–370.
Zurita, G., & Nussbaum, M. (2004). Computer supported collaborative learning using wirelessly interconnected handheld computers. *Computers & Education, 42*, 289–314.

**Dr. Nigel Calder** is an Associate Professor at the University of Waikato, Tauranga. He has taught in both primary and secondary schools. His research interests are predominantly in the use of digital technologies in mathematics education. His present and recent research projects are: The influence of apps on primary-school mathematics teaching and learning; The influence of mobile technologies on the engagement of 16–18 year-old reluctant learners; Connecting science and mathematics instruction through exploratory talk and inquiry; and How mathematical thinking emerges through student-centred inquiry learning in secondary schools. He wrote the book: *Processing Mathematics Through Digital Technologies: ThePrimaryYears* and has published numerous resources (including Figure It Out books), book chapters and journal articles.

**Carol Murphy** previously at the University of Waikato, is currently a Senior Lecturer in Mathematics Education at the University of Tasmania, Australia. Her interests in student learning in mathematics stemmed from her background as a primary school teacher in England. Now, as a primarymathematics education lecturer and researcher, her interest into children's mathematical thinking has led her to focus on the teaching of calculation strategies as well as on collaboration and use of talk. Carol has also written on teacher subject knowledge and changes in teacher practice. These research interests have informed her recent investigations into teachers' practices and student learning in the use of iPads in New Zealand primary mathematics classrooms.

# Mobile Technologies in the Primary Mathematics Classroom: Engaging or Not?

**Catherine Attard**

**Abstract** Many schools invest in mobile technologies or actively promote their use through Bring Your Own Device (BYOD) programs with the expectation that the use of such devices will improve student engagement and, as a result, improve student learning outcomes. However, there is little research to date that explores teacher and student perceptions of whether and how the use of mobile technologies within mathematics classrooms does indeed improve engagement with mathematics. This chapter draws on data from a small range of research projects investigating the use of mobile technologies and associated applications in the primary mathematics classroom. It uses a multidimensional view of engagement and the Framework for Engagement with Mathematics as a lens to re-analyse existing and new data. Issues relating to engagement and the use of mobile technologies will be explored within the context of classrooms where students and many of their teachers are now considered to be 'digital natives', and Information and Communication Technologies are an integral and ubiquitous part of their daily lives.

**Keywords** Student engagement · Mobile devices · iPads · Mathematics Primary

## Introduction

Over the past decade, the range of mobile devices in contemporary classrooms have fast become the standard learning tools, replacing the once prominent desktop computer (Meletiou-Mavrotheris, Mavrou, & Paparistodemou, 2015). Mobile technologies are "ubiquitous in nature, highly portable and endowed with multimedia capabilities offering a new dimension to curriculum, making learning

C. Attard (✉)
Centre for Educational Research, Western Sydney University, Sydney, Australia
e-mail: C.Attard@westernsydney.edu.au



N. Calder et al. (eds.), *Using Mobile Technologies in the Teaching and Learning of Mathematics*, Mathematics Education in the Digital Era 12,
https://doi.org/10.1007/978-3-319-90179-4_4

accessible 'anywhere, anytime'" (Handal, Campbell, Cavanagh, & Petocz 2016, p. 200). An illustration of this phenomenon is evidenced in the increase in popularity of computer tablets. The Apple iPad, for example, quickly became one of the most popular devices in schools when it was released in 2011 alongside smartphones and other mobile devices.

The relatively low cost and the vast range of affordances offered by mobile devices has made them attractive to schools. Devices are often purchased in the hope of improving student engagement, leading to improved learning outcomes. This is based upon the assumption that when students are deeply engaged with tasks, they are more likely to develop positive attitudes. Positive attitudes are more likely to promote learning. Expectations that students would be more engaged when using such technologies, and eventually resulting in improved learning outcomes are reflected widely in literature (e.g., Beavis, Muspratt, & Thompson, 2015; Bray & Tangney, 2015; Ke, 2008, as cited in Chang, Evans, Kim, Norton, & Samur, 2015; Pierce & Ball, 2009). There is also an emerging body of literature from studies that have shown students are, in fact, more engaged with mathematics as a result of experiencing the use of mobile technologies in their mathematics lessons (Attard & Curry, 2012; Bray & Tangney 2015; Hilton, 2016; Ingram, Williamson-Leadley, & Pratt, 2016: Muir & Geiger, 2016).

Although we have emerging evidence that students are more engaged, there is a gap in the literature that explores student engagement on a deeper level to investigate what specific aspects and uses of mobile devices do, in fact, improve student engagement. Is it the device itself, the use of specific types of applications (apps), or is it the pedagogical practices of the teacher? Are students engaged simply because of the novelty of devices or activities that are introduced?

This chapter explores more deeply what it is to be engaged with technology within the primary mathematics classroom using the Framework for Engagement with Mathematics (FEM) (Attard, 2014), as a theoretical lens. First, a definition of engagement will be provided before the FEM is introduced. Next, to contextualise the discussion a brief exploration of literature pertaining to how mobile technology is being used in mathematics classrooms is provided. Examples of mobile technology use from three studies conducted in Australian primary classrooms will then be aligned to the FEM using voices from the classroom to explore how mobile technologies may or may not lead to engagement with mathematics.

## Theorising Engagement

Within an educational context, the construct of engagement can be characterised as meaningful participation in a context where knowledge and learning are valued and used. An important element of this level of engagement is the maintenance of interpersonal relationships and identities within the classroom community, in addition to positive interactions within the environments in which the individual has significant personal investment (Hickey, 2003). Consistent with this socio-cultural view of engagement is the definition provided by Fredricks, Blumenfeld and Paris (2004), who define engagement as a deeper student relationship with classroom work, multi-faceted and operating at cognitive, affective and behavioural levels. It is argued that viewing engagement as the combination of behaviour, emotion and cognition provides a characterisation of children that is much more valuable than researching individual components. It is this view that informs the multidimensional view of engagement for this chapter. Engagement is the coming together of cognitive, operative, and affective facets (Fair Go Team NSW Department of Education and Training, 2006; Munns & Martins, 2005), leading to students valuing and enjoying school mathematics, and seeing connections between school mathematics and their own lives.

In this definition, engagement includes individual thoughts that are projected outwards in terms of a person's investment and effort towards learning, as well as those relational behaviours that occur within the mathematics classroom (Attard, 2014). This definition forms the theoretical foundation for the FEM (Fig. 1), introduced by Attard (2014) as a tool devised to assist teachers in planning engaging learning experiences in mathematics. The FEM is used in this chapter as a

**FRAMEWORK FOR ENGAGEMENT WITH MATHEMATICS**

**In an engaging mathematics classroom, positive pedagogical relationships exist where:**

- students' **backgrounds** and pre-existing knowledge are acknowledged and contribute to the learning of others;
- the teacher is **aware** of each student's mathematical abilities and learning needs;
- **interaction** amongst students and between teacher and students is continuous;
- the teacher models **enthusiasm** and an enjoyment of mathematics and has a strong **pedagogical content knowledge; and**
- **feedback** to students is constructive, purposeful and timely.

**In an engaging mathematics classroom, engaging pedagogical repertoires mean:**

- there is **substantive conversation** about mathematical concepts and their applications to life;
- tasks are positive, provide opportunity for all students to achieve a level of **success** and are **challenging** for all;
- students are provided an element of **choice;**
- technology is embedded and used to enhance mathematical understanding through a **student-centred** approach to learning;
- the **relevance** of the mathematics curriculum is explicitly linked to students' lives outside the classroom and empowers students with the capacity to transform and reform their lives; and
- mathematics lessons regularly include a **variety** of tasks that cater to the diverse needs of learners.

**Students are engaged with mathematics when:**

- mathematics is a subject they enjoy learning,
- they value mathematics learning and see its relevance in their current and future lives, and
- they see connections between the mathematics learnt at school and the mathematics used beyond the classroom.

**Fig. 1** The Framework for engagement with mathematics (Attard, 2014) lens to assist in determining how the use of mobile technologies assists in increasing (or decreasing) students' engagement with mathematics.

The framework was derived as part of a qualitative, longitudinal study of the influences on student engagement during the middle years of schooling (Grades 5–8 in Australia) (Attard, 2014). The FEM emerged as an outcome of the research and an ongoing review of literature. Importantly, the framework takes student voice seriously in its consideration of what engages learners in mathematics classrooms, hence student voice features in this chapter. Although this study overwhelmingly indicated the teacher was the strongest influence on these students' engagement, this influence is complex, consisting of two separate, yet inter-related elements: pedagogical relationships and pedagogical repertoires. For the purpose of the FEM, pedagogical relationships refer to the interpersonal teaching and learning relationships between teachers and students that optimise the learning of and engagement with mathematics. Pedagogical repertoires refer to the day-to-day teaching practices employed by the teacher.

Pedagogical repertoires, as referred to in the FEM, assume aspects of more traditionally recognised frameworks and constructs such as Shulman's pedagogical content knowledge (1986), technological pedagogical content knowledge (Koehler & Mishra, 2009) and mathematical knowledge for teaching as described by Hill, Ball and Schilling (2008). However, they also encompass non-content specific practices that have been found to directly influence student engagement with mathematics such as opportunities for substantive conversation, provision of choice, and task variety.

It is suggested that it is difficult for students to engage with mathematics without a foundation of strong pedagogical relationships. It can also be argued that it is through engaging pedagogies such as the effective use of mobile technologies, that positive pedagogical relationships are developed, highlighting the connections between pedagogical relationships and engaging pedagogical repertoires. Although technology is specifically mentioned in only one statement within the FEM, in this chapter the framework is applied to the way technology is embedded in teaching and learning: the pedagogical relationships that inform the use of technology, and the pedagogical repertoires that embed technology.

Just as there are a range of influences on student engagement with mathematics through the use of mobile technologies, there are also a range of influences on how teachers use or envisage the use of these technologies.

## How Are Teachers Using Mobile Technologies in Primary Mathematics Classrooms?

One of the biggest benefits of mobile technologies is the wide range of affordances available to teachers and students. One mobile device can give a student a tool to access all the processes and contents from the mathematics curriculum, through tasks that range from low level, fluency building activities to tasks that require the student to analyse, evaluate and create. Unlike a traditional text book, mobile devices provide opportunities for interaction in many different formats such as individual or collaborative gamification of activities. They also allow for various software applications to be used simultaneously. One example of this is the ability to capture audio and written work, allowing students to show and explain the solution to a mathematical problem. Another example, is the dynamic and instantaneous response to input the mobile technologies afford (Calder & Campbell, 2016). The range of affordances has resulted in a wide variety of ways that these technologies are being implemented in primary mathematics classrooms.

Arguably a one device per student (1:1) program is the ultimate resourcing goal of many schools. However, this is not the reality in many classrooms. Although mobile devices are more cost effective than the traditional desktop computers, few schools can afford to purchase one device per student, and many are reluctant to enforce a mandate that all students purchase a specific device. To combat the financial burden, schools are beginning to change direction in relation to the purchase of devices, and rather than providing or prescribing a specific device for students, 'Bring Your Own Device' (BYOD) programs, where a range of devices and platforms are used and students provide their own devices, are gaining momentum (Cristol & Gimbert, 2014; Hu & Garimella, 2014).

This change brings about further significant challenges for teachers, particularly in the area of mathematics. Many teachers find it challenging to design technology integrated tasks that move beyond requiring students to act as consumers, through the use of drill and practice (apps) that build fluency, to producing, authoring and problem solving through the use of more generic productivity apps (Attard, 2013; Zhang, Trussell, Gallegos, & Asam, 2015). Issues beginning to emerge in classrooms relate to the range of devices and operating platforms being used at any one time, and often, the disparity that develops when the number of devices brought to school is inconsistent from one day to the next. However, allowing students to bring their own devices to school does have the potential to promote engagement by enhancing the links between students' home lives and school.

Regardless of whether students bring their own devices or have devices supplied by schools, their incorporation into mathematics learning varies widely from, for example, teachers using one device per group of students, one per student within a group, or whole class, 1:1 use. These variations alone can influence student engagement, particularly if the devices are shared. Even more influential on student engagement is the software, or apps, and the way they are used in mathematics lessons. Tasks can use targeted, mathematics-based apps, or they can use productivity apps such as *Explain Everything* or perhaps the generic inbuilt apps such as a still or video camera. Other options include accessing subscription-based apps or programs that include a wide range of activities and allow the teacher to track student achievement such as *Mathletics, Maths Online* or *Matific*.

Mobile technologies lend themselves well to game playing and often this is the default pedagogy. Given that almost all young people are actively involved in game playing in either a concrete or digital form, it makes sense to expect that some use of digital games in education could assist in increasing student engagement with content such as mathematics, that may otherwise feel irrelevant to students' everyday lives. The use of digital games could also assist in bridging the digital divide between how ICT is used at home and at school, as described by Selwyn, Potter and Cranmer (2009).

The terms 'game based learning' (GBL) and 'gamification' have begun to appear regularly in academic literature. GBL, defined as the use of video games for educational purposes (Kingsley & Grabner-Hagen, 2015), has been shown in some research to enhance motivation towards learning and academic performance. One concept stemming from GBL is gamification, which as Goehle and Wagaman (2016) suggest, is a natural fit for education. Interestingly, there are several interpretations of the definition of gamification, which is generally suggested to be the use of game design elements within a non-game context (Brigham, 2015). A teacher might gamify an activity or the teaching of a particular concept by adding achievement badges, rewards and levels in an attempt to increase student engagement (Goehle & Wagaman, 2016; Kingsley & Grabner-Hagen, 2015). The purpose of gamification within education is the use of game elements such as rewards and game-like activities to promote learning and engage and motivate students.

Regardless of what affordances are being utilised, if we consider the FEM, student engagement is tied closely to the pedagogical relationships and pedagogical practices within the classroom. However, the OECD claim that "we have not yet become good enough at the kind of pedagogies that make the most of technology … adding 21st century technology to 20th-century teaching practices will just dilute the effectiveness of teaching" (2015, p. 3). So how do teachers use emerging mobile technologies and adapt their pedagogies effectively in the teaching of primary mathematics? The following provides a brief description of the three studies used in this chapter to illustrate engaging (or disengaging) practices.

## The Studies

This chapter draws from data derived from three separate studies. Data from the first two studies have been presented elsewhere, but are re-purposed in this chapter for analysis against the FEM.

### *Study 1*

This study was an exploratory case study that took place in the early days of iPad integration in schools. The study explored one primary mathematics classroom and sought to understand how iPads were introduced into teaching and learning in a Grade 3 context. Further detail of the methodology involved have been reported in Attard and Curry (2012) and Attard (2015).

### *Study 2*

Study 2 was a multiple case study that investigated the pedagogies of four classroom teachers within their first six months of iPad integration. The teachers were situated at the same school and data was gathered from students and teachers in Kindergarten, Grade 2, Grade 4 and Grade 6. Further detail of this methodology has been reported in Attard (2013).

### *Study 3*

Study 3 was a multiple case study involving 16 teachers and their students from eight schools across New South Wales. This study investigated whether the use of a subscription based program, *Matific*, would improve student engagement with mathematics. *Matific* is a range of digital mathematics resources that are game-based applications, available to students on any device or operating platform. Participants in this study used a range of devices that included desktop computers, laptops and iPads. A more detailed methodology is found in Attard (2016).

All three studies used qualitative methods that included student focus group discussions and individual teacher interviews. Studies 1 and 2 also incorporated classroom observations. Study 3 included data from pre- and post-tests. Some data from Studies 1 and 2 have been presented elsewhere. However, for the purpose of this chapter they have been re-analysed for alignment with the FEM which is now presented. Each element of the FEM is presented and aligned with data derived from the three studies and existing literature as an alternate way of considering how the use of mobile technologies can contribute towards improving student engagement with mathematics.

## Mobile Technology to Enhance Pedagogical Relationships

Pedagogical relationships form the foundation for deep student engagement with mathematics. The following is a brief discussion of how each of the elements of the FEM have been evidenced in the three studies.

### *Students' Backgrounds and Pre-existing Knowledge Are Acknowledged and Contribute to the Learning of Others*

Technology enhanced tasks that are open-ended and provide opportunities for students to apply pre-existing knowledge for the benefit of other students have been evidenced. An example of one such task was observed in Study 2, in a Grade 5 classroom. The task required students to plan an itinerary and budget for a 'big day out' in the city. The students were to include the use of public transport and a trip to the cinema in their plans. Each group of three students were provided with an iPad that was used to access a range of apps relating to public transport timetables, trip planners and movie timetables. Although this task could have been conducted using standard computers and the Internet, the mobility and ubiquitous access provided by the iPads allowed the students to focus more on the mathematics embedded within the task and it promoted collaboration and discussion. The open-ended nature of the task and the real-life context allowed students to draw on personal experience and resulted in high cognitive, affective and operative engagement. This comment was typical of what was heard amongst the groups of students: "I thought about it and showed them and Luke said I should get this train because I'll have more time" (Grade 5 student, Study 2).

### *The Teacher Is Aware of Each Student's Mathematical Abilities and Learning Needs*

One of the affordances in the Matific suite of resources was the ability to allocate specific episodes for lesson time and homework to each student according to the identified need. Students were then able to access these tasks from any device at any time, using their login details. Although only seven out of the 16 teachers in this study utilised this affordance, it made a significant difference to their students' engagement with mathematics. One teacher who did differentiate the tasks talked about how she planned its use and how her strategy engaged her students in the following comment:

> *...we started off with the pre-test and I gave them feedback immediately afterwards and I had explained to them that I had grouped them so they knew they were grouped based on the pre-test. The kids were really aware of their goal for (the topic of) 'time' so because they were aware of their goal for time and because they knew that it was linked with Matific and the activities there they were conscious of their learning more...They loved having those goals and they knew that it was related to the Matific game that I had assigned to them (Grade 3/4 teacher, Study 3).*

The teachers felt that the ability to differentiate the tasks allowed their students to build confidence and ability with appropriately levelled episodes, while not appearing to be different from their more advanced peers:

> *It was perfect in a sense that we made it a point that we started at the middle and we went down for those who needed extra support, which was fabulous because they were still doing it visually, they were doing the exact same thing, and then we also gave the option that they could go up if they felt confident enough but at the same time visually, it was exactly the same for those kids that don't want to be different, that maybe do need that little bit of extra support (Grade 6 teacher, Study 3).*

Being able to experience success is an important element of student engagement, and plays a significant role in building confidence and developing a positive attitude towards mathematics.

## *Interaction Amongst Students and Between Teacher and Students Is Continuous*

A common argument against the use of 1:1 device programs is the perception that opportunities for interaction are diminished. However, there is evidence that when embedded in strong pedagogical practices, the use of mobile devices can promote important mathematical discussion. This was evident in all three studies where there was a mixture of 1:1 implementation and shared devices.

In Studies 1 and 2, students were provided with opportunities to share the work they had completed either individually or in pairs (predominantly using productivity apps such as *Show Me* or *Explain Everything*) with their peers through the use of an interactive whiteboard (IWB). The purpose of this was to reflect on learning as well as to receive constructive feedback from peers. In Study 3, there were several instances where the structure of the software in relation to the rewards system (to be discussed later) promoted discussion and collaboration despite students working on individual devices. The teachers believed the resources promoted mathematical discussion, perseverance, and 'collective encouragement', saying: "They would challenge themselves but also challenge each other, it was very good" (Grade 6 teacher, Study 3).

## *The Teacher Models Enthusiasm and an Enjoyment of Mathematics and Has a Strong Pedagogical Content Knowledge*

This section will focus on the second part of the above statement, the teachers' pedagogical content knowledge (PCK). In all three studies, PCK significantly influenced the success of mobile technology use. In Study 1, the teacher (with less than one year of experience), openly admitted his current depth of PCK limited his ability to incorporate iPads into mathematics lessons. "I could teach other things so well but my maths is always—like I have had to learn two things" (Grade 3 teacher, Study 1). In Study 2, there were several examples where PCK influenced students' learning. In a lesson on area using the *Doodle Buddy* app, students' questions took an unanticipated direction, and the teacher was unable to explain how to multiply decimal fractions even though this was a concept that appears within the primary mathematics curriculum. Conversely, in a Grade 4 classroom, the teacher used her

PCK to craft an effective lesson that used a drill and practice app to analyse student errors 'in the moment' and provide timely intervention (see Attard 2013 for a full description).

## *Feedback to Students Is Constructive, Purposeful and Timely*

Although it is feedback from teachers that is an important foundation for the development of positive pedagogical relationships, feedback derived from the use of technology can contribute to engagement. One of the most significant affordances of using mobile technologies is the provision of immediate feedback when using consumable apps such as mathematics games. Not surprisingly, this affordance featured heavily in all three studies and is considered a major contributor to the increase in student engagement when using these devices. The following are representative quotes from all three studies: "… it makes me happy because if you touch it and you make a mistake it just like takes it away … if it's on the iPad you can just go oh, that's wrong and you can take it away" (Grade 3 Student, Study 1). "… it's a lot quicker instead of just asking to go to the teacher and look at the answers" (Grade 4 student, Study 2). In Study 3, the immediate feedback was more than just an indication of a correct or incorrect answer: "Well it would like tell me it's wrong, and then it would like give an example and stuff like that, and I would try again on a different one." (Grade 6 student, Study 3).

The teachers also recognised the power of immediate feedback in terms of improving engagement: "It was very, very engaging for the kids. They found it—and I found it as well—they were learning from their mistakes based on the feedback that they were getting instantly from the program itself which was really good" (Grade 6 teacher, Study 3).

## Pedagogical Repertoires that Include Mobile Technologies

It is when strong pedagogical relationships are developed that a teacher's pedagogical repertoire becomes more influential in improving and maintaining engagement. There are strong connections between and amongst the elements of the FEM, and many of the affordances described above and below address several elements of the framework. There are also some elements that relate entirely to the teacher's pedagogical decisions rather than the technology use, so to retain the focus of this chapter and to avoid repetition, select elements that pertain directly to the use of mobile technologies are presented below.

## *There Is Substantive Conversation About Mathematical Concepts and Their Applications to Life*

The first element of pedagogical repertoires is one such example that links closely to the need for continuous interaction. As discussed already, creative use of mobile technology promotes mathematical discussion. Tasks that required students to investigate real-life contexts such as the 'big day out' task in Study 2 allowed students to see the application of mathematics to day-to-day living. Substantive conversations about mathematics were also promoted in Study 2 when the Grade 4 teacher used iPads to photograph mathematics outside the classroom. The students photographed each other as they measured various items around the school. The teacher then used the photographs they had taken in their next mathematics lesson as a focus for discussions on measurement. Another example was a Grade 5 lesson where students conducted a comparison of virtual dice and real dice. This provided a foundation for discussions of probability in real-life situations.

## *Tasks Are Positive, Provide Opportunity for All Students to Achieve a Level of Success and Are Challenging for All*

The *Matific* resources used in Study 3 were particularly effective in providing all students with the opportunity to achieve success while being challenged. Along with the ability to assign different tasks to different students, each episode, consisting of five questions, was carefully structured. The level of difficulty of each question increased gradually to ensure students were appropriately challenged. In addition, students were provided with scaffolding when answers were incorrect. This was a major benefit of the software that the students were very aware of and which, according to them, helped them learn. Students as young as Grade 2 noticed this:

> *...pretty much the best thing about Matific is because the last ones are pretty hard and it can teach you things. Like the first one gets you started with it the second one can make you like, can be a tiny bit tricky, and then the middle one is easy and hard, and then it goes quite hard. So the good thing is it is quite hard and they can teach you more (Grade 2 student, Study 3).*

All students appeared to have been challenged as a result of the structure of the resources, as evidenced in this quote: "the degree of difficulty challenged even some of my top kids" (Grade 6 teacher, Study 3). In this case, the high operative and cognitive engagement led to high affective engagement—students enjoyed and valued mathematics because they felt they were learning.

### *Technology Is Embedded and Used to Enhance Mathematical Understanding Through a Student-Centred Approach to Learning*

The very nature of mobile devices has resulted in a more student-centred approach when compared to traditional devices such as desktop computers or interactive whiteboards, that are typically positioned at the front of the classroom and often perpetuate teacher-centred practices. The significant uptake of gamification caused by the use of mobile devices within mathematics lessons has resulted in improved student engagement when the games provide appropriate challenge and are accompanied by mathematical discussion, promoting high affective, cognitive and operative engagement.

Students in all three studies confirmed the use of games enhanced their engagement and made mathematics lessons fun, and the following quotes provide an example of how powerful this was: "Because they have games but all the games are educational, most of the time, and you can use games to help with different maths skills" (Grade 3 student, Study 1). The use of games promoted discussion when students in Kindergarten compared their progress, making comments like "look how far I've gone", and "I've gone further" and encouraging them to work harder. Likewise, in Study 3, the game element made learning fun and the associated rewards were a powerful motivator to work hard and understand the mathematical concepts: "I keep on practicing. I keep on doing it again" (Grade 6 student, Study 3).

The most significant benefit of the reward system within the *Matific* resources was that it provided motivation for students to continue working hard. The simple 'super awesome' statement that appeared for five correct answers promoted perseverance amongst almost all of the students in Study 3. They spoke about how they wanted to try harder when they got answers incorrect, in order to achieve a 'super awesome' status, with comments like this one being typical: "I kept on going back and back and I finally got five stars" (Grade 6 student, Study 3).

Although the use of games does generally engage students, caution should also be taken. In some instances, during Studies 1 and 2, games did not always improve engagement due to a mismatch between student ability and the content of the game; a mismatch between the content of the game and the purpose or mathematical goal of the lesson; or a lack of reflection or discussion following interaction with the game.

## Mobile Technologies in the Primary Mathematics Classroom: Engaging or Not?

This chapter has provided a theoretical framework for engagement with mathematics and applied it to the use of mobile technologies in the primary classroom. In each of the studies explored here, there were similarities and differences between the depth of engagement. This was due to various factors, including the use of different software applications (with varied affordances), different levels of teacher confidence and experience, and the diverse ways in which the devices were used in terms of mathematical activities and the number and type of devices. There is little doubt that when used thoughtfully, mobile devices can improve engagement with mathematics, but it is more than just the device that makes the difference. Ultimately, it's the way they are used, the purpose for their use and the pedagogical practices that embed their use that determine how engaging they are.

The examples of engaging use of mobile technologies provided in this chapter give only one side of the story. There are also examples of mobile technology use that result in either no change, or a decrease in engagement. Often the disengaging uses are the result of poor pedagogical relationships or pedagogical practices that assume the devices alone (with no teacher input) will engage students and pedagogical content knowledge that requires further development. In addition, the distraction caused by logistical issues such as unreliable wifi access, or software and hardware malfunction, can lead to lowered levels of engagement and significant time wasting. Teachers need to be thoroughly prepared and have a strong awareness of the technological limitations within their individual contexts. This also includes having a thorough understanding of the affordances and limitations of the software applications they are using. However, it is important to focus on successful use of mobile technologies.

Finally, we must consider whether the use of mobile technologies is just a result of the novelty effect. Because mobile devices are developing so rapidly, there has been little time for longitudinal research to investigate whether the use of such devices result in sustained student engagement. However, for the moment, we can conclude that when used well, mobile technology does improve student engagement at operative, cognitive, and affective levels. As one seven-year-old student said: "So basically it is fun and you also get to learn stuff which is pretty good" (Grade 2 student, Study 3).

## References

Attard, C. (2013). *Introducing iPads into primary mathematics pedagogies: An exploration of two teachers' experiences.* Paper Presented at the Mathematics Education: Yesterday, Today and Tomorrow (Proceedings of the 36th Annual Conference of the Mathematics Education Research Group of Australasia), Melbourne.

Attard, C. (2014). "I don't like it, I don't love it, but I do it and I don't mind": Introducing a framework for engagement with mathematics. *Curriculum Perspectives, 34*(3), 1–14.

Attard, C. (2015). Introducing iPads into primary mathematics classrooms: Teachers' experiences and pedagogies. In M. Meletiou-Mavrotheris, K. Mavrou, & E. Paparistodemou (Eds.), *Integrating touch enabled and mobile devices into contemporary mathematics education* (pp. 197–217). Hershey, PA: IGI Global.

Attard, C. (2016). *Research evaluation of matific mathematics learning resources: Project report.* Penrith, NSW: Western Sydney University. https://doi.org/10.4225/35/57f2f391015a4.

Attard, C., & Curry, C. (2012, July). *Exploring the use of iPads to engage young students with mathematics.* Paper presented at the Mathematics education: Expanding horizons, (Proceedings of the 35th Annual Conference of the Mathematics Education Research Group of Australasia) Singapore.

Beavis, C., Muspratt, S., & Thompson, R. (2015). 'Computer games can get your brain working': Student experience and perceptions of digital games in the classroom. *Learning, Media and Technology, 40*(1), 21–42.

Bray, A., & Tangney, B. (2015). Enhancing student engagement through the affordances of mobile technology: a 21st century learning perspective on realistic mathematics education. *Mathematics Education Research Journal, 28*(1), 173–197.

Brigham, T. J. (2015). An introduction to gamification: Adding game elements for engagement. *Medical References Services Quarterly, 34*(4), 471–480.

Calder, N. S., & Campbell, A. (2016). Using mathematical apps with reluctant learners. In: *Digital experiences in mathematics education.* https://doi.org/10.1007/s40751-016-0011-y.

Chang, M., Evans, M. A., Kim, S., Norton, A., & Samur, Y. (2015). Differential effects of learning games on mathematics proficiency. *Educational Media International, 52*(1), 47–57.

Cristol, D., & Gimbert, B. (2014). Academic achievement in BYOD classrooms. *Journal of Applied Learning Technology, 4*(1), 24–30.

Fair Go Team NSW Department of Education and Training. (2006). *School is for me: Pathways to student engagement.* Sydney: NSW Department of Education and Training.

Fredricks, J. A., Blumenfeld, P. C., & Paris, A. H. (2004). School engagement: Potential of the concept, state of the evidence. *Review of Educational Research, 74*(1), 59–110.

Goehle, G., & Wagaman, J. (2016). The impact of gamification in web based homework. *Problems, Resources, and Issues in Mathematics Undergraduate Studies, 26*(6), 557–569.

Handal, B., Campbell, C., Cavanagh, M., & Petocz, P. (2016). Characterising the perceived value of mathematics educational apps in preservice teachers. *Mathematics Education Research Journal, 28*(1), 199–221.

Hickey, D. T. (2003). Engaged participation versus marginal nonparticipation: A stridently sociocultural approach to achievement motivation. *The Elementary School Journal, 103*(4), 401–429.

Hill, H. C., Ball, D. L., & Schilling, S. G. (2008). Unpacking pedagogical content knowledge: Conceptualising and measuring teachers' topic-specific knowledge of students. *Journal for Research in Mathematics Education, 39*(4), 372–400.

Hilton, A. (2016). Engaging primary school students in mathematics: Can iPads make a difference? *International Journal of Science and Mathematics Education.* https://doi.org/10.1007/s10763-016-9771-5.

Hu, H., & Garimella, U. (2014). iPads for STEM teachers: A case study on perceived usefulness, perceived proficiency, intention to adopt, and integration in K-12 instruction. *Journal of Educational Technology Development and Change, 7*(1), 49–66.

Ingram, N., Williamson-Leadley, S., & Pratt, K. (2016). Showing and telling: Using tablet technology to engage students in mathematics. *Mathematics Education Research Journal, 28*(1), 123–147.

Kingsley, T. L., & Grabner-Hagen, M. M. (2015). Gamification: Questing to integrate content knowledge, literacy, and 21st-century learning. *Journal of Adolescent and Adult Literacy, 59*(1), 51–61.

Koehler, M. J., & Mishra, P. (2009). What is technological pedagogical content knowledge?. *Contemporary Issues in Technology and Teacher Education, 9*(1), 60–70.

Meletiou-Mavrotheris, M., Mavrou, K., & Paparistodemou, E. (2015). *Preface Integrating touch-enabled and mobile devices into contemporary mathematics education* (pp. xx–xxvii). IGI Global, Hershey, PA.

Muir, T., & Geiger, V. (2016). The affordances of using a flipped classroom approach in the teaching of mathematics: A case study of a grade 10 mathematics class. *Mathematics Education Research Journal, 28*(1), 149–171.

Munns, G., & Martin, A. J. (2005). *It's all about MeE: A motivation and engagement framework.* Paper Presented at the Australian Association for Academic Research Focus Conference, Cairns. http://www.aare.edu.au/05pap/mun05400.pdf.

OECD. (2015). *Students, computers and learning: Making the connection*. Paris: pISA < OECD Publishing. https://doi.org/10.1787/9789264239555-en.

Pierce, R., & Ball, L. (2009). Perceptions that may affect teachers' intention to use technology in secondary mathematics classes. *Educational Studies in Mathematics, 71*(3), 299–317.

Selwyn, N., Potter, J., & Cranmer, S. (2009). Primary pupils' use of information and communication technologies at school and home. *British Journal of Educational Technology, 40*(5), 919–932.

Shulman, L. S. (1986). Those who understand: Knowledge growth in teaching. *American Educational Research Journal, 15*(2), 4–14.

Zhang, M., Trussell, R., Gellegos, B., & Asam, R. (2015). Using math apps for improving student learning: An exploratory study in an inclusive fourth grade classroom. *TechTrends, 59*(2), 32–39.

**Catherine Attard** is an Associate Professor in primary mathematics education at Western Sydney University. She is a multiple award-winning educator who has transformed teaching and learning in primary mathematics at Western Sydney University for over 10 years. Catherine's research is focused on student engagement with mathematics and issues surrounding the pedagogical practices that influence students' engagement. Catherine is also actively researching contemporary teaching practices through the use of digital technologies, and the use of financial literacy education as a tool to engage children with mathematics. She has won several teaching awards, including a 2016 Australian Awards for University Teaching (AAUT) Teaching Excellence Award and an Office of Learning and Teaching citation for Outstanding Contributions to Student Learning.

# Part II
# Traversing the Teaching and Learning Landscape

# When Robot A.L.E.X. Trains Teachers How to Teach Mathematics

**Andreas O. Kyriakides and Maria Meletiou-Mavrotheris**

**Abstract** In this chapter, we argue for the importance of providing teachers with the time and expertise to evaluate and use mobile devices to enhance students' learning of mathematics. The research we present here comes from a multifaceted program designed to provide a group of in-service teachers with the knowledge, skills, confidence, and practical experience required to effectively exploit tablet devices as a tool for enhancing mathematics teaching and learning. The program took place within a public primary school in Cyprus. Fifteen teachers participated in a classroom workshop, attended an academic seminar, participated in interviews and integrated the app A.L.E.X. in their own lesson plans and instruction. The type of professional learning for teachers we suggest indicates a possible context within which teachers could reshape their knowledge of, and attitudes towards the use of mobile devices.

**Keywords** Mobile mathematics learning · A.L.E.X. · Tablet PCs
TPACK · Primary school teachers

## Introduction

One pervasive challenge in mathematics education at the school level is the identification and use of instructional contexts that motivate student inquiry and learning. Recent technology advances have provided the opportunity to create entirely new, inquiry-based learning environments by significantly increasing the range and sophistication of possible classroom activities (e.g. use of dynamic

A. O. Kyriakides (✉) · M. Meletiou-Mavrotheris
European University Cyprus, Engomi, Cyprus
e-mail: ak5288@gmail.com

M. Meletiou-Mavrotheris
e-mail: M.Mavrotheris@euc.ac.cy

 
N. Calder et al. (eds.), *Using Mobile Technologies in the Teaching and Learning of Mathematics*, Mathematics Education in the Digital Era 12,
https://doi.org/10.1007/978-3-319-90179-4_5

educational software, computer simulations, augmented reality applications, etc.). Although traditional teacher-centered approaches to mathematics instruction still dominate, a shift towards the adoption of more active, technology-supported learning environments is being reflected in a number of initiatives undertaken by educational systems worldwide. In Europe, for example, the European Schoolnet is a network of 31 Ministries of Education that was formed with the aim of bringing innovation in teaching and learning across Europe through supporting schools in achieving effective integration of ICT in different subjects including mathematics.

One promising approach lately explored is the potential of hand-held tablet PCs like the Apple iPads and Android tablets routinely used in daily life, as tools for enhancing mathematics learning. Tablets, as well as smartphones and other mobile devices, are becoming standard learning tools, and adopting them in the classroom is becoming more widespread (Johnson et al., 2013). The existing literature indicates strongly the significant potential of tablets and other mobile devices as ubiquitous tools that can radically transform and enrich mathematics pedagogy by creating engaging learning environments (Clark & Luckin, 2013; Henderson & Yeow, 2012; Melhuish & Falloon, 2010). Despite, however, the widely acknowledged educational potential of mobile devices, their actual success as a tool for revitalizing mathematics instruction will ultimately depend upon the abilities of teachers to take full advantage of their affordances (Becker, 2007). Several research studies (e.g. Blackwell, 2014) have asserted that it is much more demanding for teachers to exploit the growing prominence of digital technologies in instructional settings than was originally anticipated, and that many educators remain unprepared to effectively employ Information and Communication Technology (ICT) tools in their teaching practices. Thus, to bring about the necessary changes in teaching cultures that will enable mathematics education to reap the full benefits of mobile devices, it is of utmost importance to provide high quality pre-service and in-service training opportunities that will equip teachers with the required knowledge and skills to effectively infuse them into teaching and learning.

Acknowledging the transformative potential of mobile devices, but also the crucial role of teachers in any effort to bring about change and innovation, the current exploratory study focused on in-service primary teacher training on the effective integration of tablet technologies in mathematics education. Building on the notion of technological pedagogical content knowledge (TPACK) as a conceptual framework (Mishra & Koehler, 2006), a professional development program was designed and implemented in a primary school in Cyprus. The program aimed at providing the group of teachers in this school ($n = 15$) with the knowledge, skills, confidence, and practical experience required to effectively exploit tablet devices as a tool for fostering children's motivation and their learning of mathematics. Its impact on the study participants was examined from three perspectives:

Influence on teachers' attitudes and perceptions regarding tablet technologies and mobile-based mathematics teaching and learning; Impact on the development of teachers' TPACK regarding the instructional integration of tablets; and Level of transfer and adoption of TPACK competencies acquired through the study intervention to actual teaching practice.

## Literature Review

Since the introduction of the iPad in 2010, there has been a rapid adoption of its use in educational institutions worldwide. Although the availability of research studies on the integration of tablets and other mobile devices into primary mathematics teaching and learning is still relatively limited due to the novelty of these technologies, there is considerable convergence in the existing research findings, which highlights significant benefits of mobile devices (e.g. Clark & Luckin, 2013; Heinrich, 2012; McKenna, 2012; Melhuish & Falloon, 2010). Research suggests that the adoption and use of mobile devices within and beyond the classroom confines, can lead to the creation of a more active, inclusive and engaging environment, which encourages more cooperative and collaborative forms of learning (Henderson & Yeow, 2012), creativity (Bennett, 2011) and differentiation and acceleration of learning (Milman, Carlson-Bancroft, & Vanden Boogart, 2012).

Various studies within classroom settings indicate that tablet devices attract and gain students' attention, contributing to their increased motivation and engagement with schoolwork (Henderson & Yeow, 2012; Milman, Carlson-Bancroft, & Vanden Boogart, 2012). Tablet applications, which "live in the spaces where education and entertainment overlap" (EDUCAUSE, 2011), can capture students' imagination, enticing them to learn on their own. Their tactile interface, media-friendly approach, and mobility, introduce an element of fun, making the learning experience much more relevant and authentic (Burden, Hopkins, Male, Martin, & Trala, 2012). Children, but also adults of all ages, enjoy exploring and learning in ways that are natural to them when using a touch device (Cohen, 2012; Department for Education and Communities, 2012). The portability, speed, simplicity of interface, and accessibility of mobile devices makes them very easy to use for most people, including young children and the elderly, non-tech-savvy parents and teachers, and students with special educational needs (Melhuish & Falloon, 2010). The ever-growing list of interactive mobile apps available, provide instructional designers and teachers with more options for creating personalised and seamless learning experiences that move beyond static presentation, limited interaction, and the walls and schedules of formal schooling (Henderson & Yeow, 2012; Johnson et al., 2013; Burden et al., 2012).

A promising type of mobile apps that could be utilized in the mathematics classroom are coding apps, which teach children the concepts behind programming in a playful context. (e.g. A.L.E.X., Guida, 2014; Hopscotch, Hopscotch Technologies 2014; Bee-Bot, TTS Group Limited 2012). Taking their inspiration from Logo

(Papert, 1980), educational coding apps promote a constructionist approach to tablet use, emphasing students' use of tablets to become creators instead of consumers of computer games. In addition to the provision of a highly motivational and practical approach for introducing children to computer programming and developing their computational thinking (Wilson, Hainey, & Connolly, 2012), coding game apps provide rich opportunities for the reinforcement of problem-solving, critical thinking, and logical thinking skills (e.g. sequencing, estimation, prediction, metacognition) that apply across domains. At the same time, they can also be helpful in developing subject-specific mathematics knowledge. Programming provides an ideal environment for expressing and experimenting with mathematical ideas and for making abstract mathematical ideas more concrete (Aydin, 2005). The design, coding, revision, and debugging of computer commands, helps students develop higher order mathematical problem solving skills such as deductive reasoning and metacognition (Subhi, 1999), while at the same time improving their conceptual understanding of key mathematical ideas. Researchers have found that programming using constructionist environments like Logo increases students' understanding of arithmetic and measurement processes, algebraic reasoning, and general geometry abilities (Clements, Battista, & Sarama, 2001).

While apps, conducive to constructivist approaches, present some exciting opportunities for a transformative shift in mathematics teaching and learning, their introduction into the classroom does not come without challenges. The existing literature highlights not only opportunities, but also a number of pedagogical, technical, and management issues that need to be addressed for mobile devices to be effectively integrated within existing school systems. As several studies have indicated, when tablets and other mobile devices enter classrooms and other learning environments, their impressive immersive capabilities are often overlooked or underdeveloped, and instead they are used in more didactic classroom settings (Attard, 2015; Daccord, 2012).

For mobile devices to be more effectively utilized to enhance mathematics learning opportunities for all learners, there needs to be a re-conceptualization of the design and management of learning environments. Careful strategic planning and reflective implementation, grounded in solid research, is necessary. This should focus on the broad preparation and ongoing engagement of all key stakeholders involved in the educational process (prospective and practicing teachers, teacher educators and other college faculty, adult educators, educational leaders, technical managers, etc.). The provision of high quality teacher training on the educational applications of tablets, in particular, is of paramount importance to their effective integration in classroom settings (Heinrich, 2012; Henderson & Yeow, 2012), since the change in teaching practices is always one of the most important factors in any educational change. As pointed out by Pastore and Falvo (2010), teachers are the gatekeepers of what technological tools are used in their classrooms and how.

Attard (2015) used data obtained from two studies conducted in Australian primary classrooms to describe how a small group of teachers used a (then) new technology, the iPad, to teach mathematics without the support of professional development. The practices of these teachers, including the issues and challenges they experienced and examples of their teaching with iPads, were presented against a backdrop of the SAMR model (substitution, modification, augmentation and substitution) (Puentedura, 2006), and were used in conjunction with the TPACK framework to organise and analyse the observed uses of iPads. Findings illustrated the need, prior to the introduction of tablets for teaching and learning, of appropriate professional development that addresses the combination of mathematical content, pedagogy and technology that is critical for all teachers, regardless of teaching experience.

An important consideration of any model of professional development is whether it is useful and supportive of teachers' efforts to improve their teaching practices (Whitaker, Kinzie, Kraft-Sayre, Mashburn, & Pianta, 2007). Historically, professional development efforts have largely been ineffective in producing reform-based classroom change (Templin & Bombaugh, 2005). As Robinson (1998) points out, staff development often fails to transfer to the participants' workplace situations, because it is too remote from teachers' 'real-work' needs or organizational realities. Thus, the successful integration of mobile devices necessitates careful professional development that highlights the importance of engaging teachers in regular reflection and evaluation of their current practices. Ideally, professional development should take place within natural classroom settings, so that teachers can get the chance to evaluate the effect of their experiences with mobile mathematics learning on actual classroom practices.

## Methodology

The sampling method used for this study was purposeful/selective sampling. The researchers chose a rural, public primary school in Cyprus to apply their teacher professional development program. The majority of students in this school come from low-socioeconomic-status families. High dropout rates before high school graduation constitute a usual phenomenon among the area population, and this stance is often mirrored in the parents' limited interest in their children's educational attainment. The researchers knowingly selected such a context to implement the professional development program, so as to enable the teachers in this school to explore the potential of tablet technologies as a tool for enhancing their students' motivation and learning of mathematics. The program aimed at promoting, while at the same time investigating, participants' efficacy in effectively integrating tablet devices within the mathematics curriculum.

### *Conceptual Framework*

The TPACK conceptual framework guided the design and implementation of the professional development program. TPACK is a powerful and influential framework, proposed by Mishra and Koehler (2006) in response to the absence of theory guiding the integration of technology into education. Building on Shulman's (1986) idea of Pedagogical Content Knowledge, TPACK emphasizes the importance of developing integrated and interdependent understanding of three primary forms of knowledge: technology, pedagogy, and content. TPACK is based upon the premise that effective technology integration for pedagogy around specific subject matter requires an understanding of the dynamic relationship among all three knowledge components. Thus, ICT training for teachers cannot be treated as context-free and should emphasise how technology relates to pedagogy and content. The aim is to move teachers beyond technocentric strategies that focus on technology, and to promote their critical reflection on the instructional use of technological tools.

In the current study, the adoption of TPACK served a twofold purpose: (i) a guiding theory for designing the program so as to create professional development opportunities that would better prepare teachers to effectively integrate tablet devices in the mathematics teaching and learning process; and (ii) a conceptual blueprint for investigating the impact of the intervention on participants' professional growth in the use of tablet devices in mathematics.

## *Research Design: Scope and Context of Study*

A case study design was employed in the current study. The case studied consisted of the 15 members of the school's teaching personnel, including the school principal, each of whom participated in the professional development program. Their age range spanned 35–50 years old.

The professional development program aimed at building teachers' TPACK regarding the use of tablets in mathematics learning by providing them with opportunities to develop relevant knowledge, skills, and attitudes. It was designed to offer high-quality professional development experiences to teachers that would prepare them to effectively integrate tablet devices with core curricular ideas in their mathematics classrooms. It focused on content-specific issues surrounding the educational applications of tablet devices in the teaching of primary mathematics. It provided teachers with the opportunity to develop pedagogically sound strategies for the integration of apps that foster student inquiry and higher order learning of mathematics through combining experiential learning of student-centered pedagogical approaches with extensive field experience.

Following the TPACK model and action research procedures, the program was carried out in three phases: (i) Phase I. Familiarization with Mobile Mathematics Learning; (ii) Phase II: Lesson Planning; (iii) Phase III: Lesson Implementation and Reflection. Each of the three phases, described next in more detail, supported teachers in strengthening the connections among their technological, pedagogical, and content knowledge.

### Phase I—Familiarization with Mobile Mathematics Learning

During Phase I, all staff participated in a classroom workshop and a follow-up professional development seminar, which offered a critical introduction into the potential and challenges of using mobile devices in mathematics instruction. Teachers' TPACK on mobile mathematics learning began to develop through experiencing some of the ways in which a purposefully selected app (coding app A. L.E.X.), blended with carefully constructed learning experiences, could help improve children's attitudes towards mathematics, while at the same time advancing their mathematical thinking and problem solving skills.

*Classroom Workshop*
The classroom workshop was organized and led by the first author in one of the two sixth grade classes of the school. The duration of the workshop was 80 min, that is, two consecutive teaching periods. To facilitate smooth running of the school, seven of the teachers attended the workshop during the first teaching period, and eight attended during the second. The teachers and the school principal assumed the role of learners, and were mixed with the 20 students of the participating sixth grade class, into 7 groups of 4–5 members each. Each mixed group (teachers and students) had their own iPad in which the coding game app A.L.E.X. (named after the developer's nephew) was downloaded, and worked collaboratively to complete the tasks of a given worksheet.

The reason we selected A.L.E.X. to use during the workshop is the fact that it is an excellent example of an app that could be incorporated into the mathematics curriculum to promote a constructionist approach to tablet use. In a previous study we had conducted within the same primary school in Cyprus (Kyriakides, Meletiou-Mavrotheris, & Prodromou, 2016), we had utilized A.L.E.X. as a tool for engaging a group of students from a low socioeconomic background in authentic mathematical problem solving activities with very promising results. The endeavours with A.L.E.X. of the 10–11 year old children participating in that study, had led to increased enthusiasm and higher levels of classroom participation while at the same time providing deep learning opportunities for students.

### App A.L.E.X.

A.L.E.X. is an entertaining, programming puzzle game that allows players to control a robot along a path. It is a free educational app suitable for use on iPad or Android tablets. The lower levels of the game are suitable for children as young as 6 years old, while the higher levels might be challenging even for high school students or adults. A.L.E.X. offers a basic introduction to programming concepts and logic. At the same time, it has the potential to tacitly promote a number of concepts and procedures embedded in the school mathematics curriculum. This becomes feasible by offering the user the opportunity to think and plan logically as

**Fig. 1** Robot A.L.E.X.

he or she programs the robot A.L.E.X. (see Fig. 1) with a sequence of commands, in order to complete each level.

The game has two modes, Play and Create. In the Play Mode, players complete standard puzzles using the pieces provided to them. They begin at a start point and have to 'pre-plan' the robot's path. Once they plan the path by building a sequence of instructions, they execute these instructions and watch the robot 'walkout' their plan. If they have given the right instructions, the robot will reach its destination; otherwise, it will fall into oblivion. The levels start off fairly easy and increase in difficulty as the player advances. The free version includes 25 progressively demanding levels. There is also an upgrade available, which incurs a small cost, and provides 35 additional levels. At each level, players are evaluated on whether they

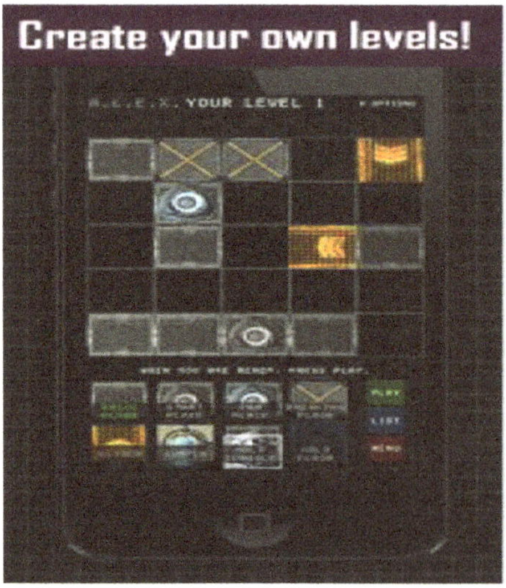


**Fig. 2** Users' potentiality to "create" their own levels

successfully complete the level, how quickly they do so, and whether they take the shortest path. The Create Mode includes features for players to create their own puzzle. In this mode, players can devise their own levels by structuring the pathways they would like A.L.E.X. to follow (see Fig. 2), and play through their own levels. In the free version, players can create up to 51 levels, which they can save and edit at any time.

The commands A.L.E.X. can follow are simple and symbolically expressed. For instance, the commands ‘turn left’, ‘turn right’, or ‘go forward’ are given when one touches the game’s screen on the particular arrow pointing to the respective direction (see Fig. 3).

The app is currently available only in English; however, the instructions are simple to follow for non-native English speakers. For the students participating in our study, language was not an issue, since they all had some command of the English language, which is taught as a second language in Cyprus throughout primary school. Moreover, most of the software that Cypriot students use has an English interface.

Despite being very simple to initially use, A.L.E.X. is a powerful educational game app that can help children improve their skills in directional language and programming through sequences of forward, backward, left, and right 90° turns. At the same time, it can help children improve their understanding of mathematical ideas related to motion, direction, and geometry. Unlike the vast majority of geometry apps currently available, which are very limited in their ability to assist students in developing geometrical conceptual understanding (Larkin, 2015), A.L.E.X. provides a user-friendly venue for children to experiment with geometrical ideas, to make and test hypotheses, and to implement corrections based on feedback received.

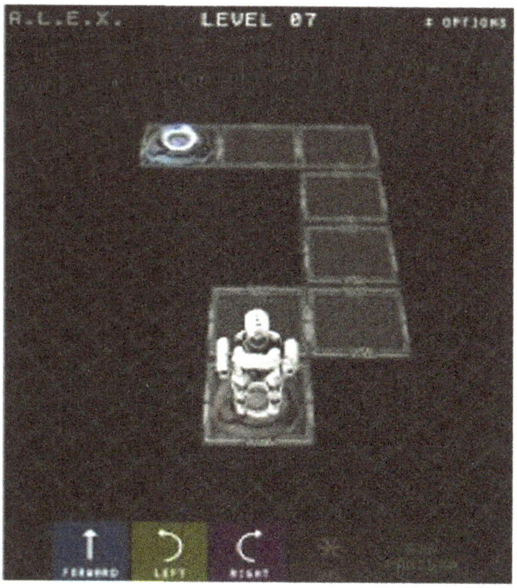


**Fig. 3** Screen’s display of commands

In the classroom workshop, we followed the same structure and content we had adopted in earlier research on tablet devices. The design of the given worksheet was such so that the employed technology could function supportively in a learning environment teaching the concept of symmetry. Activities included familiarity of learners with the first 10 levels of the game, construction of their own pathway, design, symbolic representation and implementation of a symmetrical (to a given) pathway. Symmetry was purposefully selected for the geometrical concepts included in sixth grade's curriculum because of its close relationship to the constructionist approach of using A.L.E.X. (Kyriakides, Meletiou-Mavrotheris, & Prodromou, 2015).

*Seminar*

At the end of the same school day on which the classroom workshop was held, teachers also attended a seminar organized by the researchers. In this seminar, the methodology and key findings of the classroom-based research project on mobile mathematics learning previously conducted by the authors Kyriakides et al. (2016) were presented to the school's teaching personnel. The presentation lasted for one hour. After providing an overview of existing literature relevant to mobile mathematics education, the researchers presented to the participating teachers an analysis of the data they had collected during the teaching intervention that took place in their previous study. It was clarified that the illustrated findings referred to a student population having similar characteristics to those of the students participating in the workshop the teachers had attended earlier that day. A discussion was then initiated, which focused on children and what is required to involve them in learning about mathematics though the use of game apps like A.L.E.X. It provided the venue for discussing the affordances and limitations of tablet devices, and for identifying design considerations that promote the incorporation of apps in ways that motivate children, while at the same time advancing their mathematical thinking and problem solving skills. Discussion highlighted the benefits of introducing tablets in the mathematics classroom, possible challenges and pitfalls that educators may face and tips to minimise these issues, as well as the importance of teacher training in app selection and management. We examine these issues more fully later in this chapter.

## Phase II: Lesson Planning

In Phase II of the program, teachers' TPACK was enhanced through their engagement in lesson planning. They worked in groups of five (three groups in total), to develop lesson plans and accompanying teaching materials incorporating the use of A.L.E.X.. They were instructed to develop their lesson plans based on the TPACK model in order to ensure good fit across content, pedagogy, and technology. The groups selected a topic from the primary mathematics national curriculum and prepared a lesson unit for a specific grade-level of students that was aligned with the learning objectives specified in the curriculum. At the same time, they also adhered to important pedagogical principles associated with technology enhanced educational environments such as facilitation of an authentic learning experience, active participation, collaboration, and promotion of higher order thinking skills (e.g. problem-solving, creative-innovative thinking). The lesson plans were shared with the researchers for comments and suggestions, and were revised based upon received feedback.

#### Phase III: Lesson Implementation and Reflection

Next, teachers conducted a follow-up classroom intervention, in order to apply what they had learned about mobile mathematics learning into an actual classroom setting through action research. One member from each group taught the lesson that their team had designed in his/her class, while the other members conducted peer observation. Observers took field notes of how students reacted to the delivery of the lesson plan. They commented on the ways in which the integration of the game app affected the teaching-learning process, and impacted student engagement and learning. In addition to taking field notes, observers also produced videotaped segments of the lesson implementation.

Once the classroom research was completed, a two hours session was held for teachers to reflect upon their teaching experimentation. Participants reported on their experiences, and provided videotaped teaching episodes and samples of students' work for reflection and evaluation. They shared their observations on students' reactions during the lesson, noting what went well and what difficulties they faced, and made suggestions for improvement. Following the reflection session, each group worked collaboratively to write and submit a reflection paper on their classroom research projects.

### *Instruments, Data Collection, and Analysis Procedure*

Multiple forms of data were collected to document changes in teachers' TPACK regarding mobile mathematics learning, and to their perceptions and attitudes towards the integration of tablet devices in mathematics instruction as a result of participating in the program:

*Pre-survey*: This open-ended survey gathered information on the participants' use of tablet devices in daily life (e.g. What is your level of your familiarity with mobile devices? Do you own an iPad or an Android tablet device at home? If yes, how often do you use your tablet device? For what purposes do you use your tablet?), and on their prior experiences, attitudes, and perceptions surrounding the application of ICT in general, and tablets specifically, in the mathematics classroom (e.g. To what extent is technology integrated into your classes? What types of technological tools do you use with your students? What do you perceive as the main obstacles to technology integration in your classroom? Do you feel you are properly trained in the use and integration of technology in general, and of tablets in particular, into your classroom? If not, in what professional development activities would you be interested in participating? If you had access to tablets at school, would you use them in your mathematics classroom? Explain your response).

*Individual open-ended interviews*: Upon completion of the teacher training workshop and the seminar (Phase I), the researchers conducted individual interviews with each of the 15 participants. These semi-structured interviews recorded participants' attitudes and considerations about the use of tablets in learning mathematics and in classroom management in general. They gave researchers the opportunity to trace possible shifts in teachers' attitudes and perceptions regarding mobile learning, as well as their willingness to practically apply the TPACK they had progressively acquired through the professional development program (see Phases II and III).

*Observations and artifacts collected during Phases I and II*: Researchers' observations and field notes, teachers' submitted lesson plans.

*Observations and artifacts collected during Phase III*: Teachers' field notes and reflection papers, video episodes and samples of student work provided by teachers.

Findings presented in the current chapter are mainly based on data collected during Phases I and II of the study (i.e. items i. to iii. above). Analysis of the data collected during Phase III (i.e. item iv. above) is still ongoing. Data collected during Phases I and II were transcribed (interviews), coded, and analyzed to guide the investigation on the impact of the intervention on participants' attitudes and TPACK development. We did not use an analytical framework with predetermined categories to assess how teachers' perceptions and TPACK developed after going through the program due to the lack of well-established frameworks and methodological insights for studying mobile mathematics education in the context of in-service mathematics teacher training. What we did instead was to identify, through careful reviewing of the different sources of data, recurring themes or patterns in the data. To increase the reliability of the findings, the activities were analyzed and categorized by both researchers. Inter-rater discrepancies were resolved through discussion.

## Results

Findings provide strong indications that our TPACK-guided professional development program had a positive impact on at least the first two perspectives of the participants' experiences examined: (i) Attitudes and perceptions regarding mobile mathematics teaching and learning; and (ii) TPACK development regarding the instructional integration of tablet devices. Preliminary analysis of some of the data collected during Phase III also suggests positive gains on the third perspective examined: (iii) Level of transfer and adoption of acquired TPACK to actual teaching practice.

## *Changes in Attitudes and Perceptions Regarding Mobile Mathematics Learning*

### Prior Experiences and Attitudes towards tablet PCs as learning tools

The pre-survey completed by the participants provided baseline information about the teachers' prior experiences and attitudes towards technology-enhanced mathematics education in general, and mobile mathematics learning in particular. Teachers expressed positive attitudes towards the instructional use of technology but, at the same time, acknowledged that ICT tools were not adequately used in their classrooms. They reported that technology use was limited mainly to Word Processing, PowerPoint, and Internet browsing. In mathematics, their students used technology mainly to perform routine calculations, practice skills and procedures, and check answers. Teachers reported that students rarely or never used technology to solve complex problems or discover mathematics principles and concepts in their mathematics class.

Teachers cited various factors as obstacles to technology integration: limited resources, limited time, technical issues, lack of support, and an oversized curriculum. The participants particularly emphasized the need for additional training in new trends in ICT learning across the curriculum. They pointed out the need for professional development activities focused on embedding new technologies, and especially tablet devices, in their classroom. They were particularly interested in professional activities that focus on the use of innovative educational technologies for building children's critical reasoning and problem solving skills. Ifigenia, for example, stated (all illustrative quotes from the teachers are labelled with pseudonyms to protect participants' anonymity): *My students lack critical thinking... they like being spoon-fed and get lost whenever I ask them something which is non-routine. I would like to be familiarized with technologies that can help children build their critical thinking and problem solving skills ... not applets of drill-and-practice type.*

Teachers were also positive regarding the possibility of integrating tablets in their teaching. They agreed that tablets should be viewed as worthy of consideration in the classroom. Despite, however, their generally positive attitudes toward the use of tablets in education, these teachers also had no prior knowledge and experience with mobile learning, and lacked appreciation of the potential of tablet technologies to transform the nature of mathematics education provided to students. The most commonly cited reasons for considering using tablets in the classroom were for increasing students' motivation and engagement: *Using tablets in the classroom will provide a strong incentive for children to actively participate in the learning process* (Anna). When prompted to identify opportunities for introducing tablets in the mathematics classroom, it became obvious that all of the teachers had a very limited notion of their potential to support learning. They viewed tablets as tools for practicing and/or evaluating acquired skills, but not as a powerful means of creating immersive, problem-solving learning experiences: *I believe that apps should be used at the completion of a lesson for students to practice what they have been taught and for the teacher to see whether his learning objectives having been achieved* (Achilleas). This limited vision of the educational potential of tablets is understandable given that teachers' past exposure to mathematics education applications had largely been limited to drill-and-practice ones. None had ever been exposed to a challenging and complex education app designed to help students build higher order problem solving skills. Moreover, no teacher had ever received training on how to effectively utilize apps in a classroom environment.

## Post-study Perceptions Towards Tablet PCs as Learning Tools

Findings suggest that the professional development program was quite successful in helping teachers to move beyond their restricted views of apps as educational tools. Their TPACK development led to a parallel change in their perceptions regarding teaching mathematics using tablets. Their participation in the program helped them realize tablet devices' true potential for supporting learning of the mathematics curriculum in educationally powerful and interactive ways.

In the interviews conducted upon completion of Phase I, all teachers expressed very positive attitudes and strong intention to incorporate tablets into their teaching practices. Unlike, however, the pre-survey stage, their focus was not on the playfulness of apps. They developed more sophisticated views and emphasized the fact that the integration of tablets into the instructional process can act not only as a strong motivational tool, but can also offer an effective learning context that can promote the construction of powerful mathematics knowledge and skills. Using the insights gained from their participation in the classroom workshop, teachers listed several advantages that can make using educationally sound apps like A.L.E.X. a more meaningful, engaging, and effective learning experience for students, namely: ease of use, portability, promotion of active learning and experimentation, promotion of student autonomy and self-directed learning, promotion of problem solving, building of transversal skills and competences. Unanimously, the sample in this research indicated the catalytical role that the on-going professional development program had in assisting them to overcome negative prior conceptions (if any) and to form a brand new, more positive and informed stance towards the use of tablet devices in the teaching of mathematics:

> *The professional development program we attended helped us all to realize how easy it is to use touchscreens in the teaching of mathematics and, thus, we now have a more positive stance towards tablet devices (Aristodemos)*

> *Whenever I happened to participate in a teacher professional development program, everything was theoretical and at the end, you gained nothing. However, if you see how they [tablet devices] could actually be used in the classroom, as we had the chance to do here, you get convinced to use them yourself. And the reason is that you see that someone tried it and it worked and it's not as difficult and unattainable as you could have imagined (Penelope).*

*If it wasn't someone like you [researchers], I would have never thought that I could use this thing [A.L.E.X.] in such a manner (School Principal).*

Worthy of note is the fact that the participating teachers did not reflect in a narrow manner, but articulated proposals that underscore the necessity of educating the broader community of teachers about mobile devices. Ariadne's suggestion regarding the inclusion of mobile devices in the national mathematics curriculum could serve as an indicative example:

*While observing the lesson, because if you don't see it yourself my sense is that you will not understand what it means to do mathematics with touchscreens, I thought that it would have been wise to make some suggestions to those who deal with the writing of national mathematics curricula to include in the daily lesson syllabi the use of touchscreens. It is something that could contribute in an interesting way towards the learning of mathematical concepts. Children will love it! (Ariadne)*

## Changes in Participants' TPACK Competency for Tablet Devices

During Phase I, the classroom workshop and seminar supported teachers' development of TPACK and extended their thinking regarding how students learn with tablet devices. It assisted them in moving beyond their narrow view of apps as a drill-and-practice resource, to a view of apps as a powerful exploratory tool for acquiring new knowledge and skills. Teachers came to the realization that educational apps ought to combine playfulness with instructional soundness, and developed skills in properly evaluating them, and in selecting apps with pedagogically sound design features (e.g. authenticity, interactivity, added value to the educational process, multiple levels to support differentiation of instruction, etc.). Their experimentation with A.L.E.X., helped them gain better understanding of how educationally sound apps could be integrated into the mathematics curriculum. Their endeavours with A.L.E.X. familiarized them with the design principles for constructivist mobile learning environments, and promoted their critical reflection on the use of tablets in mathematics education. Thus, participants improved their knowledge regarding the selection of appropriate apps based on appropriate pedagogy and content, an important tenet associated with TPACK.

A close examination of the interview transcripts illustrates that the participating teachers acknowledged the advantages of the employed technology (A.L.E.X.) and interpreted them through the lens of maximizing learning outcomes:

*It was easy for the children to use the app because there were not many commands. And when A.L.E.X. had to turn, the user did not need to determine the angle's size in degrees. There were just three tools, straight, right and left. Even the weakest students found it easy to deal with (Loukia).*

*They didn't have the keyboard, the mouse and the screen. They simply had something in front of them, which they touched with their fingers. I think the children of this age, of this generation don't have any problem to use this tool because they are very familiar with touch screens (Charalambos).*

*I sensed the directness the children experienced. Because A.L.E.X. was like a three-dimensional figure, I think this was the reason that made students to stand up and walk the steps in the classroom. It was like observing a human being moving around, that's why they identified with him and walked, to assure themselves that the given commands were the right ones (Melpomeni).*

*I believe in technology but only if you use it in the right way, that is, not technology for the sake of technology. Teachers must implement it in such a manner to make it beneficial for the lesson, as you did in the workshop you carried out (School Principal).*

Of note is the unavoidable comparison made with currently used school technologies:

*For me, to carry laptops in the classroom and set them up is very time consuming, it is a whole process. Touchscreens are much more easily used, they are smaller and, given that children are very familiar with this kind of technology, it is more convenient for the teacher to teach with tablets or iPads rather than with PCs or laptops (Ourania).*

## Transfer and Adoption of TPACK Competencies to Teaching Practice

Teachers' eagerness to transfer what they had gained through the program was apparent in the individual interviews conducted after the completion of Phase I. The will to extend the acquired TPACK competencies was widespread among participants. This will was not vaguely expressed, but was well thought out and included specific suggestions. For example, one teacher noted:

*I suppose that you may make a differentiation according to the grade level you work with. I, for example, who teach second graders, could do something simpler. Perhaps, I could teach squares and rectangles...I mean the students might have been asked to give commands to A.L.E.X. to go straight, to turn. And I'm saying this because the concept of 90 degrees is in our syllabus. I am thinking of planning a lesson that will include the construction of the two shapes and that will engage students in noticing the geometrical properties. For instance, students could be encouraged to observe that for the case of the square you need as many steps to go straight, as you need to go when you turn right. On the other hand, for the case of the rectangle students could see that only the opposite sides require from A.L.E.X. to walk equal numbers of steps (Maria).*

During Phases II and III, the participants transferred the knowledge acquired during Phase I into lesson planning and implementation. This was a valuable experience that helped them develop professionally in relation to mobile apps' integration into the mathematics curriculum. As the preliminary analysis of the data collected during Phases II and III indicates, teachers' hands-on teaching experiences, and sharing with peers, helped them to develop many pedagogical ideas and to apply in practice, effective instructional strategies for successful instructional integration of tablets.

Analysis of the lesson plans submitted by the three groups of teachers suggest positive gains in their ability to effectively integrate apps within the mathematics curriculum. All groups prepared high quality lesson plans that made constructive use of A.L.E.X. and aligned with their targeted grade level and curricular topic. All three lesson plans also included appropriate pre-game and post-game activities that transferred the video game mathematics experiences to other settings. In addition to a clear and focused introduction at the beginning of the lesson plan, there was also a debriefing and reflection activity, in which students had to review and analyse the events that occurred in the game, reflect on the content of the game, and share the knowledge acquired while playing. The instructional strategies used by these

WORKSHEET

Activity A
Work in your group the first 10 levels of the game.

Activity B
A.L.E.X. has to follow these 4 pathways:
Pathway 1:
Pathway 2:
Pathway 3:
Pathway 4:
Having first examined the steps of the 4 pathways, compare them and write down your observations.

Activity C
Using the images below construct the 4 pathways A.L.E.X. has to follow.
START PLACE
BASIC FLOOR
END PLACE
Follow the steps of each pathway and then play the game.
What shape did the pathways have?
Calculate the area of each shape in square units.
Pathway 1: ……………………
Pathway 2: ……………………
Pathway 3: ……………………
Pathway 4: ……………………

Activity D
What do you notice about the 4 areas? Could you write these numbers as product of factors?
Pathway 1: ……………………
Pathway 2: ……………………
Pathway 3: ……………………
Pathway 4: ……………………

Activity E
Find the common feature of the areas created by the 4 pathways A.L.E.X. followed.

**Fig. 4** Worksheet activities on square numbers

teachers are in accord with the main findings of the literature, which indicates that digital games are more effective when acting as adjuncts to a range of teaching methods rather than as stand-alone applications (Gee, 2007).

For illustration purposes, Fig. 4 presents the worksheet developed and used by one of the three groups of teachers, during their teaching intervention within a sixth grade classroom.

This group's lesson aimed at the exploration, via A.L.E.X., of the concept of square numbers. Children, in groups of four, worked with iPads on the basis of the given worksheet. The method used was the same with the one employed in the classroom workshop of Phase I. The other two groups also designed and implemented high quality lessons that made constructive use of A.L.E.X. and fitted well with their targeted grade level and curricular topic.

While the classroom experimentation further strengthened teachers' belief that appropriate use of apps can be highly motivational and can help create more conducive learning environments, it also helped participants to build more realistic expectations about what the integration of tablets in mathematics classrooms entailed in practice. Teachers took a stance of inquiry towards their teaching research experience, listing not only the benefits but also various challenges of using apps (e.g. resource limitations, issues of tablet safety, classroom management issues), and making recommendations for a productive integration of tablets into teaching practices.

## Conclusions and Implications for Teaching and Research

The success of tablets as a tool for learning mathematics in classroom situations will ultimately depend on the abilities of teachers to take full advantage of their educational potential. Findings from the pre-survey completed by teachers at the outset of the current study corroborate with the research literature, which indicates that the majority of teachers have positive attitudes towards the adoption of tablets in the mathematics classroom. However, they tend to lack appreciation of their true potential for transforming teaching and learning (Daccord, 2012) and view apps as instructional tools to be used mainly for motivational or drill-and-practice purposes.

Concurring with the literature (e.g. Niess et al., 2009), our research has illustrated the usefulness of TPACK as a means of studying and facilitating teachers' professional growth in the use of tablets in education. Key conclusions from the analysis of the data collected during the study were that the professional development program was quite successful in helping this group of educators to move beyond their restricted views of apps as educational tools, and that the program improved their confidence and ability to integrate apps within the mathematics curriculum. Although there was no pre-post assessment to formally track changes in participants' TPACK, there are strong indications in the collected data of improvement in the participants' perceptions of, and proficiency with, apps in mathematics instruction.

There are important implications of the study for the design or redesigning of pre-service or in-service teacher training curricula and programs on tablets' integration. The study design and outcomes shed light on what effective teacher training in mobile mathematics learning might entail in helping teachers learn about, adopt, and integrate tablets into their teaching. Insights from the study indicate that utilizing a conceptually based theoretical framework, such as TPACK, can enrich teachers' professional development. In developing teacher training programs, the TPACK model can serve: as a useful framework for identifying the knowledge base for teaching mathematics with tablet devices; as a guide for understanding how knowledge about mathematical content, pedagogy, and technology overlap to inform choices for curriculum and instruction; and as a model to promote and evaluate the impact of professional development on teacher TPACK. In turn, strengthening teachers' TPACK enhances their ability to purposefully and effectively integrate tablets into instruction.

Findings also indicate that the development of teachers' TPACK necessitates the provision of opportunities for both theoretical and experiential learning of technology-based pedagogical approaches. Concurring with prior research (e.g. Angeli & Valanides, 2009), this study provides evidence that teachers' involvement in professional development activities such as lesson design and field experience (e.g. conduct of action research, classroom teaching, classroom observation), can support them in developing their teaching competencies with ICT and deepen their understanding of TPACK in ways transferable into their own practice.

This study was exploratory in nature. Clearly, the presented results are only suggestive and warrant further research to better understand the TPACK development of teachers using apps to support mathematics learning. The exploratory nature of the investigation, the qualitative methodology used to research the case, the small scale of the study, and its limited geographical nature mean that the drawing of generalizations should be done very cautiously. Future iterations ought to investigate the impact of professional development on teachers' TPACK competencies and teaching skills (e.g. use of control groups, collection of pre- and post-data on the actual impact of teachers' developed TPACK on children's motivation and higher-level mathematical learning). This approach, conducted with a larger and more representative sample, could lead to the development of generalized principles and models of professional development that can help foster the expertise of mathematics teachers in incorporating apps into their pedagogy.

## References


Angeli, C., & Valanides, N. (2009). Epistemological and methodological issues for the conceptualization, development, and assessment of ICT-TPCK: Advances in technological pedagogical content knowledge (TPCK). *Computers & Education, 52*(1), 154–168.

Attard, C. (2015). Introducing iPads into primary mathematics classrooms: Teachers' experiences and pedagogies. In M. Meletiou-Mavrotheris, K. Mavrou, & E. Paparistodemou (Eds.),

*Integrating touch-enabled and mobile devices into contemporary mathematics education* (pp. 193–213). Hershey, PA: IGI Global.

Aydin, E. (2005). The use of computers in mathematics education: A paradigm shift from computer assisted instruction towards student programming. *The Turkish Online Journal of Educational Technology, 4*(2), 27–34.

Becker, K. (2007). Teaching teachers about serious games. In C. Montgomerie, & J. Seale (Eds.), *Proceedings of World Conference on Educational Multimedia, Hypermedia and Telecommunications 2007* (pp. 2389–2396). Chesapeake, VA: AACE.

Bennett, K. R. (2011). Less than a class set. *Learning and Leading with Technology, 39,* 22–25.

Blackwell, C. (2014). Teacher practices with mobile technology: Integrating tablet computers into the early childhood classroom. *Journal of Education Research, 7,* 1–25.

Burden, K., Hopkins, P., Male, T., Martin, S., & Trala, C. (2012). *iPad Scotland Evaluation.* Hull, Humberside: University of Hull.

Clark, W., & Luckin, R. (2013). *What the research says—iPads in the classroom.* London Knowledge Lab, Institute of Education: University of London.

Clements, D. H., Battista, M. T., & Sarama, J. (2001). Logo and geometry. *Journal for Research in Mathematics Education, Monograph 10.* Reston, VA: National Council of Teachers of mathematics. https://doi.org/10.2307/749924.

Cohen, S. (2012). A 1:1 IPad initiative—Vision to reality. *Library Media Connection, 30*(6), 14–16.

Daccord, T. (2012). 5 critical mistakes schools make with iPads (and how to correct them). *Edudemic.* Retrieved from http://www.edudemic.com/5-critical-mistakes-schools-ipads-and-correct-them/.

Department for Education and Communities. (2012). *Use of Tablet technology in the classroom.* Sydney, Australia: NSW.

EDUCAUSE (2011). *7 Things you should know about iPad apps for learning.* Publication of EDUCASUSE Learning Initiative. Retrieved from http://net.educause.edu/ir/library/pdf/ELI7069.pdf.

Gee, J. P. (2007). *What video games have to teach us about learning and literacy.* New York, NY: Palgrave Macmillan.

Guida, C. T. (2014). *A.L.E.X. (version 1.4)* [mobile application software]. Retrieved from https://itunes.apple.com/us/app/a.l.e.x./id597040772?mt=8.

Heinrich, P. (2012). *The iPad as a tool for education: A study of the introduction of iPads at Longfield Academy.* Kent. Nottingham: NAACE: The ICT Association.

Henderson, S., & Yeow, J. (2012). IPad in education—A case study of iPad adoption and use in a primary school. HICSS. In *Proceedings of the 2012 45th Hawaii International Conference on System Sciences* (pp. 78–87). Grand Wailea, Maui, HI, USA.

Hopscotch Technologies. (2014). *Hopscotch (version 2.13.1)* [mobile application software]. Retrieved from https://itunes.apple.com/gb/app/hopscotch-coding-for-kids/id617098629?mt=8&ign-mpt=uo%3D4

Johnson, L., Adams Becker, S., Cummins, M., Estrada, V., Freeman, A., & Ludgate, H. (2013). *NMC Horizon Report: 2013 K-12 Edition.* Austin, Texas: The New Media Consortium.

Kyriakides, A. O., Meletiou-Mavrotheris, M., & Prodromou, T. (2015). Changing children's stance towards mathematics through mobile teaching: The case of robot A.L.E.X. In M. Meletiou-Mavrotheris, K. Mavrou, & E. Paparistodemou (Eds.), *Integrating touch-enabled and mobile devices into contemporary mathematics education* (pp. 122–145). IGI Global: USA.

Kyriakides, A. O., Meletiou-Mavrotheris, M., & Prodromou, T. (2016). Mobile devices in the service of students' learning of mathematics: The example of game application A.L.E.X. in the context of a primary school in Cyprus. *Mathematics Education Research Journal, 28*(1), 53–78.

Larkin, K. (2015). The search for fidelity in geometry apps: An exercise in futility? In M. Marshman, V. Geiger, & A. Bennison (Eds.), *Mathematics Education in the Margins. Proceedings of the 38th Annual Conference of the Mathematics Education Research Group of Australasia* (pp. 341–348). Sunshine Coast: MERGA.

McKenna, C. (2012). There's an app for that: how two elementary classrooms used iPads to enhance student learning and achievement. *Education, 2*(5), 136–142.
Melhuish, K., & Falloon, G. (2010). Looking to the future: M-learning with the iPad. *Computers in New Zealand Schools: Learning, Leading, Technology, 22*(3), 1–16.
Milman, N., Carlson-Bancroft, A., & Vanden Boogart A. (2012). *iPads in a Pre K-4th independent school—Year 1—Enhancing engagement, collaboration, and differentiation across content areas*. Paper Presented at the International Society for Technology in Education Conference. San Diego, CA.
Mishra, P., & Koehler, M. J. (2006). Technological pedagogical content knowledge: A framework for teacher knowledge. *Teachers College Record, 108*(6), 1017–1054.
Niess, M. L., Ronau, R. N., Shafer, K. G., Driskell, S. O., Harper, S. R., & Johnston, C. (2009). Mathematics teacher TPACK standards and development model. *Contemporary Issues in Technology and Teacher Education, 9,* 4–24.
Papert, S. (1980). *Mind storms: Children, computers and powerful ideas*. New York: Basic Books.
Pastore, R. S., & Falvo, D. A. (2010). Video games in the classroom: Pre- and in-service teachers' perceptions of games in the K-12 classroom. *International Journal of Instructional Technology and Distance Learning, 7*(12), 49–57.
Puentedura, R. (2006). *The SAMR model: Background and exemplars*. Retrieved from http://www.hippasus.com/rrpweblog/archives/2012/08/23/SAMR_BackgroundExemplars.pdf
Robinson, B. (1998). A strategic perspective on staff development for open and distance learning. In C. Latchem & F. Lockwood (Eds.), *Staff development in open and flexible learning* (pp. 33–44). Routledge: London and New York.
Shulman, D. (1986). Those who understand: Knowledge growth in teaching. *Educational Researcher, 15*(2), 4–14.
Subhi, T. (1999). The impact of LOGO on gifted children's achievement and creativity. *Journal of Computer Assisted learning, 15*(2), 98–108.
Templin, M. A., & Bombaugh, M. (2005). An innovation in the evaluation of teacher professional development serving reform in science. *Journal of Science Teacher Education, 16,* 141–158.
TTS Group Limited. (2012). *Bee-Bot (version 1.2)* [mobile application software]. Retrieved from https://itunes.apple.com/gb/app/bee-bot-pyramid/id509207211?mt=8&ign-mpt=uo%3D4.
Whitaker, S., Kinzie, M., Kraft-Sayre, M. E., Mashburn, A., & Pianta, R. C. (2007). Use and Evaluation of web-based professional development services across participant levels of support. *Early Childhood Education Journal, 34*(6), 379–386.
Wilson, A., Hainey, T., & Connolly, T. M. (2012). *Evaluation of computer games developed by primary school children to gauge understanding of programming concepts*. Paper Presented at the 6th European Conference on Games-Based Learning (ECGBL), 4–5 October 2012. Cork, Ireland.

**Andreas O. Kyriakides** is a full-time primary school teacher and a scientific collaborator at the Department of Educational Studies of European University Cyprus. He has a Ph.D. in Mathematics Education (The Open University, UK), an M.A. in Curriculum and Teaching (Michigan State University, USA) and a B.A. in Elementary Education (University of Cyprus). In the field of Teacher Education he is interested not only in training prospective teachers to acquire the mathematics knowledge base but also in sharpening their sensitivity to the complexities of learning and directing their attention to practices and constructs which can inform preferences when teaching mathematics.

**Maria Meletiou-Mavrotheris** is a Professor of Mathematics Education at the European University Cyprus, and Director of the Research Laboratory in ICT-Enhanced Education. She has a Ph.D. in Mathematics Education (University of Texas at Austin), an M.Sc. in Statistics (University of Texas at Austin), an M.Sc. in Engineering (University of Texas at Austin), an M.A. in Open and Distance Learning (UK Open University), a B.A. in Mathematics (University of Texas at Austin), and Teacher's Diploma in Elementary Education (Pedagogical Academy of Cyprus). Her research work focuses on the study of issues related to the teaching and learning of statistics, while at the same time equally contributing to several other areas of mathematics, science, and technology education. She has established a respected research record through numerous publications in scholarly international journals and edited volumes, and has been able to attract considerable funding from national and international sources. Her research record includes the coordination of multiple EU funded projects.

# Supporting Teachers' Orchestration of Mobile Learning Activities

**Håkan Sollervall, Didac Gil de la Iglesia and Janosch Zbick**

**Abstract** In this chapter we explore how an innovative mobile learning activity, designed by the authors can be implemented by mathematics classroom teachers. The focal part of the activity involves spatial orientation tasks that are executed with the support of customized mobile technologies in an outdoor setting. In this chapter, we present a comprehensive account of our research efforts spanning a five-year period and focus on providing didactical and technological support for teachers' informed orchestration of the technology enabled learning activity.

## Introduction

In any educational context, teaching and learning are closely connected. Teachers plan and implement teaching activities that stimulate learning activities among their students. Although researchers sometimes disagree about what the key factors for promoting good learning conditions are, they usually agree that learning processes involve thinking as well as action (Bruner, 1966). Teachers need to provide support for cognitive as well as social processes when planning and implementing teaching activities. The connection between social processes and learning effects has been extensively studied. For example, constructive behavior involving self-regulated explorations of problem situations results in better learning effects than "active" behavior involving only the utilization of known methods to solve routine problems (Chi, 2009). Further improvements to learning may be achieved if constructive

H. Sollervall (✉) · D. Gil de la Iglesia · J. Zbick
Linnaeus University, Växjö, Sweden
e-mail: hakan.sollervall@lnu.se

D. Gil de la Iglesia
e-mail: didacgil9@gmail.com

J. Zbick
e-mail: janosch.zbick@lnu.se

 
N. Calder et al. (eds.), *Using Mobile Technologies in the Teaching and Learning of Mathematics*, Mathematics Education in the Digital Era 12,
https://doi.org/10.1007/978-3-319-90179-4_6

students interact with each other and explore problem situations together (Chi, 2009). However, it is a challenging task for any teacher to arrange didactical situations that promote interactive behavior while maintaining individually constructive behavior. The teacher has to provide support for students who may switch from constructive to merely active behavior in a group where they are not invited to engage in strategic deliberations and decision-making. Furthermore, the teacher has to ensure that the students' explorations are directed towards achieving relevant learning objectives.

The educational research community has long been concerned with suggesting various forms for supporting goal-oriented constructive and interactive behavior. For example, the French theory of didactical situations (Brousseau, 1997) proposes a lesson structure in three phases—devolution, adidactical situations, and institutionalization—that serves to scaffold students' goal-oriented explorations and to sum up their experiences by connecting them to institutionalized learning objectives. The institutionalization phase is similarly addressed in the Japanese tradition of problem solving oriented teaching (Shimizu, 1999) where the teacher follows up on students' activities by involving them in whole-class discussion (neriage) and summing up (matome) (Asami-Johansson, 2015). In a classroom where students engage wholeheartedly in problem solving activities, the teacher is faced with the tough decision to interrupt their work in favor of engaging them in the final phase where their experiences are summed up and institutionalized.

Problem solving oriented teaching requires carefully selected or specifically designed activities and tasks whose epistemological content align well with desired learning objectives. In our case, the researchers' design tasks that are negotiated with teachers. Our research group places emphasis on designing innovative teaching activities, for the purpose of supporting learning of mathematics. We implement these activities following a "naturalistic" approach where mathematics teachers are responsible for interpreting, adapting and implementing our suggestions with their students. As a consequence, our activities are designed with some degree of freedom and often include several possible scenarios for the teacher to pursue. For example, the teacher can choose to follow up on the students' structured explorations in an outdoor setting by focusing on observed arithmetic or geometrical patterns as well as mathematical reasoning. The researchers' suggested scenarios are communicated and negotiated with the teacher. The teaching activities are based on designed tasks that are introduced by the teacher and explored by the students, usually arranged in pairs or in small groups.

The current teaching activity involves customized mobile technologies as an intrinsic feature. Since neither the students nor the teacher have previously worked with such technologies, their implementation has to be carefully considered by the researchers and the teacher. Following a naturalistic approach, where the teacher has a prominent role, the researchers' ambition has been to provide user-friendly technologies and a teaching activity that can be introduced to the teacher during a short introductory session. The technology we have developed during this research project is called TriGo LNU. It is customized with respect to the specific teaching activity and includes a web-based[1] authoring tool and support for visualization, as well as an app that students can use to download unique sets of tasks.

In this chapter, we provide an overview of the researchers' efforts over a five-year period, regarding the development of a complex mobile learning activity. We provide a brief account of the first cycle of the design process where a preliminary activity was tested with a focus on technological performance and activity flow. The second cycle of the design process will be considered at a detailed level with focus placed on the researchers' strategies for scaffolding the orchestration of a generic activity by teachers in their classrooms. The information in this chapter builds upon the work presented in previous publications in conference proceedings (Gil de la Iglesia, Andersson, Milrad, & Sollervall, 2012; Sollervall et al., 2011; Sollervall & Gil de la Iglesia, 2015) and scientific books and journals (Peng & Sollervall, 2014; Sollervall & Milrad, 2012).

## Design Methodology

Our design research efforts involve collaboration between researchers with different backgrounds. The first author of this chapter teaches mathematics at university level and pursues educational design research in mathematics education, while the second and third authors conduct research in media technology with a focus on designing mobile technologies for educational purposes. The researchers have complementary expertise regarding the design of mathematical teaching activities supported by mobile technologies. After working collaboratively over a five-year period, we have established a common understanding regarding the essential didactical underpinnings and technological features of the teaching activity explored in this chapter. Both the didactical underpinnings and the technological features, presented in the following sections, have been introduced during the design process for the purpose of supporting orchestration of the activity and achieving specific learning objectives. This common local understanding of teaching and learning in relation to a specific teaching activity is achieved through researchers collaborating in a design process that is structured in cycles.

While the first cycle activity can be regarded as testing of a prototype, the second cycle provided a matured activity where it became meaningful to analyze the teacher's performance (Fig. 1; Sollervall & Gil de la Iglesia, 2015). In a future third cycle, where several teachers implement the activity at different locations under naturalistic conditions, will be meaningful to analyze in relation to what the students actually learn from the activity. The second cycle focused on providing opportunities for learning and was analyzed from a teaching perspective.

The cyclic design model (Fig. 1) departs from goals but does not account for how these goals are identified. In our case, the researchers were invited to observe

[1]http://trigo.lnu.se.

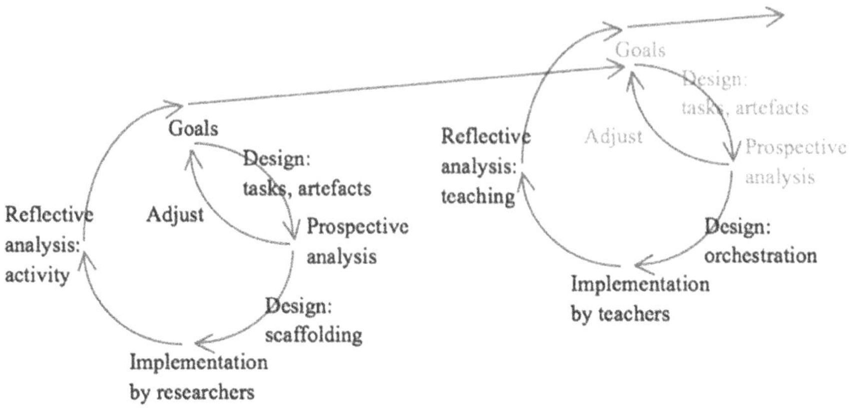


**Fig. 1** A two-cycle strategy for designing teaching activities with technologies

the demonstration of an outdoor activity involving a customized technological application that enabled students to measure the distance between two mobile devices (Spikol & Eliasson, 2010). The observed activity was designed by another group of researchers (ibid.) for students in grade 7–9 (13–16 years old). A teacher's suggestion during the demonstration stimulated a discussion about possible ways to improve the activity. The suggestion was to let the students make a personal estimate before checking the actual distance with the mobile device.

The purpose of observing the outdoor activity was to find opportunities for collaboration between researchers and teachers regarding the implementation of innovative technologies in meaningful educational settings, in our case with relevance for mathematics education.

## Didactical Underpinnings for the Teaching Activity

A few months after having observed the outdoor demonstration, the researchers in mathematics education and media technology decided to initiate a new project based on the available technological application, with students in grade 4–6 as the primary target group. In an attempt to move beyond utilizing the application as an instrument for measuring distances, the first author suggested designing tasks that required students to coordinate themselves with respect to two distances, each with respect to a given point on an open field. An example of such a task is indicated below (Fig. 2). The students stand on a large open field where they can see two physical markers (a square and a triangle). They are informed that their starting point is located 46 m away from each marker (Fig. 2a). The task "42 26" calls for the students to find the location that is 42 m away from the square and 26 m away from the triangle (Fig. 2b). Further details about tasks and the comprehensive teaching activity are provided in a later section.

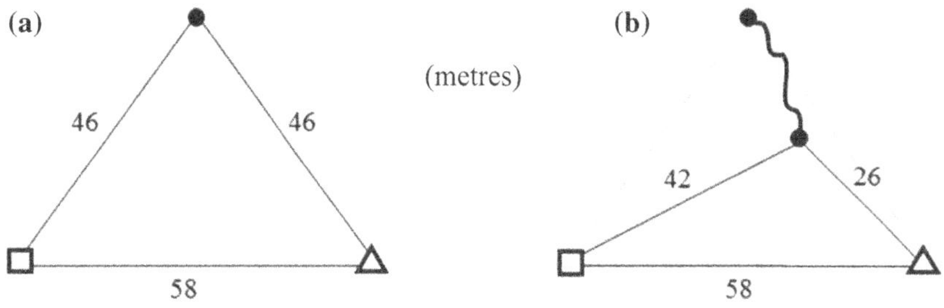


**Fig. 2** **a** The starting point. **b** The goal point

This task involves physical coordination in full-sized space and connects with the scientific notion of spatial orientation. From a mathematics education perspective, spatial orientation is a component of spatial ability that determines our ability to navigate from place to place, identify an object moving towards us, estimate quantities, understand drawings and charts, and compose various items (Patkin & Dayan, 2013). Since spatial orientation requires moving around in full-sized space and positioning self to object, it is less commonly addressed in classroom practices than object manipulation, another component of spatial ability that only requires manipulating objects. Psychometric research has established that spatial orientation is qualitatively different from object manipulation (Kozhevnikov & Hegarty, 2001) and thus needs to be addressed in the development of students' spatial ability. These findings align well with the claim by Bishop (1980, p. 260) that "insofar as we are concerned with spatial ideas in mathematics as opposed to just visual ideas, we must attend to large, full-sized space, as well as to space as it is represented in models, and in drawings on paper". Engaging in activities in full-sized space may also stimulate students' enactive mode of action and thinking, in addition to the iconic and symbolic modes that are more commonly stimulated in mathematical classroom practices (Bruner, 1966).

From a mathematical perspective the coordination of two given distances with respect to two given reference points can be interpreted as the construction of a triangle with three given sides, as treated in Euclid's Elements (Heath, 1908, p. 292; Fig. 3).

In the first design cycle, the teaching activity was implemented with students in grade 6 (12–13 years old). The activity was distributed across time (December

**Book I Proposition 22**
*Out of three straight lines, which are equal to three given straight lines, to construct a triangle: thus it is necessary that two of the straight lines taken together in any manner should be greater than the remaining one.*

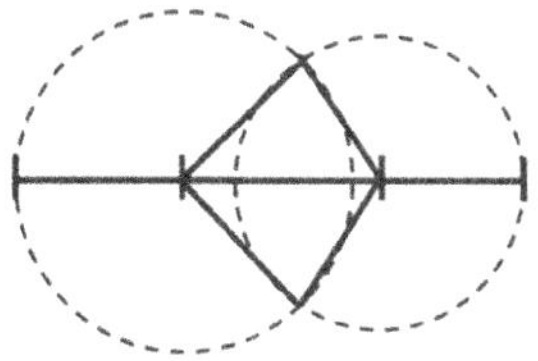

**Fig. 3** Euclidean construction of a triangle with three given sides

2010–May 2011) and across locations (indoor and outdoor), inspired by the notion of mobile-assisted seamless learning (MSL, Wong & Looi, 2011). MSL also addresses the integration of personal and social learning, combining digital and physical worlds, utilizing digital and traditional tools, features that were also addressed in the first cycle activity.

However, several challenges (didactical, organizational, as well as technological challenges) were identified after the completion of the first design cycle. The main didactical challenge concerned following up on the students' experiences. In the first cycle, the researchers let the teacher take care of the follow-up session in the classroom. The students were invited to share their experiences on the board (supported by visual web technologies) but their presentations were not clearly connected with mathematical learning objectives. Regarding organization of the follow-up session, the researchers had assumed that the teacher would support such connective work in the classroom. However, the teacher had not prepared to do this and was also not sure if he was allowed to interfere with the on-going research activities. He prioritized to confirm the students' work and chose to accept their rather superficial presentations. Regarding the use of technologies in the classroom, the students were instructed to support their presentations by displaying their specific field data on a map that was projected on the whiteboard. However, they were not able to readily access their data on the classroom computer and some students chose to present their findings without involving the intended technology enabled visualizations. These shortcomings indicated that the classroom activities needed to be more carefully designed to support the teacher's work as well as the students' presentations.

In the second design cycle, it was decided to maintain a focus on spatial orientation. In the first cycle, the teacher had not received much support from the researchers to arrange the concluding discussion in the classroom. As a consequence, the students reported on their experiences and their findings were confirmed, but not elaborated on. In an attempt to provide a supporting structure for the teaching activity, the researchers chose to involve the theory of didactical situations (Brousseau, 1997) in the second cycle. The fundamental structure of a didactical situation—devolution, adidactical situations, institutionalization (Fig. 4; adapted from Balacheff, 2013)—fits particularly well when designing across physical contexts.

While the introductory phase of devolution and the concluding phase of institutionalization are led by the teacher, the students are responsible for taking action in the adidactical situations. The three phases within a didactical situation become

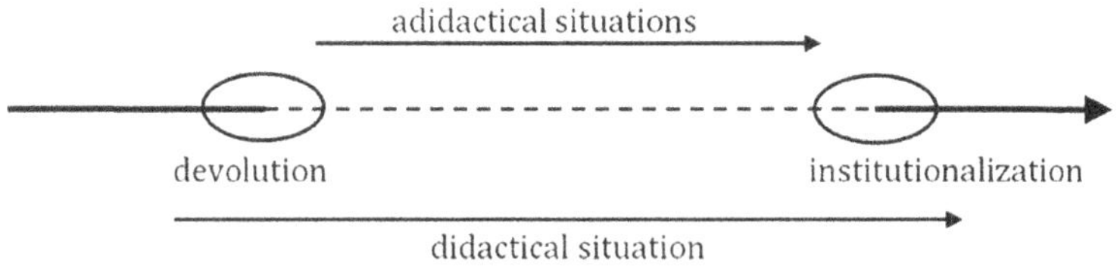


**Fig. 4** Structure of a didactical situation

(a)

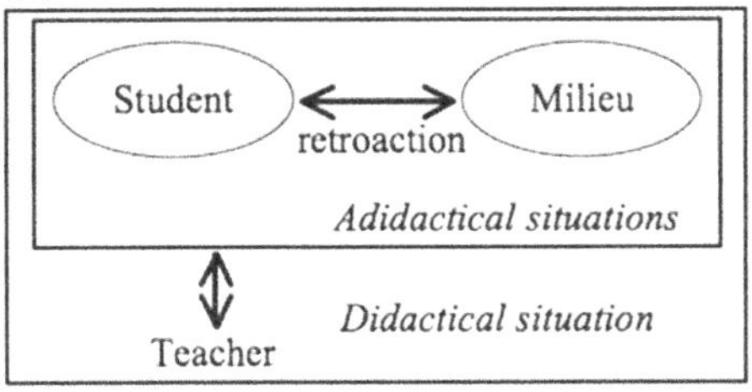


(b)

| Praxis | Logos |
|---|---|
| Tasks<br>Techniques | Technology<br>Theory |

**Fig. 5** **a** The structure of a didactical situation. **b** The elements of a praxeology

naturally separated if adidacticity is promoted by giving the students full responsibility for the technology-supported exploration of mathematical tasks by retroacting only with the milieu and not the teacher (Fig. 5a, adapted from Bessot, 2003, p. 7).

In order to capture qualitative differences of the learning opportunities that are offered during a didactical situation, we chose to utilize the notion of praxeologies (Fig. 5b).

While praxis, that is, tasks and techniques, naturally dominates within the adidactical situations, the focus may shift towards logos in the phase of institutionalization where the students are invited to reflect on their experiences by engaging in discussions about how and why the techniques work. A technology-oriented discourse may include describing techniques, explaining how they work and when they work, while a theory-oriented discourse aims at justifying the techniques and the technological claims (Rodríguez, Bosch, & Gascón, 2008). Several possible focal activities, related to technology and theory, were identified by the researchers.

**Technology**: identifying and comparing strategies for construction, distinguishing between possible and impossible constructions;

**Theory**: justifying the strategies particularly the circle strategy, formulating criteria and arguing why some constructions are possible and others are not.

The teaching activity in the second cycle was designed as a complete didactical situation, encompassing devolution, an outdoor activity, and institutionalization, with the ambition to stimulate a logos-oriented discourse during the phase of institutionalization.

## Digital and Material Infrastructure for the Teaching Activity

Implementing the coordination tasks in the first and second design cycles called for designing new technological applications that enabled each mobile device to measure either distances with respect to two fixed points or the devices held by two

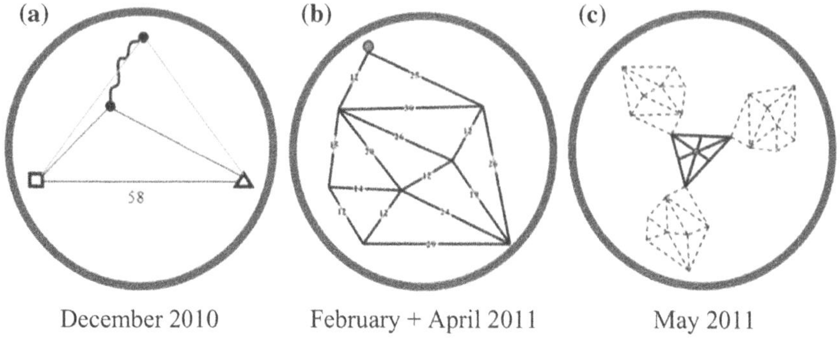


Fig. 6 a First activity. b Second activity. c Third activity

other students (Sollervall et al., 2011). Furthermore, the mobile technologies were used to document the students' attempts to find the goal points (Fig. 2). Customized web technologies enabled the students to show selected attempts on a map, in the classroom. The modified mobile technologies were deployed on Android phones that were provided by the researchers. The locations of the field markers and the students were identified through their respective GPS coordinates.

In the first cycle, the teaching activity consisted of three separate activities (Fig. 6) designed with increasing complexity regarding forms of interaction: cooperation (students working together with the same subtasks), collaboration (students working together with different subtasks), and jigsaw (student groups first working separately and then cooperating as groups) (Dillenbourg, 2009).

The three activities were implemented on four separate occasions (the second activity was implemented twice, in slightly different versions).

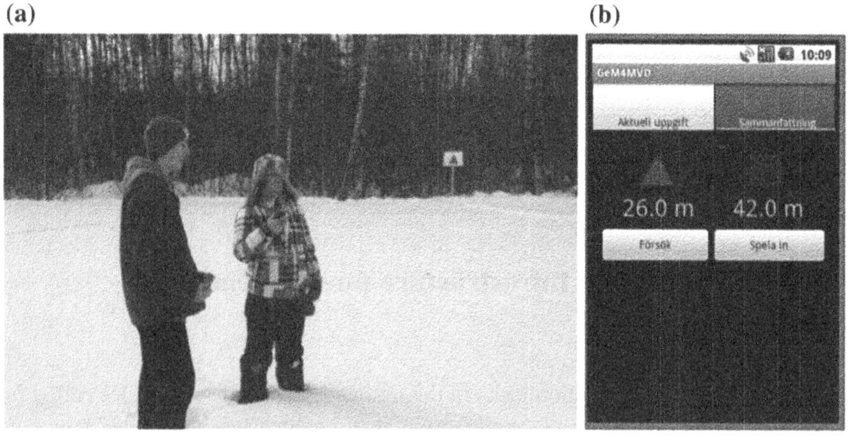


Fig. 7 a Students working on a task. b The display

The first activity (Fig. 6a) was tested with 12 grade 6 students (12–13 years old) in Sweden. They worked in pairs with 10 identical tasks, presented in a unique sequence for each group. One such sequence is shown in Fig. 8b. Each task called for coordinating distances against two fixed markers (Fig. 7; compare Fig. 2). The students' attempts were displayed and recorded on the mobile device (Fig. 8a, c). Attempts within two metres were accepted as "correct", since testing showed that a tolerance of two metres was enough to compensate for the inherent inaccuracies of the GPS technology. In the example shown below the students reached the goal point in their third attempt (Fig. 8c).

The 10 coordination tasks were planned for and executed in a specific outdoor environment, on a farmer's field in a rural area in southern Sweden (Fig. 9). The second and third activity in the first cycle (Fig. 6b, c) were planned for and executed on the same field, with further modified mobile technologies that supported measuring distances between two mobile devices. Details about the second and third activities can be found in previous publications (Sollervall et al., 2011; Sollervall & Milrad, 2012).

As mentioned earlier, several challenges were identified after the completion of the first design cycle. For example, researchers implemented the first activity in the first cycle with only 12 students. The ambition in the second cycle was to design a teaching activity that could involve a whole class (up to 30 students, a normal class size in Swedish middle school) and could be implemented with their regular teacher. The researchers also wished to pursue the goal to provide a generic activity that could be easily implemented by classroom teachers.

The design solution in the second cycle involved a total of six markers. Four of these markers were placed on the corners of a rectangular shape with dimensions 70 m by 28 m. The dimensions were chosen to fit on a sports field located directly

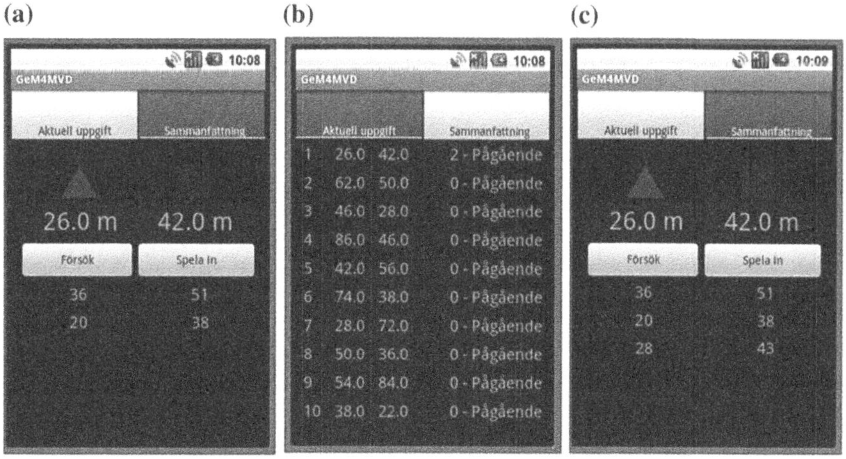


**Fig. 8** **a** After two attempts. **b** Activity log. **c** After three attempts

**Fig. 9** Birds-eye view of the farmer's field and task locations

behind a school building. The two remaining markers were placed on the midpoints of the longer edges (Fig. 10).

A total of ten tasks were designed for the activity. Eight possible constructions were mixed with two impossible constructions, such as "10 15" when the distance between the reference points was 35 m, in order to support a technology-oriented discourse during the institutionalization phase.

The activity was designed so that the students could work with their ten tasks in small groups, simultaneously and independently, in the schoolyard. The reference points were systematically varied among the six markers. For example, group 5 had their first task "20 30" against the house and the tree (Figs. 11 and 12; house-tree coded as 2–3 in the first task of group 5) while group 9 had "20 30" as their tenth

**Fig. 10** **a** The presented google map. **b** The actual field

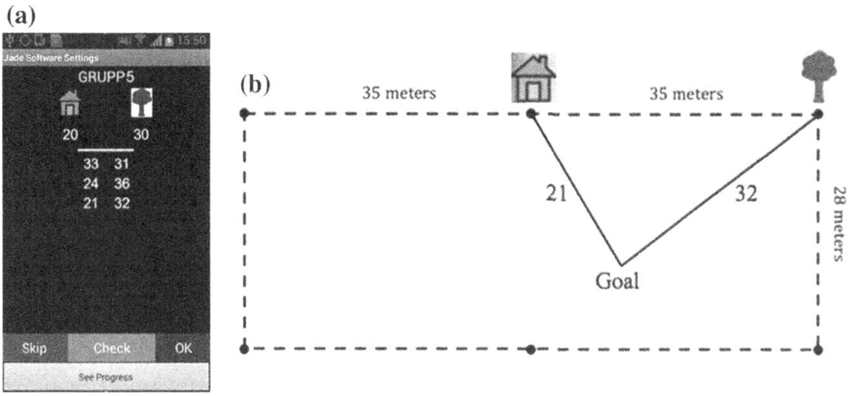


**Fig. 11** **a** The display. **b** Illustration of the solved task

| | 1 | 2 | 3 | 4 | 5 | 6 | 7 | 8 | 9 | 10 | 11 | 12 | 13 | 14 |
|---|---|---|---|---|---|---|---|---|---|---|---|---|---|---|
| 1 | 20-30<br>6-1<br>g | 20-30<br>2-5<br>g | 20-30<br>5-2<br>g | 20-30<br>1-2<br>c | 20-30<br>2-3<br>c | 20-30<br>4-5<br>c | 20-30<br>5-6<br>c | 30-20<br>2-5<br>g | 30-20<br>5-2<br>g | 30-20<br>3-4<br>g | 30-20<br>1-2<br>c | 30-20<br>2-3<br>c | 30-20<br>4-5<br>c | 30-20<br>5-6<br>c |
| 2 | 10-30<br>4-5<br>c | 10-30<br>2-3<br>c | 10-30<br>5-6<br>c | 20-15<br>3-4<br>g | 20-15<br>6-1<br>g | 20-15<br>2-5<br>g | 20-15<br>5-2<br>g | 30-10<br>1-2<br>c | 30-10<br>2-3<br>c | 30-10<br>5-6<br>c | 15-20<br>2-5<br>g | 15-20<br>5-2<br>g | 15-20<br>6-1<br>g | 15-20<br>3-4<br>g |
| 3 | 50-25<br>1-2<br>y | 50-25<br>4-5<br>y | 30-60<br>2-3<br>y | 25-25<br>1-2<br>g | [illegible]<br>[illegible]<br>r | [illegible]<br>[illegible]<br>r | 60-30<br>1-2<br>y | 25-50<br>5-6<br>y | 25-50<br>2-3<br>y | 60-30<br>4-5<br>y | 30-60<br>5-6<br>y | [illegible]<br>[illegible]<br>r | [illegible]<br>[illegible]<br>r | 25-25<br>5-6<br>c |
| 4 | 35-50<br>1-3<br>n | 35-50<br>4-6<br>n | 65-30<br>1-3<br>n | 65-30<br>4-6<br>n | 25-25<br>2-5<br>g | 25-25<br>5-2<br>g | [illegible]<br>2-3<br>r | 50-35<br>1-3<br>n | 50-35<br>4-6<br>n | 30-65<br>1-3<br>n | 30-65<br>4-6<br>n | 25-25<br>6-1<br>g | 25-25<br>3-4<br>g | [illegible]<br>[illegible]<br>r |
| 5 | 10-15<br>4-5<br>p | 10-15<br>1-2<br>p | 50-25<br>4-5<br>y | 50-25<br>1-2<br>y | 60-30<br>4-5<br>y | 60-30<br>1-2<br>y | 65-30<br>4-6<br>n | 15-10<br>5-6<br>p | 15-10<br>2-3<br>p | 25-50<br>5-6<br>y | 25-50<br>2-3<br>y | 30-60<br>5-6<br>y | 30-60<br>2-3<br>y | 30-65<br>1-3<br>n |
| 6 | 25-25<br>5-2<br>g | 25-25<br>3-4<br>g | [illegible]<br>[illegible]<br>r | 10-55<br>[illegible]<br>r | 35-50<br>1-3<br>n | 35-50<br>4-6<br>n | 10-15<br>1-2<br>p | 25-25<br>1-2<br>g | 25-25<br>2-5<br>g | [illegible]<br>1-2<br>r | [illegible]<br>[illegible]<br>r | 50-35<br>1-3<br>n | 50-35<br>4-6<br>n | 15-10<br>5-6<br>p |
| 7 | [illegible]<br>[illegible]<br>r | 55-10<br>[illegible]<br>r | 25-25<br>6-1<br>g | 30-60<br>5-6<br>y | 25-50<br>5-6<br>y | 25-50<br>2-3<br>y | 25-25<br>4-5<br>c | 10-55<br>[illegible]<br>r | [illegible]<br>[illegible]<br>r | 25-25<br>3-4<br>g | 25-25<br>5-2<br>g | 50-25<br>4-5<br>y | 50-25<br>1-2<br>y | 30-60<br>2-3<br>y |
| 8 | 30-60<br>5-6<br>y | 30-60<br>2-3<br>y | 15-10<br>5-6<br>p | 15-10<br>2-3<br>p | 30-20<br>3-4<br>g | 30-20<br>6-1<br>g | 30-20<br>5-2<br>g | 60-30<br>4-5<br>y | 60-30<br>1-2<br>y | 10-15<br>4-5<br>p | 10-15<br>1-2<br>p | 20-30<br>3-4<br>g | 20-30<br>2-5<br>g | 20-30<br>5-2<br>g |
| 9 | 65-30<br>4-6<br>n | 65-30<br>1-3<br>n | 35-50<br>4-6<br>n | 35-50<br>1-3<br>n | 10-15<br>4-5<br>p | 10-15<br>1-2<br>p | 25-50<br>5-6<br>y | 30-65<br>4-6<br>n | 30-65<br>1-3<br>n | 50-35<br>4-6<br>n | 50-35<br>1-3<br>n | 15-10<br>5-6<br>p | 15-10<br>[illegible]<br>p | 25-50<br>2-3<br>y |
| 10 | 30-20<br>1-2<br>c | 30-20<br>4-5<br>c | 30-20<br>3-4<br>g | 30-20<br>6-1<br>g | 65-30<br>4-6<br>n | 65-30<br>1-3<br>n | 35-50<br>1-3<br>n | 20-30<br>1-2<br>c | 20-30<br>4-5<br>c | 20-30<br>6-1<br>g | 20-30<br>2-5<br>g | 30-65<br>4-6<br>n | 30-65<br>1-3<br>n | 50-35<br>4-6<br>n |

**Fig. 12** Spreadsheet with 14 sets of tasks (The letters g, c, y, n, r, p represent colors)

task against the bicycle and the car (Fig. 12: bicycle-car coded as 4–5 in the tenth task of group 9).

The customized mobile application was modified to accommodate the six reference points (Fig. 11a). In the second design cycle, the phones (Androids) were provided by the researchers but were not programmed with individual task sequences. Instead, the researchers provided an app that enabled each group of students to smoothly download their unique task sequence.

The tasks were color coded on a spreadsheet according to their individual nature (grey, clear: short distances; yellow, green: long distances; red, purple: impossible distances). A black and white version of this spreadsheet is shown in Fig. 12, where these colors are indicated with the letters g, c, y, n, r, p, respectively. The didactical situation was designed for up to 14 groups of 2–3 students and targeting students in grades 4–6. The 14 sets of tasks were prepared to be downloaded to the phones (Androids) that were provided by the researchers (Fig. 11a).

## Results from the First Cycle Activity

At the implementation in December 2010, the 12 participating students were randomly organized into six groups (pairs). To avoid having the groups following each other (in order to complete their ten tasks) six variations of the initial sequence of points were constructed based on symmetry (interchanging distances to the reference points) and order of the tasks. In order to put focus on the physical orientation ability, it was decided not to provide visual references on the mobile device although this was technically possible (such as maps with marked attempts). To promote students' reflections during the activity, their new distances were shown on the display of the mobile device only when so prompted by the students (Fig. 8c). They were explicitly challenged to try to minimize the number of prompts/tries for each task. The activity took less than one hour to complete.

When the students returned to the classroom after having completed the outdoor tasks, they were asked to describe (in writing) how they solved the tasks and what they had learned from working with the tasks. When they had finished writing, their teacher initiated a follow-up session where the students shared their experiences and discussed how they solved the tasks.

Based on the students' self-written descriptions and the numerical records (digital data) from the outdoor activity, the following solution strategies were identified (Peng & Sollervall, 2014):

1. **Separate-negotiate**. The students determine two points on the field that each are supposed to fulfill one of the two distance requirements. The locations of these points are negotiated to find a third point that should fulfill both requirements.
2. **Farther-closer**. Compares current location with target points. The students attempt to move in a direction reducing the differences of both values, without considering the actual differences.
3. **Circle**. If one of the values is correct, the students attempt to move along a circular arc in order to achieve the second value while preserving the correct value.
4. **Successive adjustments**. Students react on the last obtained values and attempt to adjust either one value or both values. If they are far away from the target point they tend to correct only one of the values and if they are close they attempt to correct both values simultaneously.

The main purpose of identifying these strategies was to inform the continued design process and to support implementing the activity with new teachers.

## Summary of Design Challenges

After the completion of the first design cycle, the researchers identified several design challenges that could be addressed in a second design cycle. Some of these challenges have been addressed earlier in this chapter.

When the first cycle had been completed, the researchers discussed possible ways to improve the activity. Such informal discussions took place over several years, awaiting an opportunity to actually engage in a second design cycle. Several areas of improvement were identified for the purpose of achieving efficient implementation and large-scale dissemination of the teaching activity:

**Designing a generic activity that could be implemented elsewhere**.

Reason: The first cycle activity was adapted to the available physical field (Fig. 2) that was located within short walking distance from the school. Security of the students was not an issue. Many schools cannot offer similar conditions.

**Redesigning the activity for implementation in a whole class**.

Reason: The first cycle activity was implemented with only 12 students (six groups). Even that implementation required some variation between the tasks, but it would not suffice for whole class implementation.

**Implementing the activity with a classroom teacher**.

Reason: Researchers and technical experts implemented the activity in the first cycle activity. Hands-on efforts were needed to prepare the technologies to the physical location and to support students on the field as well as in the classroom.

These identified areas of improvement called for further technological and didactical development of the teaching activity.

**Technological development**: Preparing a user-friendly technological application that teachers can utilize for preparing and implementing the teaching activity at a location of their choice.

**Didactical development**: Preparing a teaching activity that can be implemented with a large group of students by teachers in a wide range of teaching contexts. Supporting the teacher's orchestration with focus on addressing learning objectives.

The above mentioned design challenges and identified areas of improvement were addressed in the second design cycle.

## Implementing the Teaching Activity in the Second Design Cycle

Two days before the activity was implemented with 27 students in grade 4, the researchers met with the teacher at her school. The teacher had been asked to participate by a university colleague who knew that she was interested in trying out new ways to teach. One of the goals was to evaluate the suitability of the solution for teachers external to this study. Therefore, in order to avoid personal bias, all she knew before accepting to participate was that the activity would involve mobile technologies, that it would last for at most three hours and that the students would be expected to work in the school yard as well as in the classroom.

After a short outdoor session where the mobile technologies were tested hands-on, the teacher was informed about the researchers' intention to promote a logos-oriented discourse. A summary of the information was presented on a single (two-sided) sheet of paper and included: schematic structures of a didactical situation and a praxeology (Figs. 4 and 5), the previously identified solution strategies, and possible logos-oriented learning objectives. A sample set of ten tasks was also presented and discussed. Furthermore, the teacher was informed about the possibility of showing the students' results on a Google Map. It was made clear that these ideas should serve only to inspire her implementation and that she was completely free to orchestrate the activity according to what she believed would be best for her students, not for the researchers.

The entire activity including devolution, the outdoor activity, and institutionalization, was videotaped and lasted 1 h 45 min (8.00–9.45 on a Friday morning). The students arrived to their regular classroom at 8 o'clock in the morning. They had been told in advance that they would engage in an outdoor activity and use mobile phones. Before the students arrived, the teacher had divided them into 12 groups that were displayed on the whiteboard. The three visiting researchers introduced themselves and the teacher informed the students that they were going to make use of mobile phones to solve mathematical tasks outdoors on a field where six colored markers had been placed (cat, house, tree, bicycle, car, horse).

The field and the markers were displayed on a Google map (Fig. 10a). The teacher informed the students that they were going to look for "magic points" located specific distances from two of the markers and that they were going to use the mobile phones to check the distances. The teacher asked the students what they would do if they were not satisfied with the measurements and they readily answered that they would try again. She instructed them how to download the technological application to their mobile phones. Each group received a phone from the teacher and managed to get it started after a few instructions.

When all students had opened their first tasks that were all different (an example is shown in Fig. 11a) the teacher told them to go to the field and try to solve the tasks. The time was now 8.15.

When the students arrived at the field, they showed no signs of confusion either regarding how to handle the mobile phones, or how to interpret the tasks. They even accepted the somewhat inaccurate measurements. Some students complained to the teacher that they could not achieve exact values but readily accepted the teacher's suggestion that "within two metres is fine". They were enthusiastic and engaged fully in the tasks, although some phones did not give correct measurements due to thick clouds that caused large errors in some of the GPS-values. At 8.50, the teacher told all the students to go back to the classroom.

At 8.56, everyone was back in the classroom. After a short discussion about some incorrect values and asking if the students liked the activity the teacher asked the groups to present their strategies for finding the "magic points". Most of the groups were eager to present and the teacher promised that they would all get to do so. The first group gave their mobile phone to one of the researchers who downloaded its log file to a computer that was connected to the classroom projector. Their tasks and their attempts became visible (numerically) on the board (a regular whiteboard, not interactive) to the left of the Google map (Fig. 13a). They chose the task they wanted to present and what attempts they wanted to be shown with "pins" on the Google map visualization.

For several of the groups, the teacher had to tell the students to describe the task before they started talking about how they had worked with it. During the presentations, she repeatedly asked technology-oriented questions like "How did you think when you did that?", "Why did you do walk like that?", and "How did you get those values?" Several of the previously identified solution strategies were confirmed (Peng & Sollervall, 2014) but the targeted circle strategy did not appear in the presentations.

However, when all the groups had presented, the teacher continued the discussion, focusing on the last group's presentation. They had marked a point located 20 m away from the bicycle marker (upper right corner in Fig. 13b) and had drawn

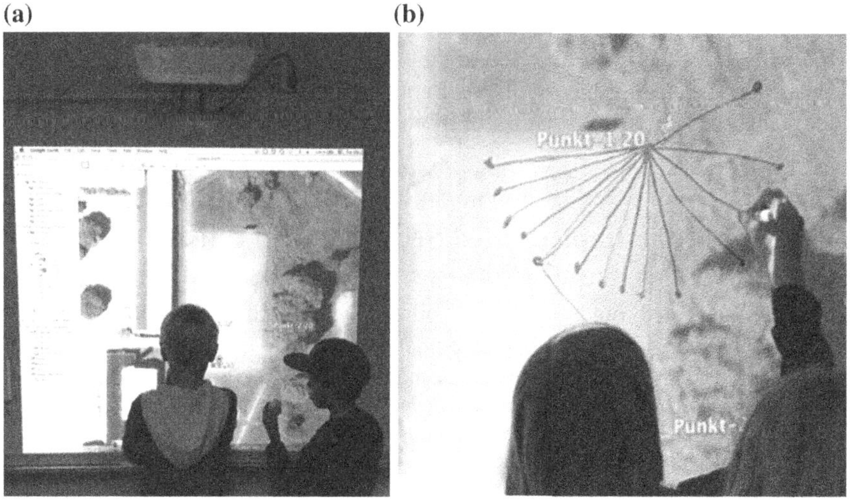


**Fig. 13** **a** One group presenting. **b** All groups contributing

a line segment from the point to the marker. The teacher asked the class if somebody could mark another point that was also located 20 m away from the bicycle. Several students tried, but failed. They seemed confused about what to do but were eager to contribute. The teacher commented on their attempts, for example "Oh that is more than 20 m", "That point is too close", "That is too far away". She tried to guide the students by asking questions: "If you stand there and it is 20 m, how can you walk to keep 20 m?", "Where else can you find 20 m?". Finally, one student managed to mark a point that seemed to be the same distance from the marker. The teacher confirmed the attempt by saying: "Yes! You found it!" and then "How did you know how to do it?". The student responded: "I just thought it out". The teacher continued with "Now I want each of you to mark a new point, that is also located 20 m away from the bicycle", and "Don't worry, there are infinitely many such points and each of you will get a chance to mark one". Most of the students caught on to the idea about keeping the distance 20 m but changing directions, and occasional mistakes were quickly corrected. When about ten points had been marked, all located on the field, the teacher commented: "Oh, nobody is being brave today". One student understood what she referred to and readily marked a point in the bushy area behind the field (Fig. 13b). A few more points were marked outside the field. A crucial scaffolding question was asked.

Teacher: Do you begin to see a pattern? You can walk in any direction.
Student 1: Oh it is a circle!
Student 2: A spider web!
Teacher: Yes! A circle! Can you all see that? [The teacher draws a circle through the points.]
Teacher: Now I have 20 here and how can I find 30 down there?

The students were invited to mark points that were initially not connected with the first circle. These points were corrected after comments from the teacher, who wrapped up the discussion at 9.45 by saying "If you can find the point where the two circles meet then you have found the magic point". Although enforced by the teacher, the concluding theory-oriented comment completed a didactical situation addressing all the four dimensions of a praxeology.

## Summary of the Implemented Second Cycle Activity

The mobile-assisted outdoor activity offered opportunities for the students to simultaneously engage in similar coordination tasks, involving the same pairs of distances but with respect to different markers. Being informed about possible logos-oriented discourses and having observed the students acting in the outdoor environment, the teacher cleverly managed to institutionalize their common experience with respect to the circle strategy. The customized mobile and web technologies inspired the students to engage in the activities and supported transitions between outdoor and indoor contexts. These supporting technologies enabled the teacher to put focus on pursuing mathematically meaningful institutionalizing activities, thus successfully finalizing a complete and complex didactical situation.

The institutionalizing discourse may be characterized as teacher-driven but student-centered. The teacher was informed about the researchers' desire to promote logos-oriented discussions and was prepared for orchestrating the session towards issues relating to technology and theory. Knowing about possible strategies for construction guided her to ask logos-oriented questions aiming particularly at the circle strategy. She patiently awaited the students to catch on to the mathematical ideas that were embedded in the didactical situation. She amplified the students' presentations by adding interpretations that led them to unfold ideas that were shared among all the students by involving them in making new constructions.

## Support for Dissemination

The findings of the activities and continuous discussions between the researchers as well as the development in technologies, in particular web technologies have lead to technical evolutions in the system. The first cycle implementation required the researchers to provide configuration of the mobile devices that were used by the students. Furthermore, location specifics of the activity had to be implemented by the researchers. To overcome these challenges, a web-based authoring tool has been added and is provided by a web-portal.[2] It allows teachers to define the activity location via a web browser, following the suggestions in Zbick, Vogel, Spikol, Jansen, and Milrad (2016). The teacher defines the location of the rectangular field where the activity will take place. The authoring tool automatically generates a configuration file that is loaded in the web-portal. This configuration file contains location-specific coordinates for the 14 configurations (14 sets of 10 tasks each) as specified in Fig. 12. Each group students download their unique configuration from the web-portal to their mobile application. Thus, the process of manually configuring the location and preparing the mobile devices has been automated and does not require involvement of the researchers.

The web-portal provides the teacher with an overview about the status of the students' preparation process. An overview about the number of downloaded configurations is presented to indicate if the students are prepared to perform the activity (Fig. 14).

[2]http://trigo.lnu.se.

**Fig. 14** The presentation of the activity on a map and the status of the preparation

Moreover, a web-based visualization has been integrated in the web-portal. In the first two design cycles, the data that was generated during the activity had to be extracted from the mobile devices by hand in order to be visualized on the Google map (Fig. 15). This process has been automated and is also available in the web-portal to enhance the seamless experience.

The mobile application is now available in the app stores of Android and Apple devices. The TriGO LNU app can be downloaded from the iTunes Store or from Google Play and a link to download the app is provided on the web portal. This enables the students to use their own devices and the teachers/researchers do not need to provide mobile devices to perform the activity.

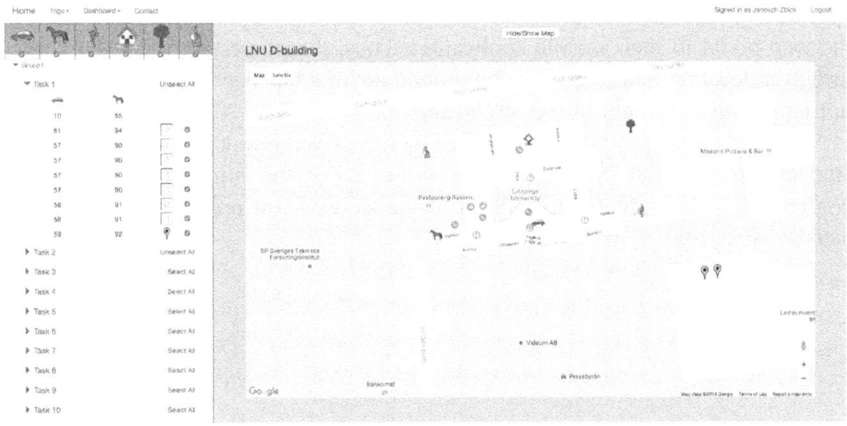


**Fig. 15** Visualization of the data collected during an activity

## Concluding Discussion

The successful orchestration of the second cycle activity can be regarded as a proof of existence regarding the feasibility of implementing the innovative mobile learning activity with classroom teachers worldwide. All that is needed is access to a rectangular open field (minimum 70 m by 28 m) and basic technological skills. Some additional efforts are needed to construct and arrange the physical outdoor markers. However, if the markers can be stored at the school they can be re-used and shared between teachers.

The research team developed the activity over a five-year period and expended substantial energy in designing suitable technologies and identifying didactical theories and research that support the structure as well as the content of the activity. The development costs can be justified with respect to educational benefits of society only when interpreted as cost per (potential) user. Furthermore, the activity may have positive educational impact beyond the activity itself, as teachers become aware of didactical theories for supporting their classroom practices and also become aware of the educational potential of mobile technologies.

A remaining challenge is how to provide support for full-scale dissemination and marketing efforts directed at teachers worldwide. The researchers have applied for funding to produce materials and videos that may improve our communication with teachers about the activity. The technologies and the didactical structure are in place, but we also have to spend efforts on making teachers aware of the activity and encouraging teachers to complete the activity with their students. Unfortunately, our proposals have been rejected since we do not have the ambition to generate profit.

## References


Asami-Johansson, Y. (2015). *Designing mathematics lessons using Japanese problem solving oriented lesson structure: A Swedish case study*. Licentiate thesis: Linköping University.

Balacheff, N. (2013). *Theory of didactical situations in mathematics. Online presentation.* Retrieved from http://www.slideshare.net/TheoRifortel/theory-of-didactical-situations.

Bessot, A. (2003). Une introduction à la théorie des situations didactiques. In *Les cahiers du laboratoire Leibniz, 91*. Grenoble, France: Laboratoire Leibniz-IMAG.

Bishop, A. J. (1980). Spatial abilities and mathematics education—A review. *Educational Studies in Mathematics, 11,* 257–269.

Brousseau, G. (1997). *Theory of didactical situations in mathematics*. Kluwer Academic Publishers.

Bruner, J. S. (1966). *Toward a theory of instruction*. Harvard University Press.

Chi, M. T. H. (2009). Active-constructive-interactive: A conceptual framework for differentiating learning activities. *Topics in Cognitive Science, 1,* 73–105.

Dillenbourg, P. (2009). Exploring neglected planes: social signals and class orchestration. In *Proceedings of CSCL 8*. Rhodes, Greece.

Gil de la Iglesia, D., Andersson, J., Milrad, M., & Sollervall, H. (2012). Towards a decentralized and self-adaptive system for m-learning applications. In *Proceedings of WMUTE 7*. Takamatsu, Kagawa, Japan.

Heath, T. L. (1908). *The thirteen books of Euclid's Elements*. Cambridge: Cambridge University Press. Retrieved from http://www.wilbourhall.org/pdfs/heath_euclid_i.pdf.

Kozhevnikov, M., & Hegarty, M. (2001). A dissociation between object manipulation, spatial ability and spatial orientation ability. *Memory and Cognition, 29*(5), 745–756.

Patkin, D., & Dayan, E. (2013). The intelligence of observation: improving high school students' spatial ability by means of intervention unit. *International Journal of Mathematical Education in Science and Technology, 44*(2), 1–17.

Peng, A., & Sollervall, H. (2014). Primary school students' spatial orientation strategies in an outdoor learning activity supported by mobile technologies. *International Journal of Education in Mathematics, Science and Technology, 2*(4), 246–256.

Rodríguez, E., Bosch, M., & Gascón, J. (2008). A networking method to compare theories: metacognition in problem solving reformulated within the Anthropological Theory of the Didactic. *ZDM Mathematics Education, 40*, 287–301.

Shimizu, Y. (1999). Aspect of mathematics teacher education in Japan: Focusing in teachers' roles. *Journal of Mathematics Teacher Education, 2*, 107–116.

Sollervall, H., & Gil de la Iglesia, D. (2015). Designing a didactical situation with mobile and web technologies. In *Proceedings of CERME 9*. Prague, Czech Republic.

Sollervall, H., & Milrad, M. (2012). Theoretical and methodological considerations regarding the design of innovative mathematical learning activities with mobile technologies. *International Journal of Mobile Learning and Organisation, 6*(2), 172–187.

Sollervall, H., Gil de la Iglesia, D., Milrad, M., Peng, A., Pettersson, O., Salavati, S., & Yau, J. (2011). Designing with mobile technologies for enacting the learning of geometry. In *Workshop proceedings of ICCE 19*. Chiang Mai, Thailand.

Spikol, D. & Eliasson, J. (2010). Lessons from designing geometry learning activities that combine mobile and 3D tools. In *Proceedings of WMUTE 6*. Kaohsiung, Taiwan.

Wong, L.-H., & Looi, C.-K. (2011). What seams do we remove in mobile-assisted seamless learning? A critical review of the literature. *Computers and Education, 57*, 2364–2381.

Zbick, J., Vogel, B., Spikol, D., Jansen, M., & Milrad, M. (2016). Toward an adaptive and adaptable architecture to support ubiquitous learning activities. In A. Peña-Ayala (Ed.), *Mobile, ubiquitous, and pervasive learning* (pp. 193–222). Springer: In collection.

**Håkan Sollervall** is an Associate Professor of Mathematics at Linnaeus University. He teaches mathematics and undertakes research in mathematics education where he makes use of educational design research methods and adaptive conceptual frameworks. He collaborates with mathematics teachers in Swedish elementary and secondary schools for the purpose of identifying appropriate types of support for designing and implementing innovative teaching activities.

**Didac Gil de la Iglesia** holds a Ph.D. in Computer science from Linnaeus University. Didac's areas of expertise are on the design of distributed systems, self-adaptive systems and machine learning, which he has combined for the design of collaborative platforms applied in Mobile Learning. Currently, Didac combines efforts in academia and industry, working as a research fellow at Linnaeus University, where he focuses on unstructured Big Data processing; and as a research member in a CyberSecurity company, where he investigates future strategies for network and enterprise security.

**Janosch Zbick** is a Ph.D. student at the Department of Media Technology. He is exploring and investigating research topics at the intersection of the fields of mobile learning, web engineering, software architectures, and ubiquitous computing. He is currently involved in various projects in the field of Technology Enhanced Learning (TEL) with the goal of defining reference architecture for a certain scenario within mobile learning.

# Collaborative Engagement Through Mobile Technology in Mathematics Learning

**Mina Sedaghatjou and Sheree Rodney**

**Abstract** When a group of students come together to engage in negotiation about mathematical ideas and activities, they draw on each other's cultural experiences for a shared understanding of mathematical meanings. This chapter considers how mobile technologies, along with children's collaborative engagements, can enhance mathematical learning. We adapted previous findings regarding touchscreen-based interactions to assess and analyse how mathematical learning occurs when learners interact with mobile technologies and with their peers. We also utilized StudioCode software to analyse children's interactions with a mathematical tool in order to better understand their collaborative practices and how they reflect using touch-screen-based devices. Our conclusions emerge from children's use of an iPad application called *TouchCounts*, which aims to develop number sense. Overall, we found that the one-to-one multimodal touch, sight, and auditory feedback via a touchscreen mobile device served to assist children's collaborative engagement and helped children develop their number sense.

**Keywords** Engagement · iPad · Numbers · Touchscreen-based device
Mathematics · Mobile technologies · Interaction · Collaborative engagement
Reflection

## Mobile Technology in Learning Mathematics

In recent years, it has become apparent that there is a shift in the way society perceives and uses technology to enhance learning, from the ancient Greek invention of the abacus, to slide rules, calculators and now more sophisticated and complex inventions such as digital technologies. Papert (1980) articulated the changing ways

M. Sedaghatjou (✉) · S. Rodney
Brock University, St. Catharines, Canada
e-mail: msedaghatjou@brocku.ca

S. Rodney
e-mail: sgilling@sfu.ca

 
N. Calder et al. (eds.), *Using Mobile Technologies in the Teaching and Learning of Mathematics*, Mathematics Education in the Digital Era 12,
https://doi.org/10.1007/978-3-319-90179-4_7

in which technology enhances learning and, for him, the invention of computers has affected "the way people think and learn," (p. 3) and occasioned debate about how these changes enhance mathematics learning. Noss and Hoyles (1996) focused on "what the computer makes possible for mathematical meaning-making" (p. 5) suggesting that interaction with computers may facilitate children's development of mathematical meanings. In keeping with Papert's idea, many researchers have sought to integrate digital technologies into mathematics learning environments (Clements, Sarama, Yelland, & Glass, 2008; Hollerbands, Laborde, & Strasser, 2008; Laborde, Kynigos, Hollebrands, & Strasser, 2006; Sinclair, Arzarello, 2010). In particular, researchers (Ainley & Ainley, 2011; Drivjers, Mariotti, Olive, & Sacristán, 2010; Hoyles & Lagrange, 2010; Sedaghatjou & Campbell, 2017; Sinclair, Chorney, & Rodney, 2016) have also suggested that the use of mobile technologies, if integrated with a suitable pedagogical structure, can make mathematics more pleasing, meaningful, practical, and engaging. That is to say, mathematics learning can be more appealing when children are provided with experiences involving the use of mobile technologies. These experiences enable children to develop strategies to better understand mathematical concepts while maintaining a sense of connection with their peers during an activity.

In this chapter, we consider ways in which touchscreen interactions with mobile technologies influenced the way a group of pre-school children engaged in mathematics learning. By mobile technologies we mean devices such as Personal Digital Assistant (PDA) Devices, iPads, cell phones, iPods, e-readers and similar handheld devices, which are increasingly being used as educational tools (Moyer-Packenham et al., 2016). We discuss how collaborative engagement—the convergence of common features and mutually constructed practices—along with mobile technologies, facilitate the development of the ways children reflect on their work and attach meanings to mathematical ideas. Furthermore, we discuss the collaborative engagement of a group of young children as they perform the task of "make 100" using *TouchCounts*.[1]

*TouchCounts* is a multimodal iPad application that provides children with the opportunity to create and represent numerical quantities. For example, by touching the screen in *TouchCounts*' "Operating World" with three fingers, a circle called a "herd" containing three small discs and labelled "3" appears, while *TouchCounts* says the number "Three" (Fig. 1a). When there are more than two herds on the screen, children can perform the gesture of *pinch-in* (Fig. 1b) for addition. *TouchCounts* represents larger quantities with larger circles. The *pinch-out* action (Fig. 1c) splits a herd. This gesture performs a subtraction operation.

[1]http://touchcounts.ca/.

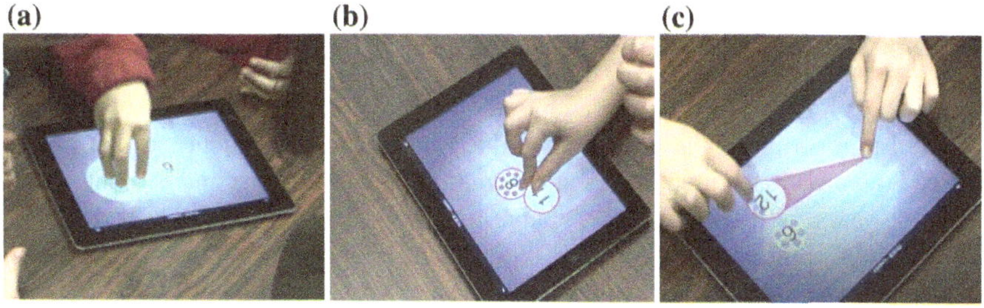

**Fig. 1** **a** In *TouchCounts'* "Operating World" each touch makes a herd of discs, as *TouchCounts* displays and speaks the number; **b** represents the action of pinch-in (gestural representation for addition) and **c** pinch-out (gestural representation for subtraction)

## Research Method

### *Participants and Data Collection*

This exploratory case study is part of a larger research project that aimed to understand the various ways that young children, aged from three to six, develop numerical abilities through embodied interactions. The study explores one way that children share mathematical meaning when they work together with *TouchCounts*. Data was collected over a four-month period in a classroom of a daycare facility located in a Canada. All 27 children registered in the daycare were free to join and leave the activities as they pleased. This means that we provided the opportunity to play with *TouchCounts* to all children, even those without parental permission to participate in the study. However, only children whose parents or guardians had signed the consent form, were video recorded. During the data collection process, children were not simultaneously exposed to formal mathematical instruction.

In the section that follows, we present evidence of collaborative engagement where a group of young children converged independently in a separate corner of the classroom engaging in their own "play" with *TouchCounts*. We noticed children's collaborative activities manifested through an "investment" (Newman et al., 1992) directed toward reaching their own goal of "making 100" or "the biggest number in the galaxy" using *TouchCounts*. Newman et al. (1992) refer to this notion as engagement. Engagement is the psychological and perhaps physical investment directed towards learning a concept, or mastering a skill that mathematical work is intended to attain. Similarly, children's engagement is exhibited through the purposeful effort they direct toward the "educationally [and mathematically] purposeful activity, which contributes directly to a desired outcome" (Hu & Kuh, 2001, p. 3).

## We Are so Good, Right John?

One day, while the senior researcher (Nathalie) was interviewing some of the children in a corner of the classroom, Mina (the research assistant and first author) noticed three boys—Sam (4 years and 8 months), Tom (5 years and 4 months), and John (4 years and 11 months)—gathering in a separate group playing with *TouchCounts*, and decided to record their activities and interactions. The names of children in this interaction are fictitious in order to conceal identity and maintain confidentiality. The children's stated overarching aim was to make big numbers. Prior to us working with them, John and Tom were able to count up to twenty starting from any given number, whereas Sam was able to count up to 20 only in a sequential form starting from one.

## Phase 1 and 2: Making a Big Number and Celebration

Sam, Tom and John are gathered around the iPad and managed to create the number 203. Tom pinches a 'one' to the herd, *TouchCounts* says "Two hundred and four". The children notice Mina.

1. Sam looks back at her, smiles and says [in a high pitch]: "We made two hundred and four! We made one hundred, and then, we made two hundred and four", with surprise.
2. Mina says, "Excellent!" Tom adds a '1'–'204'. *TouchCounts* says: "Two hundred and five." Tom says to his teammates, "Hey look at this" [smiles]. Both John and Sam happily scream: "Oh, what the heck!" Tom and Sam pinch a '3'–'205'. *TouchCounts*: "Two hundred and eight."
3. The children run towards Nathalie and other kids, jumping up and down proudly: "Look at our number, we made the biggest circle in the galaxy!" (pointing at 208 on the iPad) (Fig. 2a). Nathalie looks at the number and says: "Wow, two hundred and eight!", with an excited tone. Children scream, jump, clap, and celebrate their group work.
   Note: **a** The children show their 'big' number, '208' to Nathalie; **b** Sam on the left is upset. He asks others to work as a team. **c** The children share fingers to reach their target number, '100'; **d** They are proud that they could make 100 collaboratively.
4. Sam resets *TouchCounts* and the children make new herds on the screen. Tom says: "We are making the biggest number in the galaxy," Sam continues: "We are making one hundred and two thousand and one!" John emphasizes: "No! we are trying to make *one hundred.*"

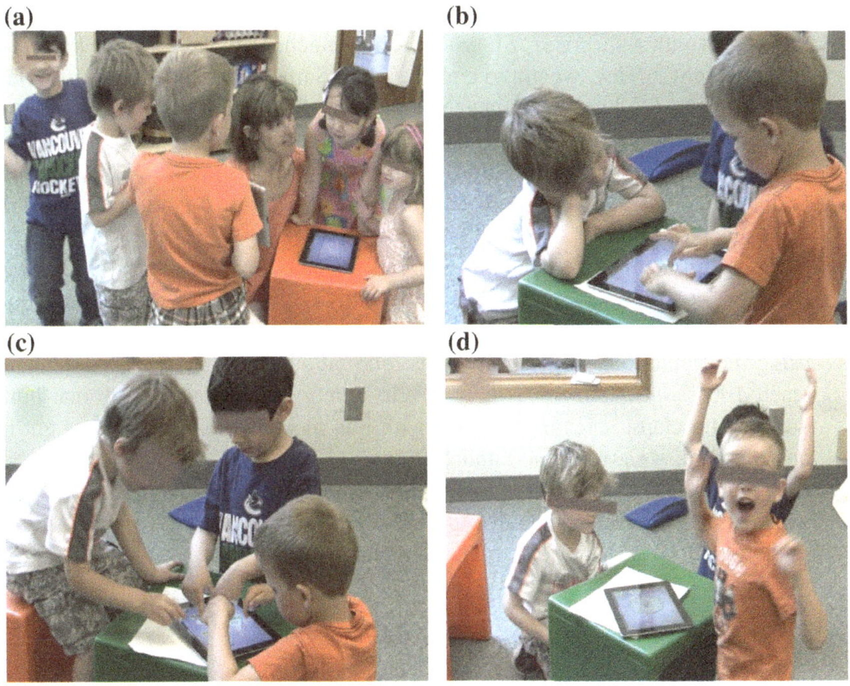

**Fig. 2** Children collaborate in making 100

## Phase 3: Making 100

5. Sam requests: "So, after that, a hundred and two thousand and one?" John rejects Sam's request and pinches herds intending to make a 100 all by himself. Sam wants to pinch herds, but John disregards this request and says, "Let me do it, Sam." *TouchCounts*: "Thirty two, forty two, forty three, fifty five …" This appears to disappoint Sam: "But John, let's make one hundred *as a team*" [very low pitch and disappointed voice] (Fig. 2b).
6. John accepts. Sam smiles, creates herds of numbers one after another with tensed fingers, "I'll make all these [numbers] and then you put them on John, okay?" says Sam excited and happily. Working together, John uses both hands to assemble the herds that Sam rapidly creates. Tom also helps John to gather herds. All fingers and eyes are on the iPad. The three children make numbers with shared fingers (Fig. 2c).
7. *TouchCounts* says: "Forty five." Sam keeps creating herds. *TouchCounts* does not allow children to both pinch herds, and create new herds, simultaneously. So, John asks Sam to wait until he puts all herds on the screen together. "Sam? No, let me first do the numbers, then you can do it again."

8. Sam says, "John look at the big circle we made." *TouchCounts*: "Sixty six." Sam says (with a surprise) "Sixty six!" John pinches 66 and 22. *TouchCounts*: "Eighty eight." "Hey, we got to eighty eight!" Sam says cheerfully. Tom: "Hey did we make the biggest number in the galaxy?" John says, "No, we got to make one hundred." Sam answers Tom, "No, *one hundred* and two thousand and *one*! That's the biggest number on the galaxy". Tom says, "No, a trillion." Sam refine his statement, "Okay, a trillion *one hundred* two thousand *and one*." John pinches 9 and 88.
9. It seems children realized that they are getting close to 100. John 'holds' the big circle of 97 to check it for few seconds. This is the first time that this action happens. Tom and Sam make some more 'ones', but no other numbers and not using more than one finger. John adds 'one' at a time to the big herd of 97 to reach 100. *TouchCounts*: "Ninety seven, ninety eight, ninety nine, one hundred." John claps, screams and says with lots of joy, "one hundred, *we* got to one hundred, *we* got to one hundred *here*!" (Fig. 2d).
10. Reaching 100 did not occur by chance, as we observed how children made and added 'ones' precisely, one at a time as a team. Tom says: "We are so good, right John?" "Yes, we are!" John responds.

## Collaborative Engagement

The previous episode demonstrates how some children "made 100" in their small group without instruction or supervision. They were *engaged* with the mathematical idea and their peers through fashioning an "educationally purposeful" (Hu & Kuh, 2001) activity and setting the goal of "making one hundred" [4 and 8]. Fredericks, Blumenfield and Paris (2004) suggest that *engagement* has three dimensions—behavioral, affective and cognitive—that provide insights into how student's collaborative engagement can be identified. According to Fredericks et al. behavioral engagement involves active contribution in learning activities; affective engagement refers to children's attitudes towards the activities; and cognitive engagement deals with the strategies used to thoughtfully involve children in understanding mathematics. These three different dimensions of engagement were evident in the observed children's group activities. For example, behavioral engagement was involved in the children's active contribution in "making 100", while affective engagement was observed in the children's attitudes towards making numbers "as a team" [2–9]. In our study children's collaborative engagement resulted in them deriving recreational value from this activity, that is, engaging in the task for their own pleasure. In addition, the boys were proud to see what they have accomplished through team work. Using *TouchCounts*, children were making numbers collaboratively, and shared their moment of joy and accomplishment with the researchers. They also commended each other for being "so good" in creating numbers [10]. The behavioral and affective engagements occurred as a consequence of the children's cognitive and mathematical engagement and understanding.

Donato (2004) offers different perspectives on the notion of collaboration. Using Gee's (2003) idea of *affinity groups,* in which "groups are continually immersed in practice and share common features" (p. 286), he argued that collaboration involves recognition of individuals engaged in the larger activity, bonding with each other, and learning mainly in cooperation with each other with knowledge usually dispersed among the members. A second perspective of collaboration comes from Petrovsky (1985), who suggested that "collaboration implies group conventionality and disregards the individual as distinctive and imitative of the social" (p. 286). Taken together, the two definitions imply that an individual within a group should not be treated as a unique or isolated agent. This also affirms the *participatory* nature of learning (Lave & Wenger, 1991). In agreement with the ideas of Donato and Petrovsky, we found children were connected with each other, learning through combined efforts, and using the mathematical tool to negotiate and achieve a group aim. The implicit cultural and social relevance of collaboration was also evident as the children influenced the construction of meanings through communal engagement.

So far, this section has explored theories of engagement and collaboration separately. To explore collaborative engagement as an integrated concept, involving the use of a mathematical tool, we adopt Gee's definition of *collaboration* and Fredericks's et al. (2004) and Hu and Kuh (2001) perspective on *engagement.* Therefore, in this chapter, *collaborative engagement* is thought of as the practice wherein children's attitudes, active contributions, and mathematical strategies promote mathematics learning using mobile technologies. It refers to harmonized learning activities and practices, within which individuals build and practice a joint mathematical engagement, using mobile technology. In the episode, *TouchCounts* provided children a collaborative engagement environment, facilitating their mathematical negotiation, contribution and exploration. The local social interaction shaped the children's goal recognition strategies, group construction and role-assignment (e.g. Tom and Sam make numbers, and John pinches them to create new numbers).

## Video Timeline: An Analytical Tool to Trace Paths of Interactions

In this section, we suggest an innovative methodology for analysing video data. This methodology helped us to assess mathematical collaborative engagement as the children interacted with *TouchCounts*. The tool we used to support our analysis is *StudioCode*: a professional video tool that captures, codes and categorizes video data for later review and analysis.

## Using Interaction Theory to Determine Codes and Categories

To analyse video data, we adapted Vogel and Jung's (2013) video-coding procedure augmented by Arzarello et al. (2014) theory of touchscreen-based interactions to develop "active" versus "basic" action categories. The theory of interaction describes basic actions as the simple ways of interacting with the touch interface. Therefore, interactions that appear to be made either randomly or with no plan are marked as basic actions. Combinations of basic actions are classified as "active" actions. For example, tapping the reset button in *TouchCounts'* "Operating World" or tapping randomly are defined as basic actions; however, tapping with multiple fingers to create a given number is categorized as an active action. In this study, active actions were identified mostly as the interaction that the individual learner (or group of learners) take(s) to attain the achieved outcome or solve the given problem (See Table 1).

During the data analysis, codes and categories were identified and modified based on different modes of interactions with the touchscreen application. The first step was to review the video data on multiple occasions to generate initial codes. This provided familiarity with the content, suggested possible codes, and reduced pre-coding bias. Next, we verified whether categories and codes were informed by

**Table 1** Basic and active actions when using *TouchCounts*

| Basic actions | | Active actions | |
|---|---|---|---|
| Tap | (a) To reset<br>(b) Single or multiple (randomly or without a plan) | Tap | (a) Single finger<br>(b) Multiple fingers |
| Hold | To know the application or by random, or without a plan | Hold | User make a contact with a herd and continues contact<br>(a) To check<br>(b) To slide |
| Slide/ Swipe | User puts finger on the screen and moves it randomly in any direction without touching a herd | Drag | (a) To organize (flick/drag): user grabs a herd and goes in a specific direction<br>(b) **Pinch-in** (user makes twin drags and brings contact together without lifting fingers- to add herds)<br>• With one hand or multiple hands<br>• Two herds or multiple herds |
| | | Pinch-out/ Spread | User performs two drags and pushes them apart without breaking contact (For subtraction —not observed in this episode)<br>(a) With one hand<br>(b) With two hands |

the chosen theory (in this case, Arzarello et al. (2014) theory of interaction). In the final step, the researcher developed and checked the integrated system of coding and created a standardized system of codes via rating accordance. Rating accordance rates and verifies the degree to which categories and codes conform to the theory being used. This can be contextually varied depending on the scope of study, theoretical framework, and/or available coding software. The process of developing the video coding system is illustrated in Fig. 3.

After identifying active and basic codes as discussed in Table 1, the researchers coded video data and captured both the frequency and duration of actions. StudioCode allows timeline analysis and connects codes to each video-segment. The [video] timeline in StudioCode provides a chronologically organized, multi-layered, graphical representation of the codes, narratives, and comments from the video recording. It enables coding of a variety of verbal and visual cues (such as gestures, body position) to precisely capture what occurred during a specific time. We used the software to comprehensively detect different interactions among the children while they interacted with *TouchCounts*.

In StudioCode's timeline, each row represents its corresponding code on the left and demonstrates the distribution, frequency, and length of each occurrence chronologically. The resulting coded timelines are summarized and synchronized below. Phases 1 and 2 correspond to the children creating a big number and celebrating, and Phase 3 to the group work resulted in creating one hundred.

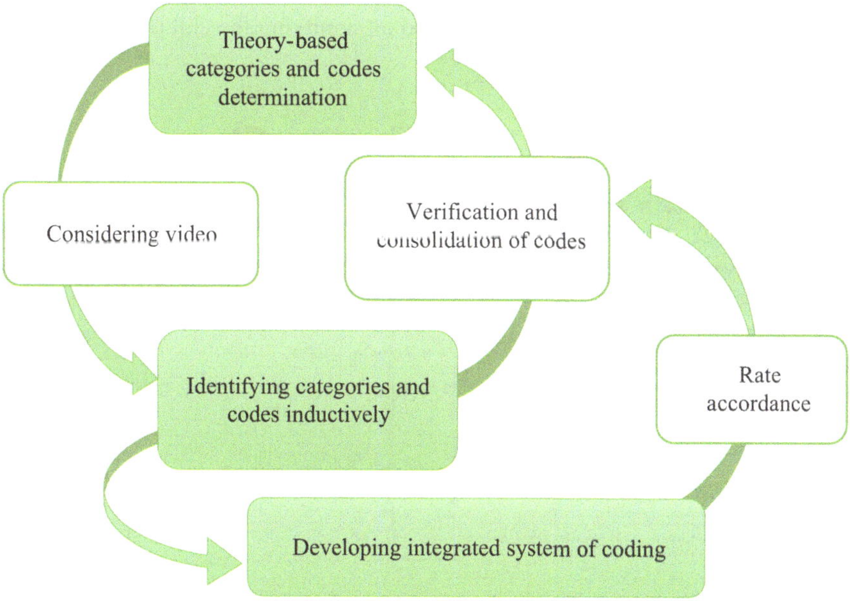


**Fig. 3** Video coding process for determination of categories and codes. (Adapted from Vogel and Jung 2013, p. 3)

Figure 4 comprises the three separate timelines for Tom, Sam, and John along with performed active actions as the codes in rows that are synchronized. As seen in Fig. 4, there are common and unshared rows in each timeline that indicate performed or unperformed active actions for each child during the reported time interval. For instance, the active action of "holding to check" was observed only once in John and Tom's interaction with *TouchCounts*.

At first glance, the timelines show a periodic form of activity (working in turns) among group members. This could be either because children took turns while they worked as a team or it might demonstrate a limitation of *TouchCounts* in not allowing a user to create new herds, while another user is simultaneously gathering existing herds. We also exported the "frequency matrix" of StudioCode to Statistical Package for the Social Sciences (SPSS) software to create a scatterplot graph that displays frequency distribution of observed active actions (Fig. 5). The frequency matrix is a table where the rows indicate active actions in the video timeline and the columns indicate different forms of a given active action that are coded in the instances of each row.

Rather than attempting to demonstrate a correlation between codes, we used a scatterplot to indicate the distribution of active actions across the video timeline. We observed higher numbers of active actions between 3:50 and 4:20 (as indicated in the horizontal box at the top of Fig. 5) when children were able to successfully reach 100 with their collaborative engagement. During this period, we observed that the children progressed from creating many herds by multi-tapping, to precisely making single-tap 'ones' after getting close to 97. That is to say, as a group, Tom and Sam made 'ones' on the screen and John used those 'ones' to reach one hundred (Fig. 5). This might be interpreted as signifying the children's awareness

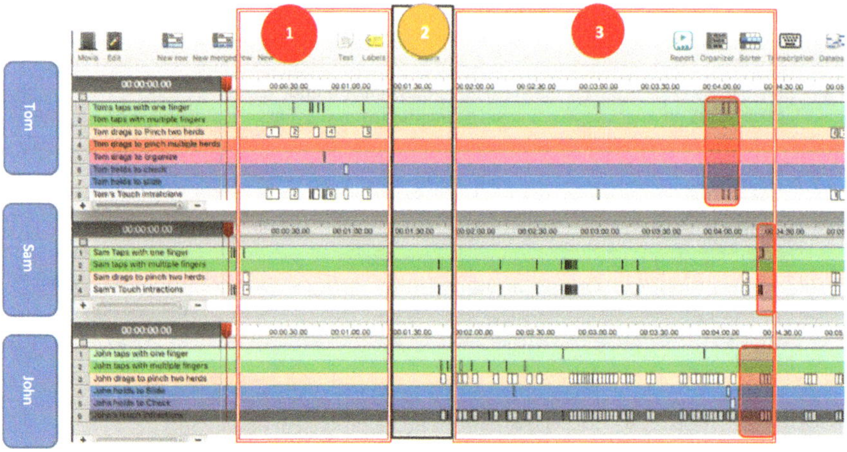


**Fig. 4** Synchronized coded timeline for children's interactions: In phase one, children make a big number; in phase two they cellebrate; in phase three they make 100. The highlighted boxes in the third phase show the children's active actions after reaching 97

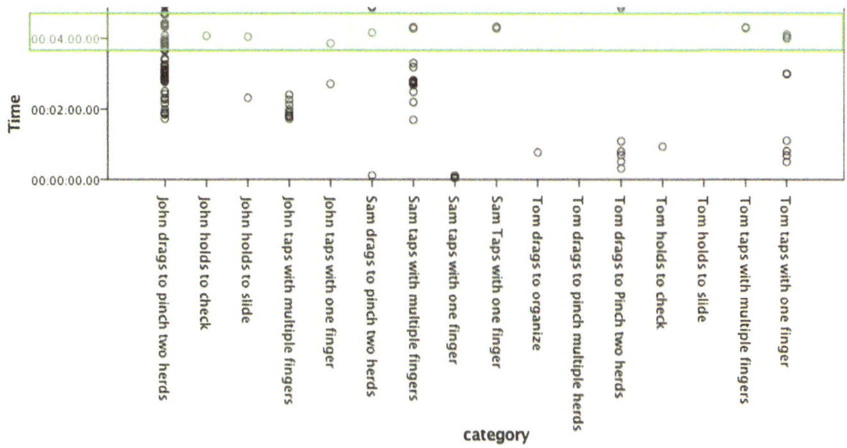


**Fig. 5** Frequency distribution of active actions over time for Tom, Sam and John

of approaching one hundred. The children's shared active actions while interacting with each other and *TouchCounts* also demonstrates their collaborative engagement. The graph also shows an increase in the distribution of active actions during the time period under investigation.

In addition, more sophisticated forms of interaction were observed just before completing the task of making 100, such as 'hold to check' by John (03:58). Children also become proficient in pinching herds, working in turns, and collaborating to make 100. This precision in using the tool cannot be separated from the precision of articulating each number by gestures (Sinclair & Heyd-Metzuyanim, 2014).

The analysis above demonstrates how collaborative engagement, assisted by mobile technologies, appear to facilitate the development of mathematics learning. In agreement with Harper and Quaye (2009), we found that engagement entails more than participation. It also incorporates "feelings and sense-making as well as [bodily] activity" (p. 5), adding to Hu and Kuh's (2001) definition of engagement. The video-timeline data demonstrates the children's bodily interactions with the mathematical tool *TouchCounts* through basic and active interactions, while they were collaboratively engaged in mathematical practice to achieve their goal of "making a hundred."

## Reflection

One of the key features of mobile technology, which makes it fruitful for adaptation in collaborative mathematical activities, is its potential to help children reflect on their work. In the context of using mobile technology, we refer to reflection as a catalyst for linking and revisiting learning. Rodgers (2002) introduced four criteria of reflection, inspired by Dewey's (1916) long-historical perspectives on the subject. The first criterion of reflection involves a meaning-making process; one that moves the learner through various levels of experience. The second criterion posits reflection as a systematic and rigorous way of thinking where the learner draws on past experiences that are similar or different from the new experience. The third criterion involves a mind-set that place emphasis on personal and intellectual growth of the individual and others who are a part of the community. Finally, reflection is embedded in a community of interaction with others, which implies that reflection is a necessary component of collaborative engagement and is, therefore, of paramount importance to our research.

Schön's (1983) model of reflection presents an alternative approach. In this model, two phases of reflection: reflection-in-action (learning through practice) and reflection-on-action (learning after the event) are described. Our primary focus here is on reflection-in-action, emphasizing that the reflection observed was temporally extended across the entire activity, rather than occurring at a discreet moment at the conclusion of the activity. This means that children's meaning-making processes can take place at any point throughout the activity. The reflection-in-action was evident when John was leading the group to reach one hundred as a milestone before making "the biggest number in the galaxy." We observed that using *TouchCounts* supported the children's reflection-in-action and also reflection-on-action. It facilitated conversations and collaborative practice among children themselves, and also between children and adults (Cochrane & Bateman, 2010). Through the experience of working with *TouchCounts*, children learned they might pass 100, and then 200 to reach 204 (See John's explanation in [1]). They also drew the conclusion that "the biggest number in the galaxy" has the "biggest circle in the galaxy," implying a relationship between circle size and number. This was because of their continuous reflection on the size of circles appearing on the screen at different intervals (reflection-in-action); the bigger the number-the larger the circle (limited to the size of the screen).

## Using the Affordances of Mobile Technologies to Enhance Engagement

According to Lai et al. (2007), the term 'affordance' originates in the work of Gibson (1977) and means "the relationship between an object's physical properties and the characteristics of a user that enables particular interactions between user and object" (p. 328). The development of this relationship has opened up diverse trajectories for learners to construct and comprehend mathematical knowledge (Sacristan et al., 2010). Mobile technology offers the ability to simultaneously connect to and explore visual, symbolic, and numerical representations in a dynamic way (Sacristan & Noss, 2008). They allow the learner the flexibility to quickly rearrange information and re-engage with activities from new perspectives (Calder, 2005; Clements, 2000). They also provide a system of networking where interaction and collaboration within a structured system is used to share and discuss issues relating to mathematics (Sinclair, 2005).

Mobile technologies also provide cognitive benefits based on their various functionalities and the purpose for which they are used. Results from studies (Gadanidis & Geiger, 2010; Pierce & Stacey, 2010) have shown that the use of mobile technological tools supports the learning of mathematics skills such as problem-solving, reasoning, computational thinking, and justifying.

For example, although we are not arguing whether or not the proposed "biggest number in the galaxy" by children is mathematically meaningful or not, we found it interesting when children shared their knowledge of "the biggest number in the galaxy", as it surprisingly almost always had a '*one*' at the end (e.g. "a trillion one hundred two thousand and *one*"). Presumably, this is the influence of one of the *TouchCounts'* affordances, which promotes creating larger numbers by adding a 'one' to any given number. We suggest two distinct explanations for this phenomenon. First, the capability of *TouchCounts* to encourage the development of an early exploration of the set of natural numbers as an infinite[2] set by young children. Second, the opportunity for children to validate their assumptions openly as far as the actuality of the designed tool allows. This is what Stone and Minocha 2005) defined as a good user interface design in that it facilitated easy, natural, and engaging interaction, which in turn allows users to carry out their required tasks or goals in a natural and logical order. In this sense, children initially proposed a "trillion", as "the biggest number in the galaxy" and their continuous adding of the aforementioned special '*one*' is a manifestation of a fundamental number theory axiom, which proves that the set of natural numbers is an infinite set. Based on set theory, natural numbers are the counting numbers that usually represent the cardinality of a (non-zero sized) set with each natural number being "built upon" the previous number add one. $A_1 = 1$, $A_2 = 1 + 1 = 2$, $A_3 = 2 + 1 = 3$, …, $A_{n+1} = A_n + 1$. So, regardless of which number is suggested as *the biggest number*, children can envision a number that is greater, by adding a 'one'.

## Conclusion and Limitation

Although smart phones, tablets and other handheld devices are increasing popular among young children (Hilda et al., 2015), research regarding the implications of mobile technology on mathematics teaching and learning is still relatively new. As

[2]Suppose $x$ is the greatest natural number, then there is $x + 1$ that $x + 1 > x$ (proof by contradiction).

mathematics education researchers, we sought to gain better understanding of the relationship among collaborative engagement, mobile technology, and mathematics learning. We initially examined collaboration and engagement as separate entities, and then examined them collectively to operationalize the idea of a group of children working collaboratively on a purposeful activity towards a shared outcome. We used a case study to show that social interactions among children engaging with mobile technology shape goal recognition and role-assignment strategies. We also showed that mathematical meanings were developed when children worked together on a task mediated by mobile technology. They were able to "make one hundred" collectively, and demonstrated early understanding of a fundamental idea of number theory: specifically, that the set of natural numbers is an infinite set. Furthermore, the children were able to envision that an amount can get greater if 'one' or more is added.

In addition, collaborative engagement using *TouchCounts* was most notable during teamwork, goal setting, negotiations, sharing knowledge, and joint construction of practices. We also discussed the ways children discarded individual practices to collaboratively achieve goals. The diverse paths that children utilized in achieving their goal were evidence that the affordances of *TouchCounts* provided cognitive benefits for children and that they were able to use their problem-solving skills to test and justify conjectures. As an analytic tool, the video timeline showed the emergence of a remarkable amount of bodily engagement as the children completed active actions (Arzarello, Bairral, & Danè, 2014). In this case study, while we specifically examined the relationship between collaborative engagement and mathematics learning through mobile technology, we suggest that StudioCode is a powerful and sophisticated data analysis tool for other mathematics research investigating different modes of interactions, communication, and gestures in educational contexts. As a result, we believe that it would be beneficial to further explore how the use of mobile technology might also contribute to older children's collaborative engagements.

## References


Ainley, M., & Ainley, J. (2011). Student engagement with science in early adolescence: The contribution of enjoyment to students' continuing interest in learning about science. *Contemporary Educational Psychology*, *36*(1), 4–12.

Arzarello, F., Bairral, M. A., & Danè, C. (2014). Moving from dragging to touchscreen: Geometrical learning with geometric dynamic software. *Teaching mathematics and its applications*, *33*(1), 39–51.

Calder, N. S. (2005). "I type what I think and try it": Children's initial approaches to investigate through spreadsheets. In P. Clarkson, A. Downton, D. Gronn, M. Horne, A. McDonough, R. Pierce & A. Roche (Eds.), *Building Connections: Theory, Research and Practice, (Proceedings of the 28th Annual Conference of the Mathematics Education Research Group of Australasia)* (pp. 185–192). Melbourne, Sydney: MERGA.

Clements, D. H. (2000). From exercises and tasks to problems and projects: Unique contributions of computers to innovative mathematics education. *Journal of Mathematical behavior, 19*, 9–47.

Clements, D. H., Sarama, J., Yelland, N. J., & Glass, B. (2008). Learning and teaching geometry with computers in the elementary and middle school. In M. K. Heid & G. Blume (Eds.), *Research on technology and the teaching learning of mathematics: Research syntheses* (Vol. 1, pp. 109–159). Greenwich, CT: Information Age.

Cochrane, T., & Bateman, R. (2010). Smartphones give you wings: Pedagogical affordances of mobile web 2.0. *Australasian Journal of Educational Technology, 26*(1), 1–14.

Dewey, J. (1916). *Democracy and education.* MacMillan, New York.

Donato, R. (2004). Aspects of collaboration in pedagogical discourse. In M. McGroarty (Ed.), *Annual review of applied linguistics: Advances in language pedagogy* (pp. 284–302). West Nyack, NY: Cambridge University Press.

Drijvers, P., Mariotti, M. A., Olive, J., & Sacristán, A. I. (2010). Introduction to section two. In C. Hoyles & J.-B. Lagrange (Eds.), *Mathematics education and technology-rethinking the terrain: The 17th ICMI study* (Online).

Fredricks, J. A., Blumenfield, P. C., & Paris, A. H. (2004). School engagement: potential of the concept, state of the evidence. *Review of Educational Research 74*(1), 59–109.

Gadanidis, G., & Geiger, V. (2010). A social perspective on technology enhanced mathematical learning—From collaboration to performance. *ZDM, 42*(1), 91–104.

Gee, J. P. (2003). What video games have to teach us about learning and literacy. *ACM Computers in Entertainment, 1(1).* New York: Palgrave McMillon.

Gibson, J. J. (1977). The theory of affordance. In R. Shaw & J. Bransford (Eds.), *Perceiving, acting and knowing perceiving, acting, and knowing: Toward an ecological psychology* (pp. 67–82). Hillsdale, NJ: Lawrence Erlbaum.

Harper, S. R., & Quaye, S. J. (Ed.) (2009). *Student engagement in higher education.* New York and London: Routledge.

Hilda, K., Kabali, M. D., Matilde, M., Irigoyen, M. D., Nunez-Davis, R., Jennifer, G., et al. (2015). Exposure and use of mobile media devices by young children. *Pediatrics, 136* (6), 1044–1050. https://doi.org/10.1542/peds.2015-2151.

Hoyles, C., & Lagrange J. B. (Ed.), (2010). *Mathematics education and technology—Rethinking the terrain: The 17th ICMI study 13*, 81–88. USA: Springer.

Hollerbands, K., Laborde, C., & Strasser, R. (2008). Technology and the learning of geometry at the secondary level. In M. K .Heid & G. Blume (Eds.), *Research on technology and the teaching learning of mathematics: Research syntheses* (Vol. 1, pp. 109–159). Greenwich, CT: Information Age.

Hu, S., & Kuh, G. D. (2001). Being (Dis) Engaged in educationally purposeful activities: the influences of student and institutional characteristics. In *Paper Presented at the American Educational Research Association Annual Conference* (pp. 10–14). Seattle, WA.

Jackiw, N., & Sinclair, N. (2014). TouchCounts. Application for the iPad. Burnaby, BC: Tangible Mathematics Project.

Laborde, C., Kynigos, C., Hollebrands, K., & Strasser, R. (2006). Teaching and learning geometry with technology. In A. Gutierrez & P. Boero (Eds.), *Handbook of research on the psychology of mathematics education: Past present and future.* (pp. 275–304). Rotterdam: Sense Publishers.

Lai, C. H., Yang, J. C., Chen, F. C., Ho, C. W., Liang, J. S. & Chan, T. W. (2007). Affordances of mobile technologies for experiential learning: interplay of technology and pedagogical practices. *Journal of Computer Assisted Learning, 23*(4), 326–337.

Lave, J., Wenger, E. (1991). *Situated learning: Legitimate peripheral participation.* Cambridge: Cambridge University Press.

Moyer-Packenham, P. S., Bullock, E. K., Shumway, J. F., Tucker, S. I., Watts, C. M., Westenskow, A., ... Jordan, K. (2016). The role of affordances in children's learning performance and efficiency when using virtual manipulative mathematics touch-screen apps. *Mathematics Education Research Journal, 28*(1), 79–105. http://doi.org/10.1007/s13394-015-0161-z.

Newman, F. W, Wehalage, G. G., & Lamborn, S. D. (1992). *The significance and sources of student engagement.* New York, NY: Teachers college press.

Noss, R. & Hoyles, C. (1996). Windows on mathematical meanings: Learning cultures and computers. In *Mathematics education library* (Vol. 17). Boston, London: Kluwer Academic publisher.

Papert, S. (1980). *Mindstorms: Children, computers and powerful ideas*. New York (NY): Basic Books.

Petrovsky, A. V. (1985). *The collective and the individual*. Moscow: Progress.

Pierce, R., & Stacey, K. (2010). Mapping pedagogical opportunities provided by mathematics analysis software. *International Journal of Computers for Mathematical Learning, 15*(1), 1–20.

Rodgers, C. (2002). Defining reflection: Another look at John Dewey and reflective thinking. *Teachers College Columbia University, 104*(4), 842–866.

Sacristan, A., & Noss, R. (2008). Computational construction as a means to coordinate representations of infinity. *International Journal of Computers for Mathematical Learning, 13* (1), 47–70. https://doi.org/10.1007/s10758-008-9127-5.

Sacristan, A. I., Calder, N., Teresa, R., Santos-Trigo, M., Friedlander, A., Hartwig, M., … Perrusquia, E. (2010). The influence of shaping of digital technologies on the learning—and learning trajectories—of mathematical concepts. In C. Hoyles, & J.-B. Lagrange (Eds.), *Mathematics education and technology—Rethinking the terrain* (Vol. 13, pp. 179–226). New York, NY: Springer.

Schön, D. A. (1983). *The reflective practitioner: How professionals think in action*. New York (NY): Basic Books.

Sedaghatjou, M., & Campbell, S. R. (2017). Exploring cardinality in the era of touchscreen-based technology. *Journal of Mathematical Education in Science and Technology* (Online). http://dx.doi.org/10.1080/0020739X.2017.1327089.

Sinclair, N., Chorney, S., & Rodney, S. (2016). Rhythm in number: exploring the affective, social and mathematical dimensions of using TouchCounts. *Mathematics Education Research Journal, 28*(1), 31–51. https://doi.org/10.1007/s13394-015-0154-y.

Sinclair, N., & Heyd-Metzuyanim, E. (2014). Learning number with TouchCounts: The role of emotions and the body in mathematical communication. T*ech Know Learn, 19*(1), 81–99. https://doi.org/10.1007/s10758-014-9212-x.

Sinclair, N., Arzarello, F. (2010). Implementing digital technologies at a national scale. In C. Hoyles & J.-B. Lagrange (Eds.), *Mathematics education and technology—Rethinking the terrain* (pp. 61–78). New York, NY: Springer.

Sinclair, N. (2005). Mathematics on the internet. In S. Johnston-Wilder & D. Pimm (Eds.), *Teaching secondary mathematics with ICT* (pp. 203–216). Berkshire, UK: Open University Press.

Stone, D., Jarrett C., Woodroffe M., & Minocha, S. (2005). *User Interface Design and Evaluation*. San Francisco, Elsevier: Morgan Kaufmann Publisher Inc.

Vogel, R., & Jung, J. (2013). Video coding—A methodological research approach to mathematical activities of kindergarten children. *In Proceedings of the Eighth Congress of the European Society for Research in Mathematics Education in Antalya* (pp. 6–10). Turkey, Ankara.

**Mina Sedaghatjou** is a mathematics educator, whose primary research interest focuses in the areas of mathematics and STEM education with the use of emerging technology. Mina engages in qualitative and quantitative research approaches to explore how using technology for teaching and learning contributes to embodied cognition and conceptual understanding. She currently serves as an assistant professor at Brock University.

**Sheree Rodney** is a Ph.D. candidate in mathematics education in the Faculty of Education at Simon Fraser University. She is a Jamaican native with experiences at various levels of the education system (Elementary, Secondary and Tertiary) and also serves as a Mathematics Specialist and National School Inspector with the Jamaican Ministry of Education. She earned both Bachelor's and Master's degrees at the University of the West Indies Mona, in Jamaica and a Teaching Diploma at the College of Agriculture Science and Education also in Jamaica. Her research interest is the integration of digital technology in mathematics classrooms with a specific focus on the relationship between embodied curiosity and digital technology tools.

# Augmenting Mathematics with Mobile Technology

**Christian Bokhove, Alison Clark-Wilson and Marios Pittalis**

**Abstract** This chapter describes two case examples of the use of mobile technology for mathematics. Building on the assumption that mobile learning has a positive effect on student attitudes and academic outcomes including STEM subjects (Hsi, 2007; Wu et al., 2012) we develop a theoretical lens for future studies for 'mobile mathematics'. The two case examples describe how mobile technology could provide opportunities for 'mathematics outside the classroom'. The first example describes a dynamic Ferris wheel, the second a static cathedral. Both examples demonstrate how 'geo-location' and 'augmented reality' features allow mobile technologies to bridge formal and informal mathematics learning (Lai et al., 2016).

**Keywords** Augmented reality · Mathematics education · Mobile learning

## Introduction

This chapter capitalizes on the potential of, and synergy with, informal learning using mobile devices (Laurillard, 2009). Whilst research is limited, evidence suggests that mobile learning has a positive effect on student attitudes and academic outcomes including STEM subjects (Hsi, 2007; Price et al., 2014; Wu et al., 2012). Lowrie (2005) argues that the technology-rich contexts that are used at school are

C. Bokhove (✉)
University of Southampton, Southampton, UK
e-mail: C.Bokhove@soton.ac.uk

A. Clark-Wilson
UCL Knowledge Lab, York, UK
e-mail: a.clark-wilson@ucl.ac.uk

M. Pittalis
University of Cyprus, Nicosia, Cyprus
e-mail: m.pittalis@ucy.ac.cy



N. Calder et al. (eds.), *Using Mobile Technologies in the Teaching and Learning of Mathematics*, Mathematics Education in the Digital Era 12,
https://doi.org/10.1007/978-3-319-90179-4_8

often different than the technology that children regularly use at home and so it is important for educational research to consider the impact that technology not commonly found in school can have on children's meaningful mathematics learning. In a study that explored opportunities for engaging children in mathematical activities through the use of a location-based game with mobile handheld technology, Wijers, Jonker, and Drijvers (2010) collected data from observations, online storage game data, an online survey and interviews and report findings that indicate enhanced student engagement. Ludwig and Jesberg (2015) explored the potential of mobile technology by provided "geo located" modelling tasks, that is, "maths trails" that were guided by the GPS options of mobile phones. Another development concerns the use of "augmented reality" in informal learning environments; a field experiment in a mathematics exhibition showed that visitors performed significantly better on knowledge acquisition (Sommerauer and Müller 2014). Despite a lack of mobile learning research in informal contexts (Wright and Parchoma 2011), we aim to better understand the concept of mobile learning and how mobile technologies can be used to bridge formal and informal mathematics learning (Lai, Khaddage, & Knezek, 2013). We hypothesize that the planned activities for different geolocations will provoke curiosity (Arnone, Small, Chauncey, & McKenna, 2011) and improve learning and interest in mathematics because:

Visitors to these locations are likely to attend in a friendship or family group. In this sociocultural context, visitors become learners and learning takes place through social interaction with others in a sociocultural context in which they act and interact in shared experiences (Vygotsky, 1978); The locations and their planned activities within them are intriguing and this physical factor may influence how visitors feel about learning in this context. Learning outcomes can be a result of the ease with which the activity can be accomplished and how well it demonstrates scientific and mathematical concepts;
Visitors are engaged in multiple ways e.g., physically, socially, emotionally and cognitively; Visitors have control over whether to engage in the activities or not; The mathematical ideas are experienced in an authentic and dynamic fashion.

In the following we further elaborate on the key dimensions of our study.

## Theoretical Perspectives

### *What Is Distinct in 'Mobile Learning'?*

Mobile technology, such as portable and handheld devices, with powerful social networking, communication and geo-location capabilities, has become ubiquitous worldwide and offers immense opportunities and new potentials in education (Dhir, Gahwaji, & Nyman, 2013; Domingo & Gargante, 2016; Larkin & Calder, 2016).

Mobile technology devices have become widely available, convenient and less expensive, with each successive generation being equipped with new features and sophisticated applications (Wu et al., 2012). The immense power of mobile technology is underlined by the fact that society and mobile technology interact with, and shape each other. Despite its ubiquitous nature, increased affordability and functionality, the integration of mobile technology devices in education is considerably limited and the effectiveness of mobile learning needs to be evidenced in a more systematic way.

To a large extent "mobile learning" or learning with mobile technology builds on the same foundations as that of technology-enhanced learning. It is not our intention to review the complete literature on the topic; we refer to the large body of literature available (e.g., Voogt & Knezek, 2008). We will focus specifically on the "mobile" aspect, mobile learning (m-learning), which differs from the broader technology topics by its ability to obtain and supply information at any time, resulting from its built-in wireless connectivity (Kukulska-Hulme & Traxler, 2005). However, it is problematic to conclude a concise definition of m-learning due to the ambiguity of the concept of m-learning itself (Kukuska-Hulme, 2009). As so often in technology-oriented literature it revolves around the question whether m-learning is about the mobility of the learning technology or the mobility of the learner his/herself? The same question is noted by Traxler (2009). To illustrate this with an extreme example, imagine a student brings his/her desktop computer outside, or a student places his/her mobile phone on the desk in a classroom and types in an essay. In other sources there seems to be a distinct emphasis on one, the other or both, without really concluding a clear definition of m-learning. Consequently, we focus on m-learning from the perspective of the mobility of the learner, which resonates with the views of O'Malley et al. (2003) suggesting that m-learning happens when the learner is not at a fixed, predetermined location and takes advantage of the learning opportunities offered by mobile technologies. Kukulska-Hulme and Traxler (2005) approached m-learning as learners' engagement in educational activities and communications with others via wireless technologies in mobile devices, without any specific location. Mobile learning takes place whenever and wherever the learners desire (Keengwe & Bhargava, 2014; Traxler, 2009). In addition, the affordances of mobile technologies offer to learners different levels of engagement and may provide inquiry-based learning activities inside the school, but also in out-of-school environments (Churchill & Churchill, 2008). What these perspectives have in common is that the learner is central and mobile.

In a review synthesis of 164 studies on m-learning from 2003 to 2010, Wu et al. (2012) revealed two major research-strands. The first strand concerns the effectiveness of mobile learning and the second one the design of mobile learning systems. A significant number of studies revealed positive, neutral and negative findings regarding the effectiveness of mobile learning. From a methodological perspective, surveys and experiments were used as the primary research methods. As the review is already somewhat older, mobile phones and PDAs were the most widely used devices for mobile learning. The authors suggest that these findings might be displaced by emerging technologies, which has become evident in the case of tablet computers. Crompton and Burke (2014) concluded similar findings to Wu et al. (2012) from a mathematics specific review. They found that: (a) most of the studies focus on effectiveness, followed by learning design, (b) mobile phones were the most widely used device, and (c) the use of mobile devices for mathematics learning was most common in elementary (5–11 years old) school settings.

There is an ongoing need to examine the pedagogies that are suitable for mobile learning from the perspective of learners' needs and not only based on the affordances of the new technological features (Traxler, 2009). Mobile learning devices have been considered as a new type of computing platform that can be used to push beyond the restrictions of traditional pedagogies, provided they are designed and implemented in a way that takes into consideration the social and cultural context of learning (Crompton & Traxler, 2015).

## *What Are the Advantages of Augmented Reality (AR)?*

Another development concerns the use of "augmented reality" in informal learning environments. When reality is augmented, technology adds an additional layer to reality. It combines real and virtual objects, has real-time interaction and three-dimensional affordances (Azuma, 1997). With AR devices users can actually see 3D objects, work with complex spatial problems and involve spatial relationships. In addition, AR technologies help learners engage in authentic exploration in the real world and conduct investigations of the real-world surroundings. As Sommerauer and Müller (2014) indicate, advances in mobile technologies (especially smartphones and tablets with built-in cameras, location options and internet access) have made augmented reality (AR) applications available for the broad public. Their pretest–posttest crossover field experiment with 101 participants at a mathematics exhibition aimed to measure the effect of AR on acquiring and retaining mathematical knowledge in an informal learning environment. The study was based on principles from the cognitive theory of multimedia learning (CTML), suggesting that people learn better from words and pictures than from words alone (Mayer, 2010). AR might, when designed correctly, address several design principles for effective multimedia instruction: firstly, the multimedia principle by overlaying pictorial content with text; secondly, the spatial and temporal contiguity principles by aligning virtual and physical information, for example in three dimensions; thirdly, the modality principle by integrating auditory elements. Finally, the signaling principle could be obtained by highlighting essential information in a learning environment through cues, for example geographic location information and triggers (Sommerauer & Müller, 2014).

## *Multiple Representations in Task Design*

Building on the dynamic nature of mobile learning and the affordances of AR technologies, we can also position AR in relation to prior research on task design in mathematics education, with a particular emphasis on the potential role of multiple representations. This was a prominent focus within the technology chapter of the 22nd ICMI study on task design (Watson & Ohtani, 2015). This study argued that often "abstract generalizations come about when critical aspects from multiple mathematical representations and discourses fuse and blend together" (p. 216). In addition, Whiteley and Mamolo (2013) used a framework of conceptual blending. It was found that teachers and students had multiple ways of reasoning about the task and created different conceptual blends for these representations. Earlier, Kaput (1986) had already argued that a multiple representational environment supported by technology might enhance high-level engagement with mathematics. So although AR might realize the potential of doing exactly that, it is important that the bridging and moving between tools and representations are key task design considerations. In moving between different representations we can also think about the distinction between a real situation, a real model as overlay on the real situation and an abstract mode, as per the modelling cycle by Blum and Leiss (2007). This cycle takes as starting point the "real situation" from which a situation or real model is inferred. The process of mathematising then transports it to a mathematical model, which is used to get mathematical results. Finally, the interpretation of these results leads to the final, real results, perhaps leading to an adjustment of the situation model.

## *Bridging Formal and Informal Learning*

Mobile learning has the potential to bring out-of-school contexts and problems into the classroom for learning mathematics and take school mathematics into out-of-school contexts because mobile technologies have the ability to work within the specific context and environment of the learning (Khaddage, Muller, & Flintoff, 2016). The importance of informal learning has been stressed in research (Cox 2013). Children can learn anywhere and anytime outside a formal learning environment resulting to an increased desire to continue interacting, playing and exploring from different perspectives. Informal learning is self-directed, has an intentional-interest, is non-assessment driven and spins-off mainly from leisure activities (Lai, Khaddage, & Knezek, 2013). Sawaya and Putnam (2015) suggested that this can be achieved by utilizing the affordances of mobile devices, such as computing input, consuming content, capturing surrounding context, communicating and collaborating with others and creating content. Thus, a suggestion could be to investigate in depth the way in which mobile technologies can be used to bridge formal and informal mathematics learning (Lai et al., 2013; Wright & Parchoma, 2011). Along with Sawaya and Putmans' framework regarding mobile devices affordances, Lai, Khaddage, and Knezek (2013) described a Mobile-Blended Collaborative Learning model that only describes three categories of mobile application tools, namely tools for collaboration, tools for coordination and tools for communication. We suggest that a category, let's call it "tools for augmentation", given the affordances described previously, also might facilitate formal and informal learning, simply because they augment reality (which we see as informal) with a virtual layer (which can be the formal content, for example provided by curriculum content). In addition, it could be suggested that children's out of school experiences might be utilized effectively to bridge the gap between home and school (primary and secondary) or home and university. Jay and Xolocotzin (2015), based on the results of an intervention program, asserted that there is enough content and motivation in children's out of school mathematics activities to be explored in ways that may help students' build their own mathematical structures. They suggest that this can be achieved by making connections between the abstract content of mathematics lessons and the multiple ways in which mathematical concepts are involved in out-of-school activities.

## *Students' Mathematical Learning Processes and Activities*

Mobile learning could provide immense pedagogical benefits when mobile technologies are used as educational tools (Keengwe & Bhargava, 2014). Research findings suggested that mobile learning is associated with autonomous learning, students' active engagement and easy-access to information through internet resources (Spector, 2015). Mobile technology devices allow students to become contributors of knowledge and co-designers of activities by posing their own real-world scenarios and utilizing the affordances of the handheld devices, such as gathering measurement data, building structures, conducting virtual/augmented experiments or creating multimedia videos. In addition, such devices encourage pupils to take control of their own learning and manage their self-directed learning and individual development (Spector, 2015). Individual development refers to the enhancement of inquiry exploration and self-regulation strategies. The virtual and augmented affordances of the devices facilitate students move from passive-reproducers of information to content creators and thus the further development of reasoning skills, such as analysis, synthesis, evaluation, decision-making, modeling, explanation and problem solving. In addition, mobile learning encourages collaborative learning and promotes social interaction and collaborative feedback.

We contend that there are numerous potentialities for m-learning that can be explored in relation to the above themes. Here, we describe two case scenarios, one for using AR for mathematics involving the London Eye attraction, next to the river Thames in central London, the second situated at a cathedral in the ancient capital of England, Winchester. We describe the scenarios from the viewpoint of the learner and other actors around him/her. We hope to show that elements of aforementioned themes, namely mobile learning, augmented reality, a combination of informal and formal learning, and multiple representations, might come together in one m-learning experience.

## Case One: The London Eye

Many cities in developed countries around the world boast an observation (or Ferris) wheel of some type that sits proudly on the landscape and inevitably captures the curiosity of onlookers. One such wheel is "The London Eye", developed to mark the new millennium. It dominates the London skyline and, as the most popular paid visitor attraction in London, it attracts over 3.75 million visitors per year. Some mathematics educators have capitalized on it to create classroom-based resources to support both an introduction to mathematical concepts (Knights, 2014) or to consolidate/assess prior learning (Thomas & Gitonga, 2013). Central to both of these approaches was the prominence of the image of The London Eye, alongside the use of technology to support the further analysis of the mathematics represented by its physical features. This case example demonstrates how, by moving the learning outside of the classroom to the venue, and combining potential functionality from mobile technology such as smartphones, new mathematical activity can be proposed and, more importantly, experienced.

On approaching The London Eye on foot, by wheeled vehicle or by boat, its position on London's South Bank and the curvature of the river Thames make it inevitable that the Eye is seen from different angles. A (mathematical) question such as, "Where does The London Eye look most like a circle?" is far from trivial as one considers the best place to stand for a circular view. Similarly, the other extreme, "Where can you view The London Eye at its thinnest?" takes you to a place on the Golden Jubilee Bridge (West) (Fig. 1), which runs alongside Hungerford Bridge.

**Fig. 1** Viewing The London Eye from the Golden Jubilee Bridge (West)

However, these static photographic images mask the most striking feature of this, and any other Ferris wheel—*it is moving at a constant speed of rotation*. In the case of The London Eye, it stops very occasionally to enable disabled visitors to embark and disembark from its "capsules". So, imagine that the observer, in our case a learner of lower secondary age, is standing with a friend or older family member at a marked location on the Golden Jubilee Brigade (or possibly, their mobile device has sent an alert to inform them that they are in an augmented reality mathematics space). As they look up at the Eye through the lens of their Smartphone, a mathematical question pops up to provoke their curiosity: "Why do the capsules look like they are closer together at the top of The London Eye when compared to the middle?"

Again, a few moments of thinking time pass before our learner is asked whether she would like a hint—a prompt to touch one of the capsules on the smartphone screen, so as to mark its changing position over time. Simultaneously, a line segment that indicates this distance is displayed—augmenting reality (Fig. 2, right side). The actual measurement can also be displayed.

Additionally, the sequence of data, the marked capsule position from the horizontal mid-line at fixed time intervals, is stored—and can be auto-displayed as either a table or a graph in response to the learners' own curiosity. Of course the same data can be collected and displayed whilst the learner is inside the capsule and experiencing The London Eye first-hand. An in-ride app, if it were to be designed, could be viewable on individual personal devices (or accessed via the many tablets provided in each capsule by the venue), could offer simultaneous screens showing

**Fig. 2** Augmented reality of The London Eye from the perspective of the Golden Jubilee Bridge (West)

the external views of the London Eye alongside the actual positional data of the individual capsules for the period of the ride. In this case, learners are prompted to make predictions in relation to the magnitude of, and relationships between, key data. By engaging learners with their personal experience of seeing how their own capsule's height varies in relation to those immediately adjacent to them and the ground below, their ride becomes a rich 2-D trigonometric experience as they experience for themselves the journey of a point on a trigonometric graph.

The transition from this early experience of (constant) circular motion as a model of height against time towards more formal trigonometric graphing could follow sequentially by going to stand at another AR mathematics spot that is facing The London Eye (Fig. 3).

The same sequence of questions still applies, but the different perspective allows for alternative approaches that involve optional AR tools. Initially, to justify or explain that the upper and lower capsules indeed *are* close to each other (in the horizontal plane) than those nearest the mid-line, a still image could be augmented as in Fig. 4.

Working from the moving image, for which you (the reader) need to know that the London Eye moves counter-clockwise when viewed from this perspective, the learner is again invited to mark a capsule, which results in an AR experience whereby the moving image is annotated with a "mid-line" and an angle measure that shows the marked capsule's position on the wheel—in this case as an angular

**Fig. 3** Viewing The London Eye from the embankment

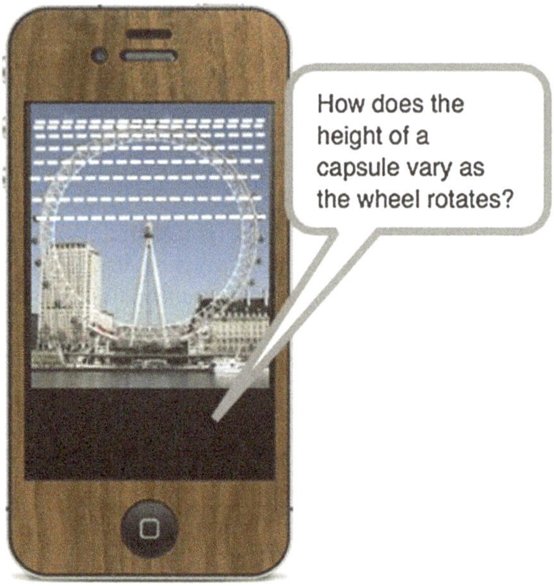


**Fig. 4** An augmented view of The London Eye from the embankment showing how the height of the capsules vary during the ride

measure relative to the "three o'clock" position to fit with the usual mathematical convention (see Fig. 5, right side).

Automated data collection from the image would then be collected and adjusted to generate a model for the capsule's motion over the journey. This could be made visible to the learner as measurement data that could be viewed and shared both in tabular form and graphically (see Fig. 6).

Any or all of our learner's explorations could be shared via social platforms—and of course hopefully with her teachers, who could use this real experience as the basis for more formal learning.

## Case Two: Augmenting a Cathedral

The second case example revolves around Winchester Cathedral, Hampshire, United Kingdom. Winchester used to be the ancient capital of England and its cathedral is one of the largest cathedrals in Europe, with the longest nave and greatest overall length of any Gothic cathedral in Europe. Upon arriving on the scene a student's mobile phone send an alert to indicate that the cathedral has some interactive features. The web-based app shows the student's geolocation and GPS coordinates and indicates that the cathedral is at the starting point of mathematical

**Fig. 5** An augmented view of The London Eye from the embankment—establishing reference points and highlighting changes in position. In the dynamic experience, the measured height would change as capsule moves during its journey

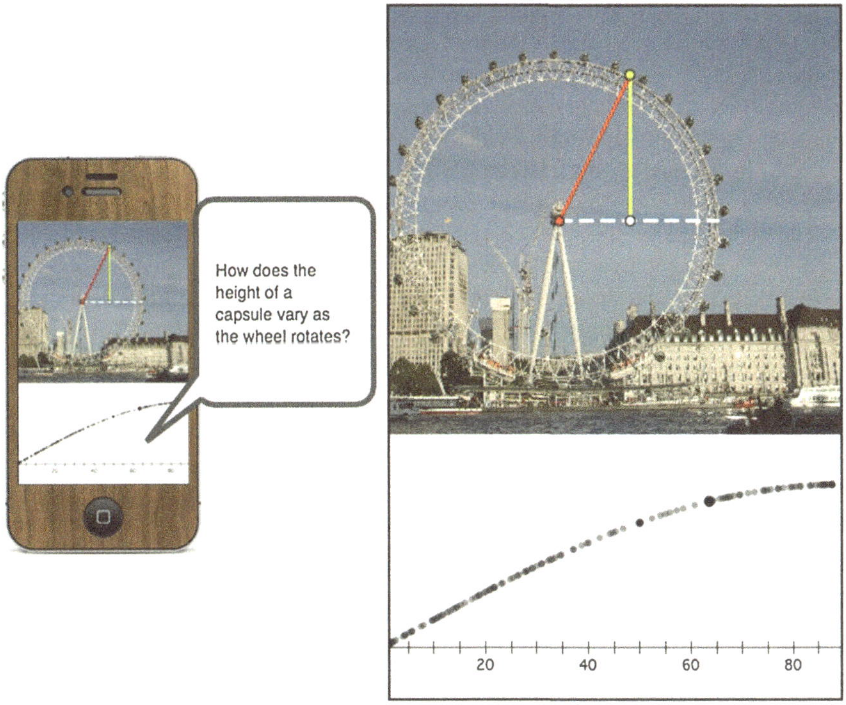


**Fig. 6** An augmented view of The London Eye from the embankment—modeling the capsule's relative position during the ride graphically

**Fig. 7** Information of Winchester Cathedral is provided

activities related to proportionality and ratios. The student can point his/her mobile device to the cathedral, after which the installed app recognizes the cathedral and provides some relevant information (see Fig. 7). An interface is provided for some further information on the web and custom information for this specific augmented location.

There is a feature to download some off-line resources such as task sheets, as part of a broader geo-located Augmented Reality package for the location. The package contains some classroom activities. Next to the information an icon also indicates that there are interactive AR activities for proportionality at this location. Clicking on the toolkit icon provides an additional 'layer' with some interactive features. In the case of the cathedral, a Dynamic Geometry System (DGS) can be used to calculate some proportions on the actual view (Fig. 8).

The platform also shows the lengths of the lines. An in-built clinometer can be used to determine the viewing angle. With some further tools, such as the geometry tool, a sketchpad and an aerial view of the area, the student can further model the situation, hinted by prompts and hints from the platform (Fig. 9).

The model the student has made is followed up by a quick pop quiz on the topic. An extension task, which the student can save for later also appears, emphasising connections between a (real) view of the cathedral in perspective and an abstract diagram, overlaying lines of the cathedral and a point on the horizon. Layers of the view can be turned on and off at will. It provides the student a means to go from reality (real situation) to an abstract model. The work is shared and commented on via the interactive, social functions of the platform (Fig. 10).

The scenario in this example can also be extended to a classroom. Students are able to experience the majesty of the cathedral but virtually. The functionality of the tools makes it possible for the teacher to make use of the location-based resources in the classroom. By pointing the device at an image of the cathedral, it can serve as a "trigger image" whereby the AR app presents a layer over the cathedral with the same functionalities as in the actual location. The location can be seen in a wider

**Fig. 8** Dynamic geometry is transposed on the view of Winchester Cathedral

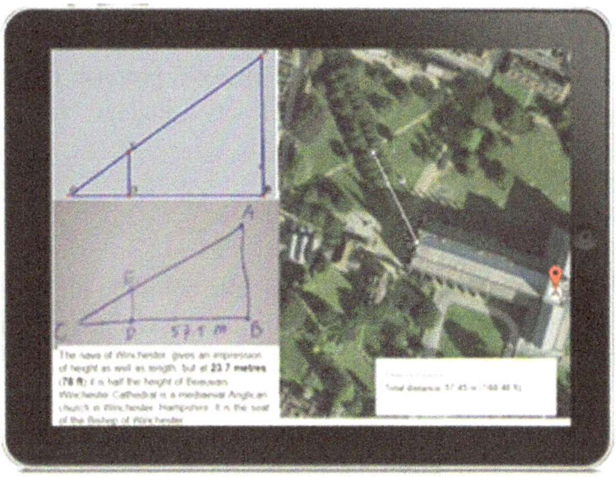


**Fig. 9** A geographical map of the surroundings of Winchester Cathedral is presented next to abstract diagrams of the situation. The top left is a geometric diagram, bottom left is a learner drawn diagram. Note that the letters in this diagram do not match those in Fig. 8; they are, however, related as the vertical AB corresponds with the height of the cathedral

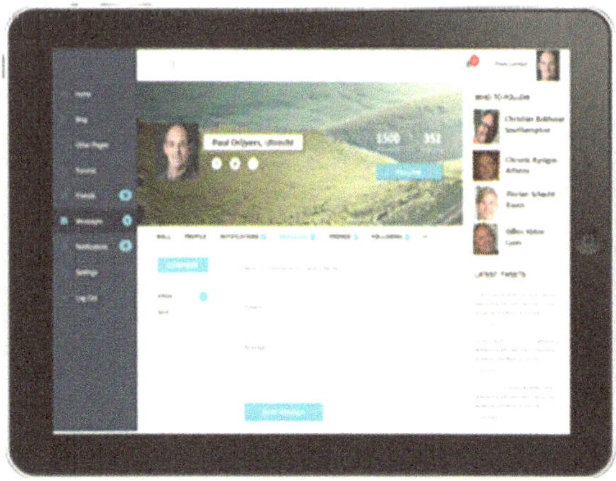


**Fig. 10** A mobile portal site provides information about the geo-location and social media functions

geographical map, disclosing that there are several other augmented spots in the area, for example at nearby Stonehenge. In addition to the same resources as the "real" location, the classroom also provides some other features that are difficult to present on mobile screens. In addition to AR, there is scope to provide a Virtual Reality (VR) experience: using a mobile device to experience the grandeur of the real cathedral, with interactive features added in. The augmented cathedral has provided a way to address proportionality "in real life" as well as to relate it to the abstract concepts.

## Towards a Theoretical Lens for Augmented Mathematics

Based on a synthesis of the literature we argued that the integration of augmented reality in mathematics teaching might "augment" learning for mathematics and contribute in developing students' reasoning skills (Spector, 2015). The above case examples made explicit the need to further investigate the role of several key dimensions, or *design decisions*. We propose that these design decisions can be grouped by three heuristics: observe, engage, and create.

### Observe Mathematics

We propose that the starting point should be the object of interest. Thus, the object of interest (the geo-location) should have interesting characteristics that can contribute in exploring salient mathematical concepts and properties. We should tap into learners' mathematical curiosity by making geo-locations the trigger: "What are the mathematical questions that might come into the learners' head?" A pertinent question related to this is, who initiates this process?—the learner, the teacher or perhaps the technology. In the theoretical section we had made clear that we see the learner as leading, but acknowledge that teacher and technology could impose constraints on their initiatives. If we indeed take the learner as starting point, this reconceptualizes the "any time, any place" assumptions of many perspectives of mobile learning, as the mathematical questions that are generated from a leisure activity are the guiding principles of one's self-directed learning process. Rather than focusing on the technology, we suggest that the focus should be on *reality* and *mathematics*. The locations where AR can be meaningfully applied is conditional on the inherent mathematics for any particular location. Luckily, mathematics is quite prevalent in most locations, whether they are *man-made*, like our two case examples or a natural phenomenon like a pattern from nature. If we make mathematics central to AR task design then in our view this also means linking the concrete reality to the mathematical abstract (and back again). AR can then serve as a tool to support the modelling cycle, by providing the means to create, apply, adapt mathematical models during the processes of interpreting and explaining real-world based problems (e.g., see Blum & Leiss, 2007; Doerr & English, 2003). Real-world based problems might arise from two-types of geo-locations, namely dynamic and static. Dynamic geo-locations are related to situations from the perspective of the user's visual and kinesthetic experience, such as roller-coaster rides or an airplane's take-off or landing. In these types of geo-locations, AR functions as a composer of the viewers' and the experiencers' perspectives. In static-geolocations, such as historic buildings, monuments, bridges and natural spots, AR facilitates mainly the in-depth study of the spot, by providing measurements. For instance, an AR experience may provide data to explore the golden ratio of measurements associated with the Parthenon or to calculate the height of the Eiffel tower based on the measures from a "selfie" picture.

## Engage in Mathematical Content: Development Issues

A second key decision, following from the mathematical content and context, concerns the appropriateness of using AR. Is it relevant, or deemed beneficial, to experience mathematics in the particular surroundings? If so, what prior mathematical knowledge/experience might be desirable? We acknowledge that this is a major prerequisite of what we should refer to as *experiential learning*. If a mathematical topic, according to the teacher or designer, is best learned without context and location-based experience, it might be difficult to make a case for AR. After all, one of the major advantages is that m-learning augmented by AR can make human experience, the surroundings, "alive" and transpose abstract mathematical concepts on the outside world. Through this AR augmented experience that integrates the real world with abstract mathematic concepts, the learners might formulate and test hypotheses, solve problems and create explanations for what they observe (Bossé, Lee, Swinson, & Faulconer, 2010). We are in no way saying that every topic *should* be experiential. In fact, there are topics where context might impede the acquisition of more abstract mathematical knowledge. Nevertheless, it should be a key consideration while thinking about the adoption of AR. The existence of mathematics in a certain geographical location does not necessarily mean that the location is suitable for augmentation. The decision of augmenting should be made on well-explicit criteria, such as whether the integration of real-world and digital-augmented learning resources has the potential to engage learners in manipulating virtual manipulatives and the underlying mathematic properties from a variety of perspectives.

## Create: Depth of Experience

A third key decision pertains to the depth of the AR experience and the extent to which the learner might assume ownership of the mathematical activity and create, share and/or communicate their productions. By exploiting different layers of AR users can be engaged in the interesting mathematical features of the geo-located spot and concretely conceptualize the problem to be explored by inspecting the spot from a variety of different perspectives that facilitate their understanding. The different layers and perspectives provided by AR provides learners with data to elaborate their thinking, seek patterns, clarify concepts, synthesize ideas, pose their own questions, and create and own mathematical models. This can be achieved by working collaboratively through the social affordances of mobile technology devices. Users can also extend their understandings to new situations and make connections (connect the characteristics of the location with the collected data and the mathematical models). We suggest that both case examples showed this: the London Eye by linking the wheel to location data and a model of the wheel, the cathedral by linking locations to geometric constructions that could aid calculations of height. The whole scenario could be completed with a reflection regarding the underlying mathematical concepts related to each spot. These considerations all reduce to decisions about how much students can manipulate or interact with the environment, and, for example, whether the technological device responds back (feedback). Mobile technology devices can offer some form of validation and opportunities to further probing and development of students' mathematical thinking.

By imagining what mathematical content students need to observe, how they need to engage with the content and how they can create their own experience, quality AR tasks can be designed more readily.

## Conclusion

In this chapter we have given an overview of how mobile learning and augmented reality might play a role in learning mathematics. After describing some relevant features of the issues involved in the study, we set out to describe two scenarios in which mobile learning, augmented reality, a combination of informal and formal learning, and multiple representations, came together. We concluded with three core aspects that need to be taken into account when designing such tasks. Firstly, it is important to reflect on the importance of the involved mathematics concepts and more importantly on how the integration of AR and the geo-location can trigger mathematical curiosity. Secondly, how appropriate it is to apply experiential learning to the topic at hand and to what extent the mathematical prerequisites of the activity meets learners' knowledge/experience. Finally, the depth of the learning experience depends on the technical functionalities of the software, and therefore the envisaged technical tool needs to be taken into account. This has less to do with technology per se but more with the learning opportunities that can be offered by the affordances of the technology and the learning design of the tasks. The above mentioned core aspects provide designers with important design parameters that they should take into account regarding what students need to observe, need to do to get engaged, and what they need to create. The scenarios presented here are practical examples of its application.

## References


Arnone, M. P., Small, R. V., Chauncey, S. A., & McKenna, H. P. (2011). Curiosity, interest and engagement in technology-pervasive learning environments: A new research agenda. *Educational Technology Research and Development, 59*(2), 181–198.

Azuma, R. T. (1997). A survey of augmented reality. *Presence: Teleoperators and Virtual Environments, 6*(4), 355–385.

Blum, W., & Leiss, D. (2007). How do students and teachers deal with mathematical modelling problems? The example "Filling up". In C. Haines, et al. (Eds.), *Mathematical modelling (ICTMA 12): Education, engineering and economics* (pp. 222–231). Chichester: Horwood Publishing.

Bossé, M. J., Lee, T. D., Swinson, M., & Faulconer, J. (2010). The NCTM process standards and the five Es of science: Connecting math and science. *School Science and Mathematics, 110*(5), 262–276.

Churchill, D., & Churchill, N. (2008). Educational affordances of PDAs: A study of a teacher's exploration of this technology. *Computers and Education, 50*(4), 1439–1450.

Cox, M. J. (2013). Formal to informal learning with IT: Research challenges and issues for e-learning. *Journal of Computer Assisted learning, 29*(1), 85–105.

Crompton, H., & Burke, D. (2014). Review of trends in mobile learning studies in mathematics: A meta-analysis. In M. Kalz, Y. Bayyurt & M. Specht (Eds.), *Mobile as a mainstream—Towards future challenges in mobile learning* (pp. 304–314). Springer.

Crompton, H., & Traxler, J. (Eds.). (2015). *Mobile learning and mathematics*. New York: Routledge.

Dhir, A., Gahwaji, N. M., & Nyman, G. (2013). The role of the iPad in the hands of the learner. *Journal of Universal Computer Science, 19*(5), 706–727.

Doerr, H. M., & English, L. D. (2003). A modeling perspective on students' mathematical reasoning about data. *Journal of Research in Mathematics Education, 34*(2), 110–136.

Domingo, M. G., & Gargante, A. B. (2016). Exploring the use of educational technology in primary education: Teachers' perception of mobile technology learning impacts and applications' use in the classroom. *Computers in Human Behavior, 56,* 21–28.

Hsi, S. (2007). Conceptualizing learning from the everyday activities of digital kids. *International Journal of Science Education, 29*(12), 1509–1529.

Jay, T., & Xolocotzin, U. (2015). Breaking barriers between out-of-school and classroom mathematics with documenting. In H. Cromptom & J. Traxler (Eds.), *Mobile learning and mathematics* (pp. 86–95). New York: Routledge.

Kaput, J. (1986). Information technology and mathematics: Opening new representational windows. *The Journal of Mathematical Behavior, 5*(2), 187–207.

Keengwe, J., & Bhargava, M. (2014). Mobile learning and integration of mobile technologies in education. *Education and Information Technologies, 19*(4), 737–746.

Khaddage, F., Müller, W., & Flintoff, K. (2016). Advancing mobile learning in formal and informal settings via Mobile App Technology: where to from here, and how? *Educational Technology and Society, 19*(3), 16–27.

Knights, C. (2014). *Introducing trigonometry*. Wiltshire: Mathematics in Education and Industry.

Kukulska-Hulme, A. (2009). Will mobile learning change language learning? *ReCALL, 21*(02), 157–165.

Kukulska-Hulme, A., & Traxler, J. (Eds.). (2005). *Mobile learning: A handbook for educators and trainers*. London: Routledge.

Lai, K. W., Khaddage, F., & Knezek, G. (2013). Blending student technology experiences in formal and informal learning. *Journal of Computer Assisted learning, 25*(5), 414–425.

Larkin, K., & Calder, N. (2016). Mathematics education and mobile technologies. *Mathematics Education Research Journal, 28*(1), 1–7.

Laurillard, D. (2009). The pedagogical challenges to collaborative technologies. *International Journal of Computer-Supported Collaborative Learning, 4*(1), 5–20.

Lowrie, T. (2005). Problem solving in technology rich contexts: Mathematics sense making in out-of-school environments. *Journal of Mathematical Behavior, 24*(3–4), 275–286.

Ludwig, M., & Jesberg, J. (2015). Using mobile technology to provide outdoor modelling tasks—The MathCityMap-project. *Procedia—Social and Behavioral Sciences, 191*, 2776–2781.

Mayer, R. E. (2010). Instruction based on visualizations. In R. E. Mayer & P. A. Alexander (Eds.), *Handbook of research on learning and instruction* (pp. 427–445). Abingdon: Routledge.

O'Malley, C., Vavoula, G., Glew, J. P., Taylor, J., Sharples, M., Lefrere, P., et al. (2005). *Guidelines for learning/teaching/tutoring in a mobile environment*. Retrieved from https://hal.archives-ouvertes.fr/hal-00696244.

Price, S., Davies, P., Farr, W., Jewitt, C., Roussos, G., & Sin, G. (2014). Fostering geospatial thinking in science education through a customisable smartphone application. *British Journal of Educational Technology, 45*(1), 160–170.

Sawaya, S. F., & Putnam, R. T. (2015). Using mobile devices to connect mathematics to out-of-school contexts. In H. Crompton & J. TRaxler (Eds.), *Mobile learning and mathematics* (pp. 9–19). New York: Routledge.

Sommerauer, P., & Müller, O. (2014). Augmented reality in informal learning environments: A field experiment in a mathematics exhibition. *Computers and Education, 79*, 59–68.

Spector, J. M. (2015). *Foundations of educational technology: Integrative approaches and interdisciplinary perspectives*. Abingdon: Routledge.

Thomas, C. D., & Gitonga, I. (2013). Mathematics in the London Eye. *The Mathematics Teacher, 106*(3), 172–177.

Traxler, J. (2009). Learning in a mobile age. *International Journal of Mobile and Blended Learning, 1*(1), 1–12.

Voogt, J., & Knezek, G. (Eds.). (2008). *International handbook of information technology in primary and secondary education*. Springer.

Vygotsky, L. (1978). *Mind in society: The development of higher psychological processes*. Cambridge, MA: Harvard University Press.

Watson, A., & Ohtani, M. (2015). *Task design in mathematics education*. Springer.

Whiteley, W., & Mamolo, A. (2013). Optimizing through geometric reasoning supported by 3-D models: Visual representations of change. In C. Margolinas (Ed.), *Task Design in Mathematics Education: Proceedings of ICMI Study 22* (pp. 129–140), Oxford, UK. Retrieved from http://hal.archives-ouvertes.fr/hal-00834054.

Wijers, M., Jonker, V., & Drijvers, P. (2010). MobileMath: Exploring mathematics outside the classroom. *ZDM Mathematics Education, 42*(7), 789–799.

Wright, S., & Parchoma, G. (2011). Technologies for learning? An actor-network theory critique of 'affordances' in research on mobile learning. *Research in Learning Technology, 19*(3), 247–258.

Wu, W. H., Wu, Y. C. J., Chen, C. Y., Kao, H. Y., Lin, C. H., & Huang, S. H. (2012). Review of trends from mobile learning studies: A meta-analysis. *Computers and Education, 59*(2), 817–827.

**Dr. Christian Bokhove** is an Associate Professor in Mathematics Education at the University of Southampton. He has been involved in many projects on mathematics education and technology, and received his PhD on this topic from the Freudenthal Institute of Utrecht University.

**Dr. Alison Clark-Wilson** is a Principal Research Associate at the UCL Knowledge Lab, University College London. A former school mathematics teacher, she has been an active researcher in the design, experimentation and evaluation of innovative technological tools and tasks for mathematics education since the late 1990s. More recently, her work has been focused on aspects of teachers' professional learning and the scaling and sustainability of technology use in mathematics classrooms over time.

**Dr. Marios Pittalis** is a researcher and educational personnel in Mathematics Education at the University of Cyprus. He participated in research projects funded by the European Union, the University of Cyprus and the Cyprus Research Foundation. He has worked for the development of the new mathematics curriculum in Cyprus and the new mathematics teaching textbooks for primary school. His work has been focused on innovative ways of integrating digital tools in mathematics teaching.

# Part III
# Navigating Content: Focussing on Particular Concepts

# Developing Mastery of Time Concepts by Integrating Lessons and Apps

**Timothy Pelton, Todd Milford and Leslee Francis Pelton**

**Abstract** Learning applications (apps) for iPads/tablets are becoming commonplace in the elementary classroom, yet there is very little research evidence to support the adoption and use of such. This chapter details our initial efforts to empirically validate the utility of a researcher-designed iPad app by integrating it into a series of collaboratively created lessons to facilitate learning of clock-reading and time concepts. A lesson study approach was used to design, refine, and improve the intervention, which included teacher-led activities, discussions, and structured use of the iPad app. Data collected included student responses to four parallel curriculum-based assessments, classroom observations, and interviews. We present our results and discuss the implications for learning time concepts, for iPad use in the classroom, for our future research efforts, and for continued app development.

## Introduction

Tablets in the elementary classroom are currently assumed to be efficient learning-support devices; particularly with respect to their potential to help students consolidate understanding and build fluency, as well as their potential to motivate student participation and engagement. This assumption has led to the rapid expansion of application (app) use in the educational context (Larkin & Milford, 2017) even though the research evidence supporting the utility of apps in this context is very thin (Goodwin, 2012; Ifenthaler & Schweinbenz, 2013). There are millions of apps available for tablets and thousands of those have ostensibly been designed specifically to help children learn mathematics; yet few studies have been

T. Pelton (✉) · T. Milford · L. F. Pelton
University of Victoria, Victoria, Canada
e-mail: tpelton@uvic.ca

T. Milford
e-mail: tmilford@uvic.ca

L. F. Pelton
e-mail: lfrancis@uvic.ca



N. Calder et al. (eds.), *Using Mobile Technologies in the Teaching and Learning of Mathematics*, Mathematics Education in the Digital Era 12,
https://doi.org/10.1007/978-3-319-90179-4_9

conducted to assess the efficiency or efficacy of these apps or of any supporting instructional protocols designed to facilitate their use in the classroom (Grant & Barbour, 2013; Moyer-Packenham et al., 2015; Pelton & Francis Pelton, 2013).

This lack of research on the efficiency and efficacy of apps in the classroom presents an opportunity for empirical investigation. Sarama and Clements (2008) have advocated for a dynamic linkage between software development and research that could potentially increase both the effectiveness of the software and its contribution to educational research. In addition, Blume and Heid (2008) suggested themes for researching the use of technology in education which include: the importance of research on students' thinking for technological tool and curriculum development; attention to technical and conceptual aspects during tool use; a need for meaningful/transferrable representations in technological environments; and the interaction of empirical research and theory to support the design and development of the tools.

Consistent with these themes, the first and third authors of this paper (in collaboration with others) have developed a series of educational apps for mathematics which are available under the general title of *MathTappers* (e.g., Pelton, Francis Pelton, & Reimer, 2010; Pelton, Francis Pelton, & Anderson, 2011). The *MathTappers* apps were explicitly designed to support students in learning mathematics by providing relevant visual models, focused content-linked games, and pedagogically sound suggestions for parents and educators (Pelton & Francis Pelton, 2010). The primary goal in the development of the *MathTappers* apps was to support learners in mastering mathematical facts and concepts by integrating and presenting interactive/dynamic models within simple games to support student sense-making and consolidation (Pelton & Francis Pelton, 2012). These apps are free and have been well received—with more than four million downloads (July, 2017). One of these *MathTappers* apps, *ClockMaster*, (Pelton, 2010) was used as a resource for this study.

*ClockMaster* is a tablet-based application designed to support students who are just beginning to master reading an analog clock face and moving toward a full understanding of the nature of time and its representation on digital and analog clocks. It encourages students to explore interrelationships between digital and analog clocks and allows them to play short games where they are challenged to represent and interpret time in digital, analog or number-word formats. Various features have been incorporated to scaffold (annotation for minutes), challenge (translate between digital and analog, missing minute hand), and provide ongoing and formative feedback (corrections, checking and reflection opportunities) to support student learning and encourage fluency in reading time on clocks.

Time is a topic that many children struggle to understand. Although it is something that we can measure easily, and talk about often, it is also something that we cannot touch or feel (Thomas, Clarke, McDonough, & Clarkson, 2016). Some aspects of time may be mastered incidentally as children experience circumstances where elapsed time and time-of-day are used to compare or plan, but this type of understanding is generally incomplete, informal and intuitive. Children need explicit opportunities to work with both digital and analog clock faces to explore and discover the workings of the system we use to tell and record time. To continue and expand this understanding of how children learn to master time, we need to gather validation evidence supporting the use of apps that teach time concepts in educational contexts.

This pilot study was designed to examine whether an integrated treatment using a curriculum-linked app, teacher-led activities, class explorations, and discussions, might have a positive impact on student learning. More specifically we focused on whether students' understanding of clock-reading (primarily analog, but also digital) and success on time related tasks improves after exposure to a teacher-led intervention integrated with the use of *ClockMaster*. The intervention consisted of five or six lessons incorporating problem solving and the use of the app as a tool to support exploration, discussion, sense-making, representation, and consolidation. The lessons were adapted and refined using a lesson study approach (Murata, 2011; Pelton, Francis Pelton, & Milford, 2015; Runeson, 2014; Shimazu, 2014). The data collected included: parallel pre- and post-assessments, video capture of student performance during interviews, teacher observations of their students, and student interviews.

## Learning About Time

Understanding the nature of time, and being able to tell time, are mathematics understandings that must be mastered to be considered a numerate person. Explicit learning outcomes are commonly found in the elementary mathematics curriculum (e.g., from grade 1 in Australia, ACARA, 2014; in grades 1–3 in the US, Common Core State Standards for Mathematics, 2010; and in Grades 1–5 in British Columbia, British Columbia Ministry of Education, 2016a). Content outcomes in British Columbia primarily focus on estimating time, clock-reading, and elapsed-time, and imply a mastery of a range of underlying concepts that we discuss below.

According to Lehrer (2003) the conceptual foundation for understanding measurement includes: unit-attribute relations, iteration, tiling, identical units, standardization, proportionality, additivity, and origin (zero-point); and the collective coordination among these components constitutes an informal theory of measure. While mastery of extensive measurement with respect to the attributes of length, mass, area, capacity/volume, and angle may be met with these foundational concepts; the mastery of the measurement of time has added complexities. Although time is also an extensive measure, it is an attribute of existence outside of 3-dimensional space and as such it is more difficult to master. In addition to the fundamental concepts of measurement, mastery of time requires students to capture an understanding of conservation of speed (i.e., that time progresses at a constant rate that is unrelated to the student's actions) (Kamii & Long, 2003) and a substantive collection of practical referents. Time is an enigmatic attribute/construct that is difficult to define. Einstein and other physicists have circularly defined time as "what clocks measure" (Mastin, 2017).

Digital clocks allow many students to be able to "tell time" before they have fully understood the nature of time and has resulted in many students facing difficulties when challenged with interpreting and setting analog clocks (Gurganus, 2007). Analog clock reading builds upon mathematical, visuo-spatial and linguistic competences and requires the development of cognitive-conceptual representations (Burny, Valcke, & Desoete, 2009). Some research shows that children develop an understanding of clocks through guided conceptualization, i.e. teachers use gestures and speech to annotate the clock face while guiding students through the process of time-telling (Williams, 2004, 2008). Kamii and Russell (2012) have suggested that children become able to deduce elapsed time qualitatively by Grade 3 and to measure time with unit iteration by Grade 6. Finally, the use of analog clocks may be a key support in developing sense-making with respect to time concepts. Despite the complexity of these important skills, few studies are available investigating the development of time-telling competency since the introduction of digital clocks (Williams, 2004) or the potential of computer technology to support the development of such competencies.

The tools and techniques used in teaching clock-reading skills have remained relatively stable over many years. However, students' limited (and continuously declining) "real-life" experiences with analog clocks, combined with traditional rote learning of time, may not be sufficient to support children in a fulsome development of all time concepts (Monroe, Orme, & Erikson, 2002). It is not clear how much of an impact this inexperience will have on students' ability to capture the complexity, and the pre-requisite skills, required to efficiently or fully master analog clocks and time, but it is hoped that new technological innovations (such as the app used in this study) provide opportunities to better address the pre-requisite skills associated with clock reading and support continued learning and mastery of analog clocks in the classroom (Masterman & Rogers, 2002). Based upon these observations, the question guiding this study was whether an integrated instructional treatment using a curriculum-linked app, teacher-led activities, and class explorations/discussions has a positive impact on student learning of time concepts, particularly those concepts associated with telling time.

## Method

Given a certificate of approval for our research protocol, consent was obtained from the school, parents, participating teachers and students prior to engaging in research. All students in two elementary classrooms, from a small independent school in British Columbia, Canada, were enlisted to assist us with two studies during regular class times in a crossed, independent-treatment approach (see Table 1). Each study treatment included a sequence of five or six 25–30-min lessons consisting of teacher-led activities developed and refined following the lesson-study model (Murata, 2011). This chapter examines the method and results for the time concepts study. The supporting iPad app, *ClockMaster*, was integrated with each of the lessons. All members of the research team participated in the review, reflection and refinement process. We chose this crossed, independent-treatment approach to ensure that students were engaged consistently over the period of the studies, and to provide a useful comparison group for each treatment (i.e., to see how each group progressed in their mastery of the corresponding topic with the intervention, while the un-treated group did not).

The participants included 21 students in a Grade 3 class (age 9) with one teacher and one pre-service teacher, and 14 students in a Grade 4 class (age 10) with one teacher. A single researcher took the lead for all of the lessons in the time concepts treatment, while a second researcher participated in, and supported, each of the lessons. Students in both classes had not yet been formally introduced to the concepts presented in the treatment, although they would have had varying degrees of informal experience with analog clocks, number-word descriptions of time and elapsed time referents. While differences in baseline understanding were anticipated between grades, it was expected that changes in mastery level would be sufficient to highlight any treatment effect. The regular classroom teachers typically remained in the classrooms during the interventions to observe and occasionally assist individual students when they were working independently. The classroom teachers were pleased to have the researchers (also certified teachers) come into their classrooms to demonstrate an appropriate integration of technology and pedagogy (Pelton, Francis Pelton, & Milford, 2015).

All lessons followed a split-delivery design. The researcher took approximately half the session to briefly lead an exploration of some of the pre-requisite knowledge and skills associated with the lesson topics (e.g., breaking the day up into segments according to student's schedules). For the second half of the lesson, pre-designated pairs or triads of students worked with the app (see Fig. 1). Students were encouraged to take turns, observe, offer feedback to peers and debrief the sessions. A final 5-min closure typically occurred in preparation for the next lesson.

Four parallel Curriculum-Based Assessment instruments (CBA1, CBA2, CBA3, and CBA4) were created to support the two studies (time concepts and number-lines). Each CBA instrument contained two independent components—component one to assess mastery of clock-reading and time concepts (see the time-related items from CBA4 in Figs. 2 through 5) and component two to assess mastery of number-line concepts (in support of the second study). Each instrument was composed of 14 selected response and short answer items (8 time-related and

**Table 1** Crossed treatment design

| Class | Pre | Treatment 1 | Post 1 | Treatment 2 | Post 2 | Interview | Follow-up |
|---|---|---|---|---|---|---|---|
| Grade 3 | CBA1 | Telling time | CBA2 | Number-line | CBA3 | Sub sample | CBA4 |
| Grade 4 | CBA1 | Number-line | CBA2 | Telling time | CBA3 | Sub sample | CBA4 |
| Time | Nov 13 | Nov 13–24 | Nov 27 | Nov 28–Dec 9 | Dec 10 | Dec 15–Jan 20 | Jan 20 |

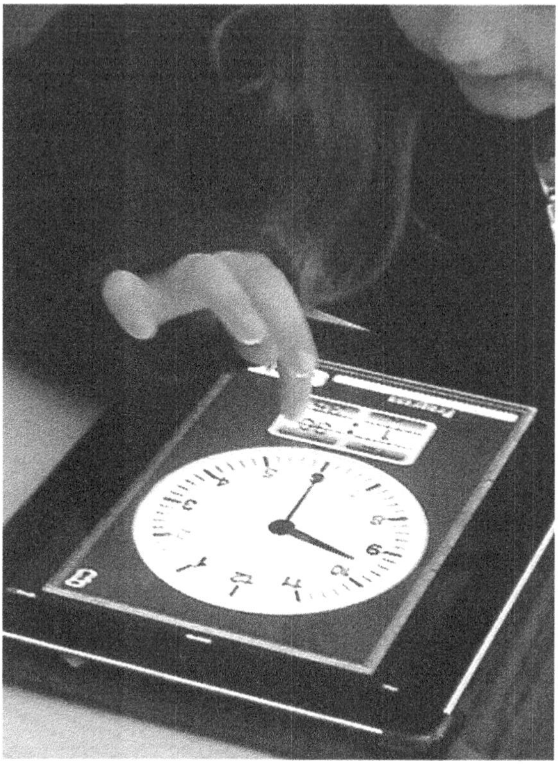

**Fig. 1** Student playing a game on the *ClockMaster* app

6 number-line items) with a maximum possible score of 19 overall (9 marks for time and 10 marks for number-line). The first instrument (CBA1) was designed and implemented as a prototype (due to time limitations and limited access to the classroom) following which the latter three instruments (CBA2, CBA3, CBA4) were created with highly parallel question formats and similar question difficulties to the first. In the analysis of CBA1, the items in the time-related component were found to be sufficiently difficult; however, instructions were adjusted to aid in assessment consistency.

To maximize construct and face validity, items were created to match British Columbia's curricular objectives with some item graphics directly echoing representations and challenges presented in the apps (see Fig. 3), and some designed to echo questions presented in the province-wide standardized assessment (Foundation Skills Assessment, FSA; British Columbia Ministry of Education, 2016b; see Fig. 5), that is administered annually to students in Grade 4.

A convenient (selected from available students at times when researchers were able to visit), roughly stratified (by performance on assessments) sample of nine students (four from grade 3 and five from grade 4) was interviewed (individually, as they were available during out-of-school programming or could be released from

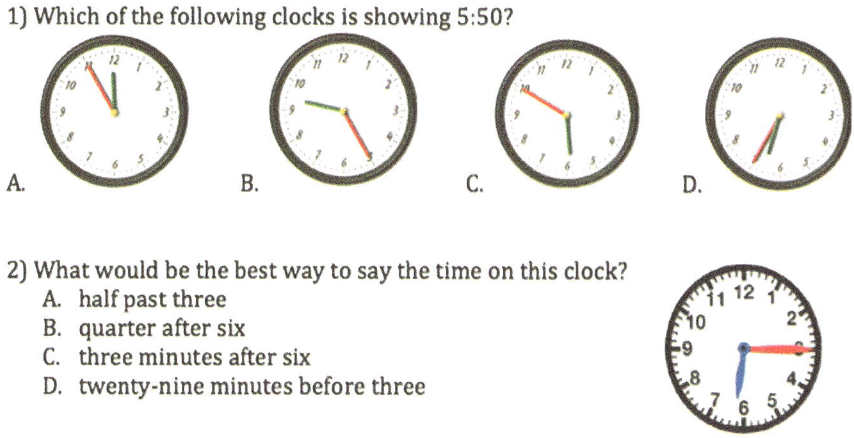


**Fig. 2** Examples of traditional questions in the time-focused component of the CBA4

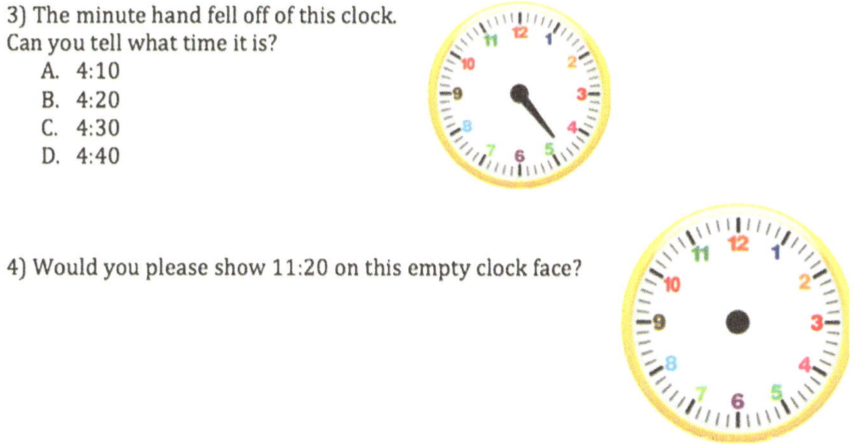


**Fig. 3** CBA4 questions linked to representations and challenges in the *ClockMaster* app. Note that 2 marks were assigned to item 4 (one mark for correctly positioning each hand)

the classroom) between one and six weeks after the treatments concluded (see Table 1). The interview process took 15–20 min for each child and consisted of two cycles, one for each study (time concepts and number-line). In each cycle, researchers first silently observed (and recorded) the student as they engaged with the related app and then gently probed their understanding via a series of semi-structured questions as they completed a second iteration on the same app to gather information on their thinking processes. For the *ClockMaster* app, each iteration consisted of a game with 10 randomly generated challenges. In each challenge the student was presented with a time in digital format and was required to manipulate the hands on the analog clock to show the equivalent representation for that time. This game was one of the app activities used regularly during the classroom lessons.

## Results

As might be expected, the Grade 4 students generally found the questions in the time-related component of the assessments less difficult than did the Grade 3 students—both in the pre-assessment(s), and on the post- and follow-up assessments. Students in both grades generally found the questions less difficult as the study progressed (i.e., item difficulty index/p-value increased across parallel items in CBA1–4). Items directly linked to the teacher-led activities (see Figs. 2, 3 and 4) had the highest mastery and retention levels. Interestingly, student responses to two of the more difficult questions (those linked to the FSA; see Fig. 5) did not follow this pattern. Rather, the apparent difficulty fluctuated erratically as some strong students failed to attempt the questions while some weaker students successfully completed the questions. We discuss why this might be the case later in this chapter.

Figure 6 presents the mean scores on the time-related components of the four CBAs for each of the classes. Some students did not complete all of the CBAs, leaving us with 19 students in our sample for the Grade 3 class and 11 students in our sample for the Grade 4 class. We can see that the Grade 3 students improved substantially immediately after the time-related intervention in CBA2 and showed additional autonomous growth in mastery weeks later in CBA3. The Grade 4

5) **About** what time is shown on this clock?

A. 5:15
B. 5:45
C. 9:30
D. 11:30

6) Mrs. Temple's class went out for a 35-minute nature walk starting at 9:40. When did the students come back to class?

A. 10:15 am B. 10:25 am C. 10:35 am D. 10:45 am

**Fig. 4** Additional examples of time focused questions in CBA4 intended to match student app experience and curricular objectives

7) The car wash started at 10:30 am, and finished at 1:15 pm. How long did it last?
A. 2 h 45 min
B. 3 h 15 min
C. 3 h 30 min
D. 3 h 45 min

8) Mrs. Ray's class took a field trip to a bird sanctuary. They left at 9:30, took 20 minutes to get to the park and then completed six 20-minute activities before lunch. What time did they start lunch?

A. 11:30 am B. 11:50 am C. 12:10 pm D. 12:30 pm

**Fig. 5** Time questions from CBA4 that most closely echoed FSA questions (British Columbia Ministry of Education, 2016b)

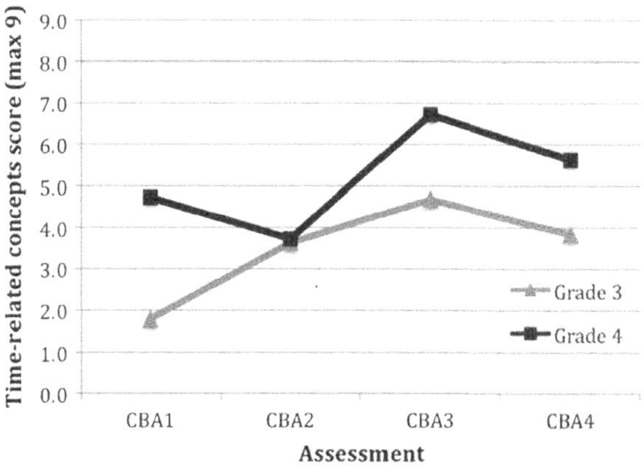


**Fig. 6** Mean student scores on time-related concepts components of the CBAs. Note that treatments occurred just before CBA2 for Grade 3 and just before CBA3 for Grade 4, and that CBA4 occurred 6 weeks after treatments concluded (interrupted by a school break)

students did not increase in their mastery until they received the time intervention after CBA2.

The follow-up interviews highlighted student mastery of the *ClockMaster* app. Almost all of the students were able to demonstrate that they had mastered the game in the app in the first (silent) round, and during the second (interview) round all students were able to explain their thinking, strategies, and actions clearly.

In Table 2 we present the results for the nine students who were interviewed—showing their scores on the four CBA assessments followed by their two performances on the *ClockMaster* game captured during the interview. Among the students interviewed, the mean scores in the clock components of the CBAs immediately preceding treatment and the mean of the scores immediately following treatment increased substantially. Interestingly, even students who had demonstrated limited proficiency on the CBAs were very successful in playing the game in the app during the follow-up interview and in communicating effective strategies with respect to representing time on an analog clock.

**Table 2** Time telling component score from each of the four CBAs presented and game results from the two games played on the *ClockMaster* app for each of the interviewed students (APP1 and APP2)

| Grade 3 students | CBA1[b] | CBA2 | CBA3 | CBA4 | APP1 | APP2 |
|---|---|---|---|---|---|---|
| Cougar A[a] (%) | 11 | 44 | 44 | 11 | 54 | 72 |
| Deer A (%) | 11 | 22 | 33 | 67 | 96 | 96 |
| Elk A (%) | 11 | 11 | 33 | 22 | 92 | 82 |
| Otter A (%) | 33 | 89 | 78 | 89 | 88 | 98 |
| Grade 4 students | CBA1[b] | CBA2[b] | CBA3 | CBA4 | APP1 | APP2 |
| Finch B (%) | 56 | 22 | 78 | 89 | 92 | 100 |
| Heron B (%) | 56 | 56 | 67 | 22 | 90 | 100 |
| Jay B (%) | – | 44 | 67 | 78 | 86 | 82 |
| Lark A (%) | 78 | 56 | 89 | 89 | 96 | 92 |
| Owl A (%) | 44 | 22 | 56 | 56 | 72 | 70 |

[a]A unique identifier was given to each student for a username on the *ClockMaster* App. They consisted of an animal name combined with the letter A, B, or C
[b]Pre-treatment assessment

# Discussion

The purpose of this pilot study was to explore the impact that an integrated treatment using a curriculum-linked app, teacher-led activities, class explorations, and discussions, might have on Grade 3 and Grade 4 student understanding of time concepts. The results briefly examined here indicate that students in both grades were able to advance in their understanding of time-related concepts and substantially improve their mastery of telling time with an analog clock over a relatively short treatment (i.e., five or six 30-min lessons over approximately three weeks).

Students in both grades generally exhibited substantial growth on their CBA scores immediately after treatment. Students were more successful on questions directly linked to the focus of the treatment (both in lesson delivery and iPad app) and generally found these questions easier following the treatments. The dip in Grade 4 student performance in CBA2 suggests that their exposure to time-related questions in CBA1 did not affect their performance on the time component in CBA2 (see Fig. 6); although the change may also have been due to student attention being focused on the number-line related items following the intervention for the number-line study. The decline in mean scores (and increased variance) for both grades on CBA4 is likely due to a combination of knowledge atrophy over the Christmas break and assessment fatigue (i.e., this was the fourth CBA they had taken in a 68 day period). Grade 4 students were also observed to spend substantially less time on the final assessment.

We expected that there might be some transfer from the content of the lessons and the app challenges to the more difficult and unfamiliar construct-linked problems (i.e., the FSA-linked items seen in Fig. 5); however, the lack of any discernible patterns in these results suggests that this did not occur, that learning was transient, or that there was too much noise in the measurement of these items.

During the follow-up semi-structured interviews, students demonstrated a strong ability to complete the app game efficiently and accurately and then to communicate their thinking and strategies clearly to the researchers. The observed facility with which students were able to communicate their understandings and processes suggests that a gentle/supportive interview might be more effective and accurate in gauging true student mastery of analog clocks than the repeated application of more traditional assessments (i.e., CBA1–4).

Using *ClockMaster* within a lesson study approach supported engagement, expanded learning opportunities, and provided more efficient learning support. Students found the app easy to use, responsive, and authentic—taking them beyond traditional and typically passive paper or chalkboard representations of clocks and closer to the haptic and kinesthetic aspects of playing with real clocks or traditional geared plastic learning clocks. The app allows for direct exploration when the hour hand is moved (weak gears and friction in physical clocks preclude this), and a "broken clock" mode allows students to discover more explicitly the direct relationship between the minute hand and the hour hand. Combining these functional differences with the available game modes expands learning opportunities by challenging students to generate correct corresponding representations of time (i.e., digital or analog) and improves efficiency by providing instant, substantive feedback with respect to clock reading and setting mastery. In addition, the game can be played individually or collaboratively, adjusted to meet student needs or abilities, and can be supported and monitored by teachers synchronously or asynchronously (through progress reports).

The results of the intervention provided here suggest that learning can be supported with the integrated use of well-aligned apps, and perhaps more importantly, the level of such learning can be captured more clearly with the use of performance observations and oral assessments. The *ClockMaster* app allowed students to take the content and skills provided in the instructional segments of the lesson intervention and reach mastery in a relatively short and enjoyable period of time.

## Conclusion

Given the small number of students involved in this pilot study, significance testing would have been neither practical nor meaningful with this data. However, the level of engagement experienced with students during the delivery of classroom lessons and iPad activities, the trends toward growth as measured by the CBAs, and the level of understanding exhibited in follow-up interviews, provides sufficient encouragement for an expansion of this investigation. Even with this limited evidence, we believe that the integration of curriculum-linked apps, in this case an app developing time-telling skills, with teacher-led activities as outlined in this chapter, may be something that the typical primary classroom teachers might reasonably apply to their classroom practice.

## References


Australian Curriculum, Assessment and Reporting Authority [ACARA]. (2014). *Foundation to year 10 curriculum: Language for interaction* (ACELA1428). Retrieved from http://www.australiancurriculum.edu.au/english/curriculum/f-10?layout=1#cdcode=ACELA1428&level=F.

Blume, G. W., & Heid, M. K. (2008). The role of research and theory in the integration of technology in mathematics teaching and learning. In G. W. Blume & M. K. Heid (Eds.), *Research on technology and the teaching and learning of mathematics: Volume 2 cases and perspectives*. Charlotte, NC: Information Age.

British Columbia Ministry of Education. (2016a). *Mathematics K-9*. Retrieved from https://curriculum.gov.bc.ca/curriculum/mathematics.

British Columbia Ministry of Education. (2016b). *Foundations Skills Assessment (FSA)*. Retrieved from http://www.bced.gov.bc.ca/assessment/fsa/.

Burny, E., Valcke, M., & Desoete, A. (2009). Towards an agenda for studying learning and instruction focusing on time-related competences in children. *Educational Studies, 35*(5), 481–492. https://doi.org/10.1080/03055690902879093.

Common Core State Standards Initiative. (2010). *Common core state standards for mathematics*. Washington, DC: National Governors Association Center for Best Practices and the Council of Chief State School Officers.

Goodwin, K. (2012). *Use of tablet technology in the classroom*. Retrieved from http://fad.teluq.ca/teluqDownload.php?file=2013/11/iPad_Evaluation_Sydney_Region_v2.pdf.

Grant, M. M., & Barbour, M. K. (2013). Mobile teaching and learning in the classroom and online: Case studies in K-12. In Z. Berge & L. Muilenberg (Eds.), *Handbook of mobile learning*. New York: Routledge.

Gurganus, S. P. (2007). *Math instruction for students with learning problems*. Boston, MA: Allyn & Bacon.

Ifenthaler, D., & Schweinbenz, V. (2013). The acceptance of Tablet-PCs in classroom instruction: The teachers' perspective. *Computers in Human Behavior, 29*(3), 525–534. https://doi.org/10.1016/j.chb.2012.11.004.

Kamii, C., & Long, K. (2003). The measurement of time: Transitivity, unit iteration, and the conservation of speed. In D. H. Clements & G. Bright (Eds.), *Learning and teaching measurement*. Reston, VA: NCTM.

Kamii, C., & Russell, K. A. (2012). Elapsed time: Why is it so difficult to teach? *Journal for Research in Mathematics Education, 43*(3), 296–315.

Larkin, K., & Milford, T. (2017). Using cluster analysis to enhance student learning when using geometry mathematics apps. In S. Ladel & C. Vale (Eds.), *ICMI—TSG41-7 the usage of tablet-applications by students with special learning needs in mathematics education*. Springer.

Lehrer, R. (2003). Developing understanding of measurement. In J. Kilpatrick, W. G. Martin & D. Schifter (Eds.), *A research companion to principles and standards for school mathematics*. Reston, VA: NCTM.

Masterman, E., & Rogers, Y. (2002). A framework for designing interactive multimedia to scaffold young children's understanding of historical chronology. *Instructional Science, 30*, 221–241.

Mastin, L. (2017). *Definition of time—Exactly what is time? Exactlywhatistime.com*. Retrieved from http://www.exactlywhatistime.com/definition-of-time/.

Monroe, E. E., Orme, M. P., & Erikson, L. B. (2002). Working cotton: Towards an understanding of time. *Teaching Children Mathematics* 8, 475–479.

Moyer-Packenham, P. S., Shumway, J. F., Bullock, E., Tucker, S. I., Anderson-Pence, K. L., Westenskow, A., … Jordan, K. (2015). Young children's learning performance and efficiency when using virtual manipulative mathematics iPad apps. *Journal of Computers in Mathematics and Science Teaching, 34*(1), 41–69.

Murata, A. (2011). Introduction: Conceptual overview of lesson study. In L. C. Hart, A. S. Alston & A. Murata (Eds.), *Lesson study research and practice in mathematics education*. Netherlands: Springer.

Pelton, T., & Francis Pelton, L. (2010). Creating handheld applications to support the development of conceptual mastery. In D. Gibson & B. Dodge (Eds.), *Proceedings of Society for Information Technology and Teacher Education International Conference 2010* (pp. 2029–2034). Chesapeake, VA: Association for the Advancement of Computing in Education (AACE).

Pelton, T., & Francis Pelton, L. (2012). Building mobile apps to support sense-making in mathematics. In P. Resta (Ed.), *Proceedings of Society for Information Technology and Teacher Education International Conference 2012* (pp. 4426–4431). Chesapeake, VA: Association for the Advancement of Computing in Education (AACE).

Pelton, T., & Francis Pelton, L. (2013). 1:1 iPad Adoption—Preparing middle school teachers to teach math. In R. McBride & M. Searson (Eds.), *Proceedings of Society for Information Technology and Teacher Education International Conference 2013* (pp. 4837–4842). Chesapeake, VA: Association for the Advancement of Computing in Education (AACE).

Pelton, T., Francis Pelton, L., & Anderson, M. (2011). MathTappers: Numberline. iOSApp. Victoria, BC: HeavyLifters Network Ltd. App Retreived from http://itunes.apple.com/us/app/mathtappers-numberline-math/id463632109?mt = 8.

Pelton, T., Francis Pelton, L., & Milford, T. (2015). Teacher reflections on using iPads and apps to support mathematics education. In *Proceedings of Society for Information Technology and Teacher Education International Conference 2015* (pp. 2813–2818). Chesapeake, VA: Association for the Advancement of Computing in Education (AACE).

Pelton, T, Francis Pelton, L., & Reimer, G. (2010). MathTappers: ClockMaster v. 2.0. Victoria, BC: HeavyLifters Network Ltd. App Retreived from http://itunes.apple.com/ca/app/mathtappers-clockmaster-math/id336932114?mt=8.

Runeson, U. (2014). Learning study in mathematics education. In S. Lerman (Ed.) *Encyclopedia of mathematics education*. Dordrecht: Springer.

Sarama, J., & Clements, D. H. (2008). Linking research and software development. In G. W. Blume & M. K. Heid (Eds.), *Research on technology and the teaching and learning of mathematics: Volume 2 cases and perspectives*. Charlotte, NC: Information Age.

Shimizu, Y. (2014). Lesson study in mathematics education. In *Encyclopedia of mathematics education* (pp. 358–360). Springer: Netherlands.

Thomas, M., Clarke, D., McDonough, A., & Clarkson, P. (2016). Understanding time: A research based framework. *Mathematics education research group of Australia*. Retrieved from: https://eric.ed.gov/?id=ED572342.

Williams, R. F. (2004). *Making meaning from a clock: Material artifacts and conceptual blending in time-telling instruction*. PhD diss., University of California.

Williams, R. F. (2008). Guided conceptualization? Mental spaces in instructional discourse. In T. Oakley & A. Hougaard, (Eds.), *Mental spaces in discourse and interaction*. Amsterdam: John Benjamins.

**Timothy Pelton** is an Associate Professor in the Faculty of Education at the University of Victoria. He teaches classes on mathematics pedagogy and educational technology. His research interests include: the creation and evaluation of educational apps; the application of appropriate technologies to enhance the teaching and learning process (e.g., clickers, iPads, GPS, 3D printing etc.); and the use of creative apps on iPads to engage students in problem solving, communicating, reasoning and representing their understandings through virtual manipulatives, games, comics and video artifacts.

**Todd Milford** is an assistant professor in science education at the University of Victoria. He was a lecturer in the Art, Law, and Education Group at Griffith University in Brisbane Australia. He has mathematics, science and special education teaching experience at the elementary and secondary level (Victoria and Vancouver, BC) as well as in the on-line environment (SIDES, Saanich, BC). He has been teaching at the postsecondary level since 2005 primarily in the areas of science education, mathematics education, and classroom assessment. His research has been and continues to be varied; however, the constant theme is on using data and data analysis to help teachers and students in the classroom.

**Leslee Francis Pelton** is an Associate Professor in the Faculty of Education at the University of Victoria where she teaches classes on mathematics content and pedagogy. Her research interests include: design and evaluation of educational apps; integration of technology in the teaching and learning of mathematics; and use of technology in problem-based learning (problem solving, reasoning, representing, communicating and making connections).

# Heatmap and Hierarchical Clustering Analysis to Highlight Changes in Young Children's Developmental Progressions Using Virtual Manipulative Mathematics Apps

**Christina W. Lommatsch, Stephen I. Tucker, Patricia S. Moyer-Packenham and Jürgen Symanzik**

**Abstract** The purpose of this study was to examine what patterns were revealed using heatmaps with hierarchical clustering to examine preschooler's performance, speed, and developmental progressions in counting and seriation. The chapter describes a study conducted with 35 preschoolers who used six touchscreen virtual manipulative mathematics apps in two different learning sequences: counting and seriation. The analysis employed heatmaps coupled with hierarchical clustering to highlight changes in children's performance, speed, and developmental progressions, between a pre- and post- assessment app after using two learning apps. This method allowed for analysis of individual and whole group data examining several tasks within each app and also several apps within each learning sequence. The analysis revealed different clusters of children grouped according to their developmental progressions which were related to incremental changes in performance and speed from the Pre to Post App use.

**Keywords** Virtual manipulative · Touchscreen app · Developmental progression Performance · Speed · Heatmap · Hierarchical clustering · Preschool Mathematics app · Seriation · Counting

C. W. Lommatsch · P. S. Moyer-Packenham (✉) · J. Symanzik
The Virtual Manipulatives Research Group, Utah State University, Logan, UT, USA
e-mail: Patricia.Moyer-Packenham@usu.edu

C. W. Lommatsch
e-mail: Christina.Lommatsch@gmail.com

J. Symanzik
e-mail: Juergen.Symanzik@usu.edu

S. I. Tucker
University of Louisville, Louisville, KY, USA
e-mail: S.Tucker@louisville.edu



N. Calder et al. (eds.), *Using Mobile Technologies in the Teaching and Learning of Mathematics*, Mathematics Education in the Digital Era 12,
https://doi.org/10.1007/978-3-319-90179-4_10

## Introduction

With the widespread introduction of mobile devices for personal and educational use, young children are increasingly encountering touchscreen mathematics apps in their daily learning experiences. Vidiksis, Jo, Hupert and Llortente (2013) reported that technology can be a useful tool in the preschool classroom, if supported by thoughtful interventions by teachers. While there have been some initial studies on young children's use of mobile apps, there is still much to be learned. In two previous papers, we examined the use of touchscreen virtual manipulative mobile apps with children aged 3–8, focusing on the affordances of the apps for children (Moyer-Packenham et al., 2016) and how children's learning was impacted by the apps (Moyer-Packenham et al., 2015). The purpose of this study was to focus more specifically on preschoolers' performance, speed, and developmental progressions when using touchscreen apps through an examination that used heatmaps with hierarchical clustering.

In this study, we selected apps classified as *virtual manipulatives*, defined as, "an interactive, technology-enabled visual representation of a dynamic mathematical object, including all of the programmable features that allow it to be manipulated, that presents opportunities for constructing mathematical knowledge" (Moyer-Packenham & Bolyard, 2016, p. 13). The positive effects of virtual manipulatives have been reported in two meta-analyses (Moyer-Packenham & Westenskow, 2013, 2016).

### *Theoretical Perspective*

The use of technology-enabled visual representations for mathematics teaching and learning is grounded in representation theory (Goldin, 2003; Goldin & Kaput, 1996) and the notion that internal and external representations are foundational to interpreting and expressing mathematical thinking (Manches & O'Malley, 2012). These notions about representations are central to Artifact-Centric Activity Theory (ACAT) (Ladel & Kortenkamp, 2013), which serves as an important framework for understanding young children's interactions with touchscreen virtual manipulative mathematics apps.

In ACAT, there are interactions among the subject, the artifact, and the object (see Fig. 1). The subject is the child, the object is the mathematics, and the artifact mediates internalizations and externalizations between the two. For example, a virtual manipulative mathematics app (artifact) is an external visualization of the mathematics (object). The child can use the artifact to externalize her understanding of the mathematics. As Ladel and Kortenkamp (2013) note about the touch-screen environment: "The direct manipulation enables children to work with virtual

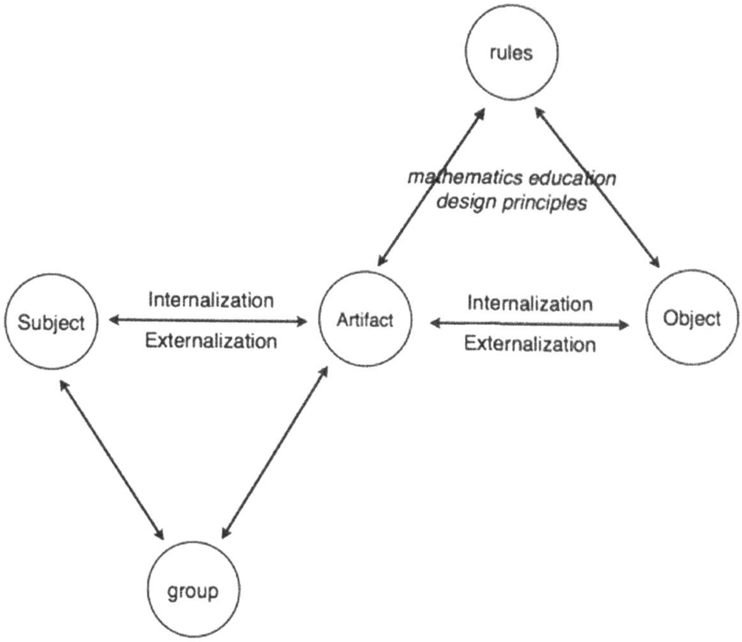


**Fig. 1** Artifact-centric activity theory (Ladel & Kortenkamp, 2016, p. 30)

manipulatives directly instead of being mediated through another input device" (p. 1). ACAT, therefore, helps to explain how the child's interactions with the artifact (i.e., virtual manipulative mathematics app) become a link to the child's mathematics (the object) and the potential for the child to learn the mathematics. As the child interacts with the artifact and the object, there are continual, gradual shifts in learning and understanding. Over time, these gradual shifts accumulate as the child increasingly connects different mathematical understandings and meanings. Eventually, the accumulation of different mathematical understandings and meanings may form a knowledge package (Ma, 1999) that helps the child to, for example, understand the counting process.

An influential construct in technology-mediated activities is technological distance, which is the degree of difficulty in physically interacting with the technology (Tucker, 2016). In ACAT, technological distance is the difference between what the subject enacts in relation to what the artifact (in this case, the virtual manipulative) recognizes as part of the intended activity. For example, during a task that requires a child to answer by tapping three fingers on the touch screen, a high degree of technological distance is present when the taps are too light to register with the app. The child might decrease technological distance by understanding that applying sufficient pressure is necessary for the app to recognize the input. Struggling with input gestures can lead children to focus on physical elements of the activity instead of the mathematics (Rick, 2012). In the context of representation and ACAT, a high degree of technological distance, during a subject's attempts to internalize and externalize representations via the artifact, may detract from the subject's access to the object (i.e., the mathematics).

## *Research on Early Mathematics Learning*

In this study, we examined two foundational topics for preschoolers: counting and seriation. Between the ages of 3 and 5, children begin to form the foundations for counting which involves mapping symbolic numbers (i.e., number words and numerals) with basic understandings of quantities (Le Corre & Carey, 2007; Pica, Lemer, Izard, & Dehaene, 2004; Siegler & Booth, 2004). Connecting their understanding of symbolic numbers with quantity forms the basis for counting and later understandings of the formal and abstract nature of the base-10 number system. Making connections among these representational systems is critical for number development. Seriation is the ability to sort and order objects according to a defined characteristic (e.g., length, area) (Inhelder, 2013). This logical reasoning task is commonly used in pre-number experiences for young children. Most seriation is interdependent with counting and understanding the base-10 place value system (Sarama & Clements, 2009b).

Foundations for counting and seriation do not happen instantaneously; rather, they form along a *developmental progression.* Clements and Sarama (2010) describe developmental progressions as a natural process, similar to children first learning to crawl, then walk, then run and skip with increasing dexterity. They write that children "follow natural developmental progressions in learning math; they learn mathematical ideas and skills in their own way" (p. 1). The terms "developmental progressions" and "learning progressions" have both been used to describe the gradual changes in children's increasing mathematical understanding. In this study, we use the terms developmental progression and domain specific progression as described by Clements and Sarama (2007):

*Developmental progression.* Most content knowledge is acquired along developmental progressions of levels of thinking. These progressions play a special role in children's cognition and learning because they are particularly consistent with children's intuitive knowledge and patterns of thinking and learning at various levels of development… with each level characterized by specific mental objects (e.g., concepts) and actions (processes)… (p. 464)

*Domain specific progression.* These developmental progressions often are most propitiously characterized within a specific mathematical domain or topic… Children's knowledge (i.e., the objects and actions they have developed with that domain) are the main determinant of the thinking within each progression, although hierarchic interactions occur at multiple levels within and between topics, as well as with general cognitive processes… (p. 464).

For example, let us examine the domain specific, or mathematical progression for counting. When children begin to learn the counting process, they may know that there are numbers that can be used to name quantities, but they cannot match the quantities with their number names. Next, children may be able to say the number names in order, but they cannot match the number names with their quantities. Then children may be able to match number names with objects, showing one-to-one correspondence, but be unable to tell how many objects there are in all. Finally, children count out collections to five and to ten and have an understanding of cardinality (Van de Walle, Karp, & Bay-Williams, 2010).

Studies on children learning counting with technology (e.g., Zaranis, Kalogiannakis & Papadakis, 2013) show that children progress through levels—pre-existing level, context-bound counting and calculation, object-bound counting and calculation, pure counting and calculating—and that these progressions can be mediated by technology. In one study involving 160 preschool children, Spencer (2013) found that using the *Know Number Free* app led to statistically significant growth in performance relative to traditional instruction. Holgersson, Barendregt, Rietz-Lepannen, Ottosson, and Lindstrom (2013) followed 87 children (ages 5–7) for eight weeks while the children played the *Fingu* app. Pre-test, post-test, and delayed post-test video interviews showed increases in children's computation abilities. Other research has focused on app affordances that support learning. Baccaglini-Frank and Maracci (2015) reported on 25 preschool children who collaboratively interacted with three multi-touch mobile apps: *TouchCounts*, *Ladybug Count*, and *Fingu*. Results showed that multi-touch affordances influenced preschoolers' development of number-sense. Numerous studies have shown that there are helping and hindering affordances of touchscreen mathematics apps. In summary, these studies demonstrate a foundation of research on the generally positive use of educationally appropriate touchscreen mathematics apps for young children.

## Methods

The design of this inquiry was an exploratory mixed methods study (Creswell & Plano Clark, 2011) that investigated learning as preschoolers interacted with touchscreen virtual manipulative mathematics apps. This study used video coding, developmental progressions, performance measures, speed measures, heatmaps, and hierarchical clustering with a subset of participants from a study that involved multiple grade levels (preschool, kindergarten, and Grade 2), apps, and content areas (e.g., Moyer-Packenham et al., 2015). The research question that guided this study was: What do patterns in heatmaps with hierarchical clustering reveal about preschoolers' performance, speed, and developmental progression levels in counting and seriation when using touchscreen virtual manipulative mathematics apps?

## Participants

The participants in this study were 35 preschool children, ages 3–5 (18 females, 17 males). Researchers distributed informational letters and brochures to parents through local schools. On demographic surveys completed by parents, most children were Caucasian (94%) and 20% of parents reported low socio-economic status. Most parents reported that their children used touchscreen devices at home at least once per week (88%), including nearly one-third of the children using the devices daily (31%). For confidentiality and data analysis, researchers assigned each child a code from 1 to 35.

## Procedures

Researchers began the study by selecting apps, designing interview protocols, and designing data collection procedures. We chose the iPad because of its touchscreen capabilities, mobility, and range of available apps. We use the term "app" to indicate the application or section of an application used to present a set of artifact, or researcher-generated, tasks. Piloting with children in local preschool programs informed the refinement of the research tools and procedures (see Moyer-Packenham et al., 2015). Researchers chose apps and tasks featuring developmentally appropriate mathematics content related to counting and seriation (e.g., Sarama & Clements, 2009a), and used virtual manipulatives featuring categories of affordances linked to positive mathematics learning outcomes (Moyer-Packenham & Westenskow, 2013, 2016).

Each child participated in a 30-min, individual, clinical interview. One researcher conducted the interview while another researcher observed from an observation room, connected by a two-way mirror with audio and video feeds. Each interview had two parts: Counting Sequence and Seriation Sequence. Each sequence involved a Pre App, two Learning Apps, and a Post App. Each child interacted with the same apps, but the order of the Learning Apps varied (see app details in Table 1).

## Data Sources

Data sources included video recordings and observer notes of the clinical interviews. Researchers video recorded every interview using a wall-mounted camera to provide an over-the-shoulder view, and each child wore a GoPro camera to provide a child's-eye view. Observers noted occurrences that might be unclear in the

**Table 1** Descriptions of apps, tasks, and measures used in the counting and seriation sequences

| App | Screenshot | Task(s) | Performance (P-Score) | Speed (S-Score) |
|---|---|---|---|---|
| Counting Pre and Post App: *Montessori Numbers: Quantity 1-9* | 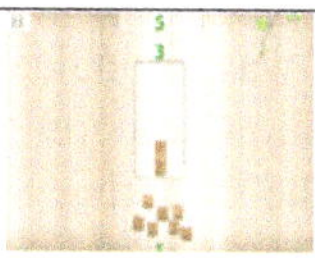 | Build a number between 1 and 5 Build a number between 6 and 10 | Quantity of blocks chosen vs. correct quantity of blocks | Total seconds to complete task/ number of target blocks |
| Counting Learning App 1: *Montessori Numbers: 1 to 20: 1-5* |  | Use the blocks to build the numbers 3, 4, and 5 | N/A | N/A |
| Counting Learning App 2: *Montessori Numbers: Numerals: 1-9* |  | Count the number of blocks and choose the corresponding numeral | N/A | N/A |
| Seriation Pre and Post App: *Pink Tower: A Montessori Sensorial Exercise: Free Moving* | 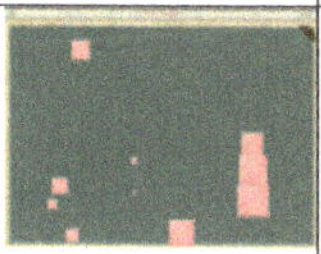 | Drag blocks to make a tower | A: Correct number of blocks until an error is made B: Correct moves vs. total moves C: How far from chosen block to correct block | Seconds per block |
| Seriation Learning App 1: *Pink Tower: #12 Tap* | 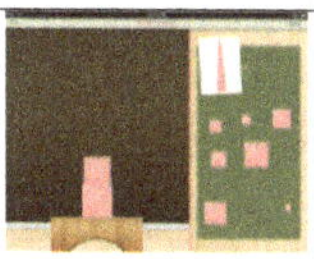 | Tap the blocks to create the pink tower; blocks only move when correct | N/A | N/A |
| Seriation Learning App 2: *Intro to Math: Red Rods* | 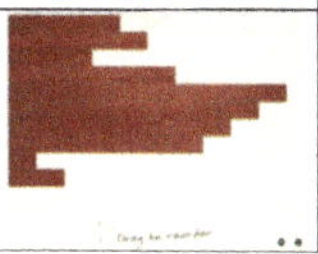 | Drag red rods to order them from longest to shortest | N/A | N/A |

*Note* Some apps have been updated since this study was conducted

recordings, including affective responses, interviewer actions, and counting and seriation strategies that occurred outside of the camera views, such as counting with fingers below the table. The combination of multiple video-recordings, with observation notes, supported triangulation and complementarity of the data. The rigorous practices of collecting and analyzing data from multiple sources strengthened the validity of the results (Creswell & Plano-Clark, 2011).

## *Data Analysis*

Researchers analyzed the video data to obtain three different scores. The first score was a performance score (P-Score). The P-Score was an indication of the child's accuracy when completing the counting and seriation tasks on the pre-assessment and the post-assessment. There was one task in the Counting Sequence and three tasks in the Seriation Sequence. The second score was a speed score (S-Score). The purpose of the S-Score was to track any changes in the speed with which the child completed each of the tasks on the pre-assessment and the post-assessment. Often a change in speed of completion can indicate more confidence in completing the tasks or more comfort with the features in the apps. Table 1 includes the apps, descriptions of the tasks, and observations recorded to determine the P-Scores and S-Scores.

The third score was a developmental progression score that represented the child's mathematical progression along a learning continuum (D-Score). The purpose of the D-Score was to determine the position of the child's knowledge along continuums of counting and seriation understanding at the pre-assessment, during the use of the first Learning App, during the use of the second Learning App, and at the post-assessment. Developmental progressions are research-based hierarchical sequences of increasingly sophisticated levels of reasoning related to a particular mathematical concept (Smith, Wiser, Anderson, & Krajcik, 2006). Based on children's development of counting and seriation understanding, we developed domain specific progressions for the counting and seriation apps. The D-Scores helped us to determine if the child made any shifts in their development progression along that continuum. Sarama, Clements, Barrett, Van Dine and McDonel (2011) describe this saying: "a critical mass of ideas from each level must be constructed before thinking… becomes ascendant in the child's mental actions and behavior" (p. 668). Table 2 shows the benchmarks along the counting and seriation progression continuum to obtain the D-Scores. As Table 2 shows, there were six levels along the counting continuum and five levels along the seriation continuum when children used the counting and seriation apps. For example, these ranged from guessing or not responding (level 1), to accurately counting a collection of 6–10 objects (level 6) in the counting progression.

Researchers used the following process to code the video data and obtain the three scores. We examined and viewed the videos multiple times, with multiple researchers viewing the videos independently to achieve data triangulation. At the beginning of this process, pairs of researchers examined 10 interview videos to develop and clarify the scoring tools. Then additional research team members were trained to code the entire data set using the three scoring tools (Corbin & Strauss, 2015; Saldaña, 2013; Stebbins, 2001). All of the video data were double-coded by two independent researchers to ensure interrater reliability.

**Heatmap and hierarchical clustering analysis**. After using the video data to obtain the three different scores, we used the mathematical progression scores (D-Scores) in a heatmaps analysis. Heatmaps with hierarchical clustering is a

**Table 2** Domain specific progression (D-Score) rubrics for counting and seriation sequences pre-post app

| Level | Counting sequence description of mathematical domain specific progression |
|---|---|
| 1 | Child guesses; no response |
| 2 | Moving blocks as app counts: child knows to move blocks to build an amount, but does not count aloud or exhibit cardinality |
| 3 | Pre-counting: child says number names, but does not match to objects (no one-to-one correspondence) |
| 4 | One-to-one correspondence (for 3+ objects): child says standard list of counting words in order; matches spoken number with one and only one object, cannot tell how many (e.g., does not stop at target number; cued by sparkles) |
| 5 | Counting a collection up to five: child understands cardinality; child can count the items in a set to five; knows that last number counted tells amount (e.g., stops at target number before sparkles feedback) |
| 6 | Counting out a collection from six to ten: child understands cardinality; child can count the items in a set to ten; knows that last number counted tells amount (e.g., stops at target number before sparkles feedback) |
| Level | Seriation sequence description of mathematical domain specific progression |
| 1 | Child guesses; no response |
| 2 | Moving block to group in some way (no understanding of order) |
| 3 | Ordered pairs—Recognition that one block in the pair is larger |
| 4 | Recognition of ordered sequence of 3 or more blocks but not the entire sequence |
| 5 | Ability to sequence; ability to self-correct |

method of analysis that is useful in analyzing multi-variate data sets. *Heatmaps* are color-shaded matrices where the color of each cell indicates its value. Data are in intervals along a continuum with each interval given a color from a divergent color scheme (e.g., blue to red). Heatmaps are popular in anthropology, bioinformatics and genetics (Wilkinson & Friendly, 2009). Their use is becoming more prevalent in educational research (e.g., Moyer-Packenham, Tucker, Westenskow, & Symanzik, 2015b). To enhance the readability of the heatmaps, hierarchical clustering was applied to the columns/rows of our heatmaps.

*Hierarchical clustering* is grouping data based on similarities by the permutation of columns/rows of a data matrix (Wilkinson & Friendly, 2009). *Dendrograms* are tree-based visual representations of the hierarchical clustering. The dendrograms depict the amount of similarity between each node, with the shorter heights indicating greater similarity and taller heights indicating greater dissimilarity. For example, a data set may include scores of several different tasks that each participant in a study performs. In the heatmap, a high score is colored with the darkest red color and a low score is colored with the darkest blue color. This allows researchers to quickly see which participants have high, low, or mixed scores. Additionally, the hierarchical clustering on the heatmap groups together participants with similar scores across the tasks allowing for a quick visual analysis of any patterns (e.g., two distinct groups of high and low performers). The dendrograms allow for a deeper analysis in comparing the strengths of these groupings forming the patterns. Dendrograms are also becoming a more prevalent tool and have been used in research evaluating apps (e.g., Larkin & Milford, 2018).

## Results

The results are organized by the two learning sequences: Counting and Seriation. Each sequence included a Pre App, two Learning Apps, and a Post App. Some apps had a fixed number of tasks while others had an open-ended number of tasks.

### *Counting Sequence*

There are three heatmaps for the Counting Sequence: one heatmap combining the Pre and Post App data and one for each Learning App. The heatmaps were generated using the children's mathematical progressions (D-Scores) for each task in each app.

Figure 2 shows the Pre-Post App results for the Counting Sequence. Four tasks are displayed in this heatmap: two from the Pre App and two from the Post App (as labeled along the bottom of the figure). The mathematical progressions (D-Scores) for each child show the child's progression in developing the ability to count from one to ten. This heatmap shows four main clusters of children. Researchers used the dendrograms and the overall patterns in children's D-scores to determine these clusters. The high cluster includes 21 children who attained the highest D-Scores on at least three of the four tasks. The mid cluster includes six children who attained middling D-Scores and improved or stayed constant from Pre to Post. The low cluster of six children attained low D-Scores. The regress (Reg.) cluster includes two children who regressed to lower levels between the Pre and Post Apps. Overall, eight children increased their average mathematical progression (D-Score), 24 stayed the same, and three decreased.

For a more in-depth analysis, we next examined the D-Score Clusters in relation to children's performance (P-Scores) and speed (S-Scores). This analysis showed that most children's performance (P-Scores) stayed the same in the Counting Sequence. Only two children's P-Scores increased on Task 1 and three increased on Task 2. In contrast, five children's P-Scores decreased on Task 1 and six decreased on Task 2. Most children in the high, mid and low clusters improved their speed (S-Score), that is, they completed tasks more quickly ($N = 31$ for Task 1; $N = 27$ for Task 2). Only four children decreased their speed on Task 1 and seven children decreased their speed on Task 2.

Interestingly, there was a relationship between children's mathematical progressions (D-Scores) and their performance (P-Score) and speed (S-Score). For the

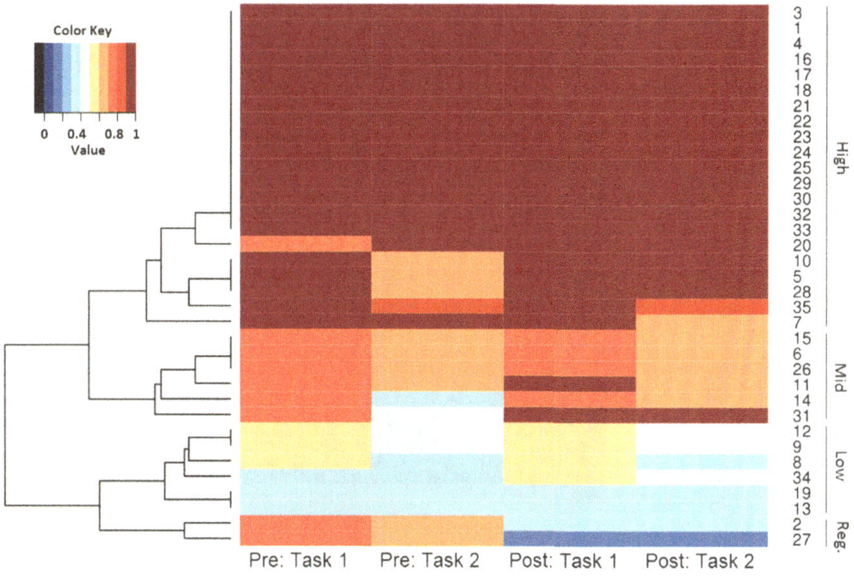


**Fig. 2** Heatmap for the counting sequence using D-scores from the pre and post apps. Reg. = regress. The ordering of the clusters (right) in this heatmap is due to the algorithm used to generate the heatmap. It does not indicate the amount of change exhibited by the children in those clusters. D-scores are indicated by the color of each of the blocks. The color key is at the top left of the figure. The dendrogram at the left of the figure depicts the clusters found in this data set and the relative similarity of each cluster. Children's ID numbers are labeled on the right side of the heatmap

high cluster, an increase in speed (S-Score) suggested that children became more technologically proficient while maintaining their performance (P-Score). The mid cluster improved their speed (S-Score) and maintained performance (P-Score). One exception was one quarter of the children who decreased in performance (P-Score) when they got faster (S-Score). In the low cluster, when one-half of children got faster (S-Score), they also made more errors and decreased in performance (P-Score). Thus, there was a relationship between the cluster groups and speed in terms of how children's performance was affected when the speed changed.

The two Learning Apps used in the Counting Sequence were also examined via heatmaps and hierarchical clustering. Figure 3 displays the mathematical progressions (D-Scores) in a heatmap for the Quantity 1–5 App and the Numerals App. On the Quantity 1–5 App (Fig. 3, top), three distinct clusters can be seen. The high cluster, labeled on the right, is the largest cluster and includes 31 children who attained the highest D-Scores on all three tasks. Because their mathematical progressions (D-Scores) were high, there was little room for growth. The mid cluster includes two children who attained mid level D-Scores. The low cluster includes two children who attained low level D-Scores. As this heatmap shows, there was little to no variation on this Learning App, with most children attaining high mathematical progression scores (D-Scores).

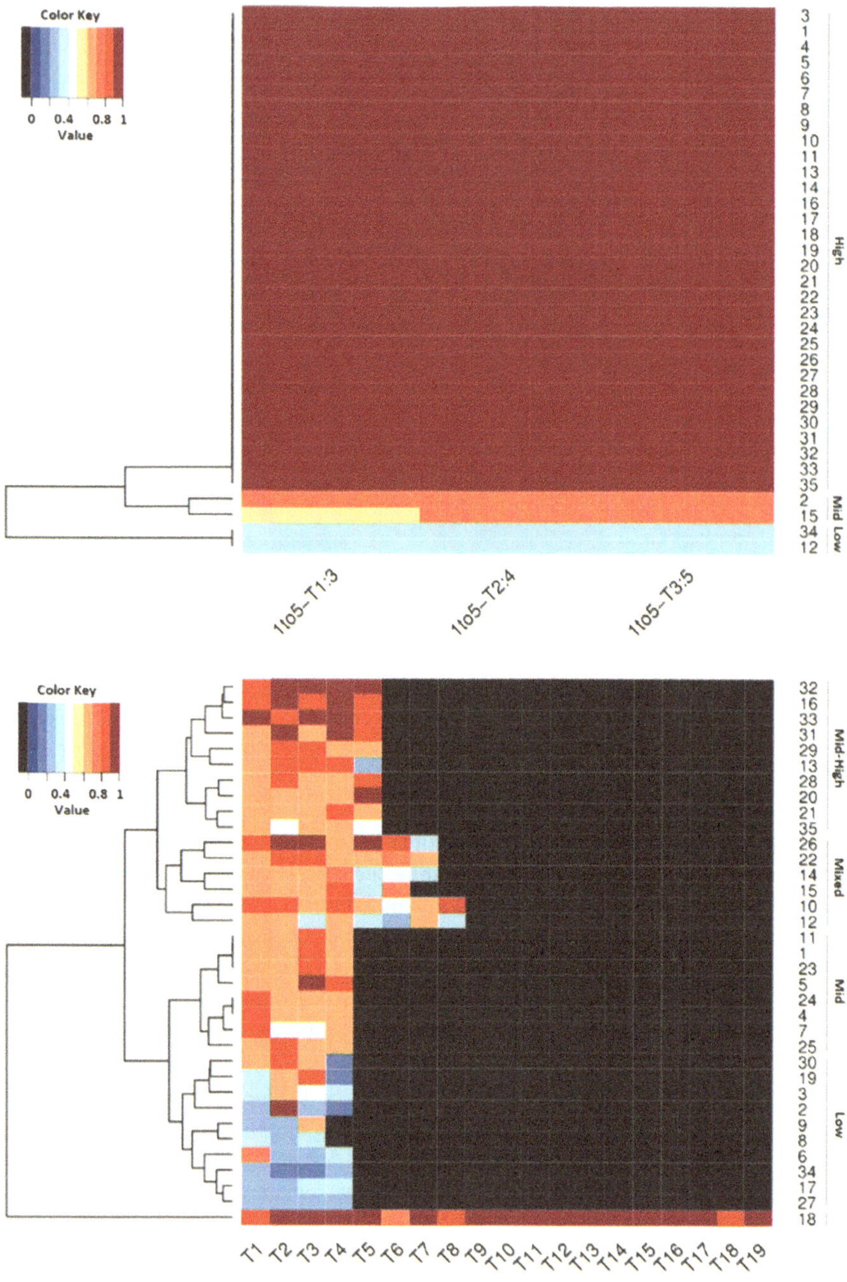


**Fig. 3** Heatmaps for the counting sequence using D-scores from the quantity 1–5 learning app (top) and the numerals learning app (bottom). Black cells indicate that the child only completed a limited number of tasks. The ordering of the clusters (right) in this heatmap is due to the algorithm used to generate the heatmap. It does not indicate the amount of change exhibited by the children in those clusters

On the Numerals Learning App (Fig. 3, bottom), children completed an open-ended number of tasks. The heatmap for the Numerals Learning App shows one color block for each task completed, with black blocks showing where the child stopped. There are five main clusters in this heatmap. In the mid-high cluster of children (labeled on the right), each child completed five tasks at a mid to high mathematical progression (D-Score). The mixed cluster completed at least six tasks and attained a range of D-Scores. Children in the mixed cluster completed more tasks than those in the mid-high or mid cluster, although they attained some lower D-Scores. In the mid cluster, each child completed four tasks and attained mid D-Scores. In the low cluster, children completed three or four tasks and attained low D-Scores. The fifth cluster contains only one child who completed a large number of tasks with nearly all of them at the highest D-Score. This heatmap shows substantial variation in children's mathematical progression (D-Scores) while they used the Numerals Learning App.

## *Seriation Sequence*

This section presents three heatmaps for the Seriation Sequence: one with the Pre-Post App data and one with each Learning App. Researchers generated these heatmaps using children's mathematical progressions (D-Scores). In this sequence, each App had only one task.

Figure 4 shows the Pre and Post App results for the Seriation Sequence. Two tasks are displayed in this heatmap: one for the Pre App and one for the Post App (labeled at the bottom of the figure). This heatmap shows five distinct clusters. The high cluster includes eight children who maintained the highest mathematical progressions (D-Score). The mid cluster includes 14 children who started at a mid-level score and stayed constant or improved to the highest level on the Post App. The progress (Pro.) cluster includes five children who attained lower scores on the Pre App and progressed to higher levels on the Post App. Children in this cluster progressed the most out of all children. The regress (Reg.) cluster includes four children who attained the highest scores on the Pre App, but regressed to a lower level on the Post App. The low cluster includes four children who attained low D-Scores on both Pre and Post apps. From the Pre to Post, 10 children increased their mathematical progression (D-Score), 21 children stayed the same, and four children decreased.

For a more in-depth analysis, we next examined these clusters (D-score) in relation to children's performance (P-Scores) and speed (S-Scores). There were three separate measures of performance (P-Score A, P-Score B, P-Score C) to capture different dimensions of each child's performance on the task. Across all clusters, most children improved their speed (S-Score), that is, they completed the tasks faster (23 became faster, four stayed the same, eight became slower). Like the Counting Sequence, children's performance (P-Scores) during the Seriation Sequence was related to the mathematical progression clusters (D-Score Clusters).

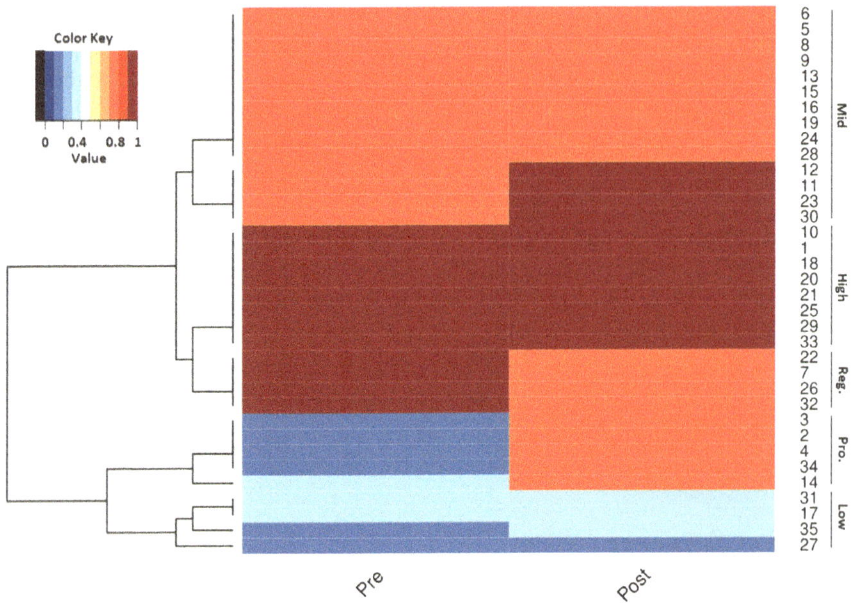


**Fig. 4** Heatmap for the Seriation Sequence using D-scores from the pre and post apps. Reg. = regress, pro. = progress. The ordering of the clusters (right) in this heatmap is due to the algorithm used to generate the heatmap. It does not indicate the amount of change exhibited by the children in those clusters. Each color block indicates the D-score that each child attained on the tasks. The dendrogram at the left of the figure depicts the clusters found in this data set and the relative similarity of each cluster. Children's ID numbers are labeled on the right side of the heatmap

Some children increased their P-Scores (14 in P-Score A, 13 in P-Score B, 14 children in P-Score C), while some children stayed the same (nine in P-Score A, 10 in P-Score B, and 13 children in P-Score C), and some children decreased their P-Scores (12 in P-Score A, 12 in P-Score B, and eight in P-Score C). Similar to the first analysis, children in the high cluster improved their speed (S-Score) and, at the same time, maintained their performance (P-Scores). Children in the progress cluster increased on two performance scores (P-Score B and P-Score C) and decreased on one (A) as they became faster. Even the children in the low cluster increased or stayed the same on their performance as they became faster. However, children in the mid cluster had a mix of performance scores as they got faster, with some improving and others getting worse. Every child in the regress cluster decreased in their performance (P-Scores) while most increased their speed (S-Score). Once again, the mathematical progression clusters (D-Scores) were related to performance (P-Score) and speed (S-Score), in terms of how the performance was affected when the speed changed.

Another analysis was performed to examine the children's mathematical progressions (D-Scores) on the two Learning Apps of the Seriation Sequence. Figure 5 shows heatmaps for the two Learning Apps in the Seriation Sequence: Pink Tower

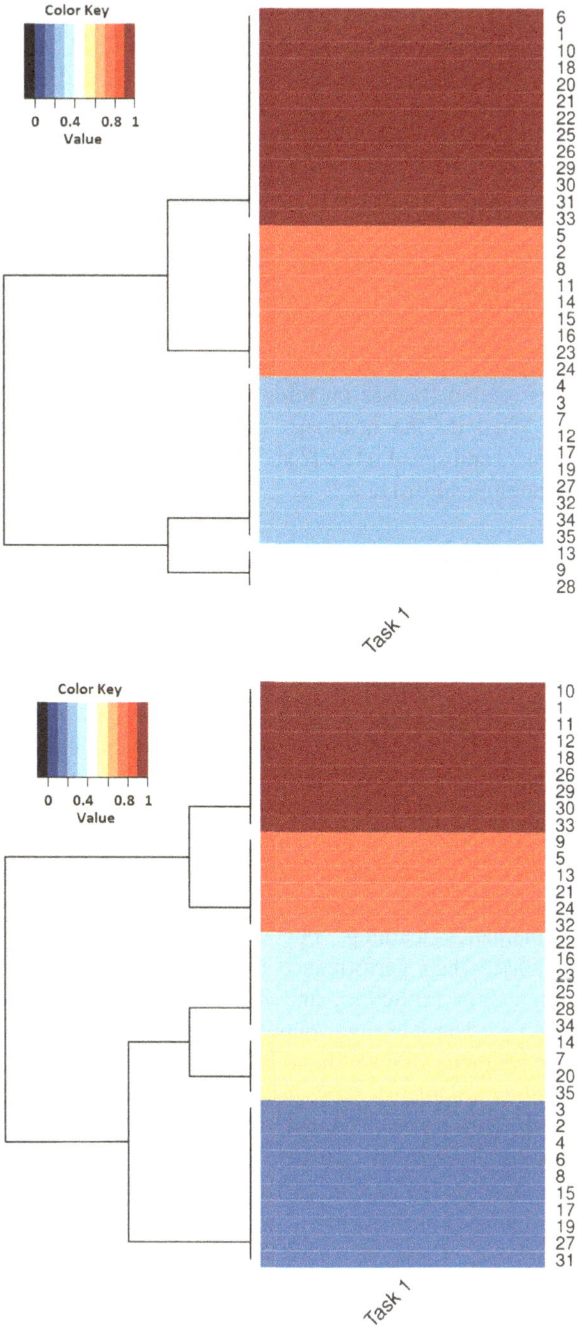


**Fig. 5** Heatmaps for seriation sequence using D-scores from the pink tower tap learning app (top) and the red rods learning app (bottom)

Tap (top) and Red Rods (bottom). Children completed one task per App, therefore the heatmaps each have a single column of D-Scores. On the Pink Tower Tap Learning App, many children (22 of 35) attained mid or higher D-Scores. On the Red Rods Learning App, children were mostly split between mid or higher D-Scores (15) and lower D-Scores (16). For both Learning Apps, there was a mixture of children who attained high and low mathematical progressions (D-Scores).

## Discussion

This section focuses on three important questions that emerged from our results: (1) What do the heatmaps reveal? (2) How are the mathematical progression clusters and speed related? and (3) What do the results mean in terms of ACAT and technological distance?

### *What Do the Heatmaps Reveal?*

The use of heatmaps with hierarchical clustering provided a different way of looking at the data and a more comprehensive picture of children's performance, speed, and mathematical progression levels. This method allowed us to simultaneously examine whole group and individual learning patterns. The mathematical progression clusters (D-Scores) generated from the hierarchical clustering allowed the examination of relationships to performance (P-Scores) and speed (S-Scores). The combination of the three scores (performance, speed, and mathematical progression levels) and the analysis method of heatmaps revealed small changes in mathematics learning. For example, in the Seriation Sequence, 14 children increased their performance (P-Score), 10 children increased their mathematical progression (D-Score), and seven of those increased on both the P-Score and D-Score. These changes may have been obscured if analyzed using only one score. The heatmaps allowed us to see that, in apps with an open-ended number of tasks, children completed many tasks and obtained a range of mathematical progression levels (D-Scores). The range of mathematical progression levels illustrates the children's productive struggle as they attempt to internalize the mathematics object. The open-ended number of tasks allowed for multiple interactions with the virtual manipulative artifact and consequently multiple refinements of the child's internalization. These increased developmental progression outcomes are consistent with findings in other research (Watts et al., 2016).

## How Are the Developmental Progression Clusters and Speed Related?

In both app sequences (Counting and Seriation), children across all mathematical progression clusters changed their speed. The videos show that children got faster at manipulating the objects and completing tasks in the apps by the end of the interviews. As the results indicated, for some children, this meant that they became faster while they maintained or improved their performance. These improvements may be due to children's increasing understanding of the task or possibly due to their reduced technological distance with the app. Or, improvements may indicate that children in the high clusters had an improved internalized understanding of the mathematics concept allowing them to focus on improving speed (S-Score), while children in other clusters had less developed internalization. In some clusters there was a mix of performance when speed increased, while in other clusters, there was a decline in performance when speed increased. While children in these clusters were similar in that they became faster at manipulating the apps, the increase in speed came at an expense for some children, who perhaps lacked a solid foundation in the mathematical concepts. In this case, getting faster was not a positive outcome as it came at the cost of children's performance. This observation has important implications for teaching and learning. When children quickly complete tasks on a touch-screen device using a mathematical app, educators must examine how the tasks were completed and with what levels of accuracy. Educators must be careful not to communicate that being "faster" is better when it comes to mathematical learning. Many children need time and multiple experiences with mathematics concepts before those understandings are solidified.

## What Do the Results Mean in Terms of ACAT and Technological Distance?

The results of this study fit within ACAT and technological distance theories, with implications for technology-mediated mathematics learning. Using the lens of ACAT, changes in the subjects' (children's) performance and speed resulted from activity involving internalizing and externalizing representations related to the object (mathematics) as mediated by the artifact (app). Children may have developed and connected their internal representations as they accessed various mathematical representations from the virtual manipulative apps, thus supporting their ability to reason with, and externalize, the representations more quickly and accurately during activity. Increased familiarity with representations in the apps may have contributed to improvements in speed. This was especially evident in the Counting sequence, where all tasks involved blocks with similar characteristics, and most children improved their speed, regardless of their performance or developmental progression levels.

Reduction of technological distance may have contributed to improvements in speed and performance. As children decreased technological distance by becoming more adept at fluently performing the physical aspects of interacting with the apps, they could increase their speed. Decreasing technological distance also may have permitted children to focus their attention on the mathematics, potentially contributing to improvements in performance. For example, the Seriation Pre-Post task involved coordinating a finger to drag a block across the screen and align it with other blocks to build a tower. Fluent use of this gesture allowed children to attend less to the act of moving the blocks and instead focus on block size and order. This analysis indicates that technological distance fits within ACAT, and together they support the examination of children's technology-mediated mathematics learning experiences.

## Conclusion

Touchscreen technology-mediated mathematics experiences can influence children's learning on mathematics tasks, but capturing these small changes requires new ways of analyzing and examining learning data. In this chapter, we attempted to understand how the subject (i.e., child) learns the object (i.e., mathematics) as the artifact (i.e., app) mediates internalizations and externalizations between the two. We offered one such method for this inquiry: heatmaps with hierarchical clustering. This process allowed us to make comparisons among different measures and use a multifaceted view for interpreting whole group and individual data on performance, speed, and mathematical progression levels.

## References


Baccaglini-Frank, A., & Maracci, M. (2015). Multi-technology and preschoolers' development of number-sense. *Digital Experiences in Mathematics Education*, 1–21.

Clements, D. H., & Sarama, J. (2007). Early childhood mathematics learning. In F. Lester (Ed.), *Second handbook of research on mathematics teaching and learning* (pp. 461–555). Reston, VA: National Council of Teachers of Mathematics.

Clements, D. H., & Sarama, J. (2010). Learning trajectories in early mathematics—sequences of acquisition and teaching. *Encyclopedia of Early Childhood Development: Numeracy*, 1–6.

Corbin, J., & Strauss, A. (2015). *Basics of qualitative research: Techniques and procedures for developing grounded theory*. Thousand Oaks, CA: Sage publications.

Creswell, J. W., & Plano Clark, V. L. (2011). *Designing and conducting mixed methods research* (2nd ed.), Thousand Oaks, CA: SAGE.

Falloon, G. (2013). Young students using ipads: App design and content influences on their learning pathways. *Computers & Education, 68*, 505–521.

Goldin, G. A. (2003). Representation in school mathematics: A unifying research perspective. In J. Kilpatrick, W. G. Martin, & D. Schifter (Eds.), *A research companion to principles and standards for school mathematics* (pp. 275–285). Reston, VA: NCTM.

Goldin, G. A., & Kaput, J. M. (1996). A joint perspective on the idea of representation in learning and doing mathematics. In L. P. Steffe, P. Nesher, P. Cobb, G. A. Goldin, & B. Greer (Eds.), *Theories of mathematical learning* (pp. 397–430). Hillsdale, NJ: Erlbaum.

Holgersson, I., Barendregt, W., Rietz-Lepannen, E., Ottosson, T., & Linstrom, B. (2013). Can children enhance their arithmetic competence by playing an especially designed computer game? *Proceedings from NORSMA 7: The Seventh Conference of the Nordic Research network on Special Needs Education in Mathematics*. Copenhagen. Retrieved from http://www.diva-portal.org/smash/record.jsf?pid=diva2%3A693178&dswid=1175.

Inhelder, B. (2013). *The early growth of logic in the child: Classification and seriation* (Vol. 83). Routledge.

Ladel, S., & Kortenkamp, U. (2013). An activity-theoretic approach to multi-touch tools in early maths learning. *The International Journal for Technology in Mathematics Education, 20*(1), 3–8.

Ladel, S., & Kortenkamp, U. (2016). Artifact-centric activity theory—A framework for the analysis of the design and use of virtual manipulatives. In P. Moyer-Packenham (Ed.), *International perspectives on teaching and learning mathematics with virtual manipulatives* (pp. 25–40). New York: Springer.

Larkin, K., & Milford, T. (2018). Mathematics apps—Stormy with the weather clearing: Using cluster analysis to enhance app use in mathematics classrooms. In N. Calder, K. Larkin, & N. Sinclair (Eds.), *Using mobile technologies in the teaching and learning of mathematics*. Mathematics Education in the Digital Era: Springer.

Le Corre, M., & Carey, S. (2007). One, two, three, four, nothing more: An investigation of the conceptual sources of the verbal counting principles. *Cognition, 105*, 395–438.

Ma, L. (1999). *Knowing and teaching elementary mathematics*. Mahwah, NJ: Lawrence Erlbaum Associates.

Manches, A., & O'Malley, C. (2012). Tangibles for learning: A representational analysis of physical manipulation. *Personal and Ubiquitous Computing, 16*, 405–419.

Moyer-Packenham, P. S., & Bolyard, J. J. (2016). Revisiting the definition of a virtual manipulative. In P. Moyer-Packenham (Ed.), *International perspectives on teaching and learning mathematics with virtual manipulatives* (pp. 5–16). New York: Springer.

Moyer-Packenham, P. S., Bullock, E. P., Shumway, J. F., Tucker, S. I., Watts, C., Westenskow, A., Anderson-Pence, K. L., Maahs-Fladung, C., … Jordan, K. (2016). The role of affordances in children's learning performance and efficiency when using virtual manipulative mathematics touch-screen apps. *Mathematics Education Research Journal, 28*(1), 79–105.

Moyer-Packenham, P. S., Shumway, J. F., Bullock, E., Tucker, S. I., Anderson-Pence, K. L., Westenskow, A., Boyer-Thurgood, J., Maahs-Fladung, C., … Jordan, K. (2015). Young children's learning performance and efficiency when using virtual manipulative mathematics iPad apps. *Journal of Computers in Mathematics and Science Teaching, 34*(1), 41–69.

Moyer-Packenham, P. S., Tucker, S. I., Westenskow, A., & Symanzik, J. (2015b). Examining patterns in second graders' use of virtual manipulative mathematics apps through heatmap analysis. *International Journal of Educational Studies in Mathematics, 2*(2), 1–16.

Moyer-Packenham, P. S., & Westenskow, A. (2013). Effects of virtual manipulatives on student achievement and mathematics learning. *International Journal of Virtual and Personal Learning Environments, 4*(3), 35–50.

Moyer-Packenham, P. S., & Westenskow, A. (2016). Revisiting the effects and affordances of virtual manipulatives for mathematics learning. In K. Terry & A. Cheney (Eds.), *Utilizing Virtual and Personal Learning Environments for Optimal Learning* (pp. 186–215). Hershey, PA: IGI Global.

Pica, P., Lemer, C., Izard, V., & Dehaene, S. (2004). Exact and approximate arithmetic in an Amazonian Indigene group. *Science, 306*(5695), 499–503.

Rick, J. (2012). Proportion: A tablet app for collaborative learning. In *Proceedings of the 11th International Conference on Interaction Design and Children* (pp. 316–319). New York, NY, USA: ACM.

Sarama, J., & Clements, D. H. (2009a). "Concrete" computer manipulatives in mathematics education. *Child Development Perspectives, 3*(3), 145–150.
Sarama, J., & Clements, D. H. (2009b). *Early childhood mathematics education research: Learning trajectories for young children*. New York, NY: Routledge.
Sarama, J., Clements, D. H., Barrett, J., Van Dine, D. W., & McDonel, J. S. (2011). Evaluation of a learning trajectory for length in the early years. *ZDM Mathematics Education, 43*, 667–680.
Siegler, R. S., & Booth, J. L. (2004). Development of numerical estimation in young children. *Child Development, 75*(2), 428–444.
Smith, C. L., Wiser, M., Anderson, C. W., & Krajcik, J. (2006). Implications of research on children's learning for standards and assessment: A proposed learning progression for matter and the atomic-molecular theory. *Measurement, 4*(1/2), 1–98.
Spencer, P. (2013). iPads: Improving numeracy learning in the early years. In V. Steinle, L. Ball, & C. Bardini (Eds.), *Mathematics education: Yesterday, today, and tomorrow* (pp. 610–617). Melbourne, Australia: MERGA.
Stebbins, R. A. (2001). *Exploratory research in the social sciences* (Vol. 48). Thousand Oaks, CA: Sage publications.
Tucker, S. I. (2016). The modification of attributes, affordances, abilities, and distance for learning framework and its applications to interactions with mathematics virtual manipulatives. In P. S. Moyer-Packenham (Ed.), *International perspectives on teaching and learning mathematics with virtual manipulatives* (pp. 41–69). Springer International Publishing.
Van de Walle, J. A., Karp, K. S., & Bay-Williams, J. M. (2010). *Elementary and middle school mathematics: Teaching developmentally*. Boston: Allyn & Bacon.
Vidiksis, R., Jo, I. Y., Hupert, N., & Llorente, C. (2013). All hands on tech: math and media in the preschool classroom. In R. McBride, & M. Searson (Eds.), *Society for Information Technology & Teacher Education International Conference 2013* (pp. 4453–4457). Chesapeake, VA: AACE.
Watts, C. M., Moyer-Packenham, P. S., Tucker, S. I., Bullock, E. P., Shumway, J. F., Westenskow, A., et al. (2016). An examination of children's learning progression shifts while using touch screen virtual manipulative mathematics apps. *Computers in Human Behavior, 64*, 814–828.
Wilkinson, L., & Friendly, M. (2009). The history of the cluster heat map. *The American Statistician, 63*(2), 179–184.
Zaranis, N., Kalogiannakis, M., & Papadakis, S. (2013). Using mobile devices for teaching realistic mathematics in kindergarten education. *Creative Education, 4*(7A1), 1–10.

**Christina W. Lommatsch** is a Doctoral Candidate at Utah State University in the Mathematics Education and Leadership Program. Lommatsch's research interests include the use of technology in teaching post-secondary mathematics topics, such as logic and Calculus, and educational app development. She has contributed to multiple funded research grants, such as App Camp for girls and STARS! GEAR UP at Utah State University. Her scholarly contributions include several publications, for both researchers and practitioners, and research presentations at local, state, and national conferences.

**Stephen I. Tucker** is Assistant Professor of Mathematics Education at the University of Louisville. Dr. Tucker's research focuses on children's mathematical interactions with technology and physically embodied mathematical practices. He has more than 20 peer-reviewed publications, including journal articles, book chapters, monographs, and refereed proceedings, reaching both researchers and practitioners. Dr. Tucker has contributed to multiple funded research grants, including as PI.

**Patricia S. Moyer-Packenham** is Professor and Director of the Mathematics Education and Leadership Programs at Utah State University. Moyer-Packenham's research focuses on uses of mathematics representations and tools (including virtual, physical, pictorial, and symbolic). She is noted for her definition of *virtualmanipulatives*, and her research on the use of manipulatives for mathematics teaching and learning. Her publications include three books and over 90 scholarly contributions including numerous journal articles, book chapters, refereed proceedings, monographs and contributions to mathematics methods textbooks. Moyer-Packenham has served as a PI on numerous grants totaling over $16 million dollars in funding for research and mathematics teacher development.

**Jürgen Symanzik** Ph.D. with a co-major in Statistics and Computer Science, Iowa State University, Ames, Iowa. M.S. in Computer Science and M.S. in Statistics, University of Dortmund, Germany. Dr. Symanzik is an Elected Member of the *International Statistical Institute* (ISI) and a Fellow of the *American Statistical Association* (ASA). He is the current editor-in-chief of *Computational Statistics* and the President-elect of the *International Association for Statistical Computing* (IASC). His research interests include dynamic statistical graphics, micromaps, visual data mining & exploratory (spatial) data analysis, and applications of statistical computing and statistical graphics to natural resources, education/teaching, medical research, and eye-tracking research.

# A Better Story: An Embodied-Design Argument for Generic Manipulatives

**Dana Rosen, Alik Palatnik and Dor Abrahamson**

**Abstract** Mathematics education practitioners and researchers have long debated best pedagogical practices for introducing to students new concepts. We report on results from analyzing the behaviors of 25 Grade 4–6 students who participated individually in tutorial activities designed to compare the pedagogical effect of manipulating objects that are either generic (non-representational, not signifying specific contexts, e.g., a circle) or situated (representational, signifying specific contexts, e.g., a hot-air balloon). The situated objects gave rise to richer stories than the generic objects, presumably because the students could bring to bear their everyday knowledge of these objects' properties, scenarios, and typical behaviors. However, in so doing, the students treated the objects' only as framed by those particular stories rather than considering other possible interpretations. Consequently, these students did not experience key struggles and insights that the designers believe to be pivotal to their conceptual development in this particular content (proportionality). Drawing on enactivist theory, we analyze several case studies qualitatively to explicate how rich situativity filters out critical opportunities for conceptually pivotal sensorimotor engagement. We caution that designers and teachers should be aware of the double-edged sword of rich situativity: Familiar objects are perhaps more engaging but can also limit the scope of learning. We advocate for our instructional methodology of entering mathematical concepts through the action level.

D. Rosen · A. Palatnik · D. Abrahamson (✉)
University of California, Berkeley, CA, USA
e-mail: dor@berkeley.edu

D. Rosen
e-mail: danarosen@berkeley.edu

A. Palatnik
e-mail: palatnik@berkeley.edu



N. Calder et al. (eds.), *Using Mobile Technologies in the Teaching and Learning of Mathematics*, Mathematics Education in the Digital Era 12,
https://doi.org/10.1007/978-3-319-90179-4_11

**Keywords** Attentional anchor · Educational design · Embodiment
Formalization · Manipulatives · Mathematics · Sensorimotor scheme
Technology

## Introduction: Dilemmas of Designing Manipulatives for Mathematics Learning

Imagine you are designing a mathematics lesson to introduce the notion of proportional equivalence, such as 2:3 = 4:6. Now further imagine that you wished for your students to understand proportionality as a multiplicative concept that builds on yet departs from additive forms of reasoning, so that the students can draw on what they already know yet expand this knowledge. In particular, you wished your students would consider a proportional equivalence sequence, such as 2:3 = 4:6 = 6:9 = 8:12, and so on, by focusing on the arithmetic difference between numbers in each ratio pair, that is, a difference of 1 in 2:3, a difference of 2 in 4:6, a difference of 3 in 6:9, a difference of 4 in 8:12, and so on. You would like the students to be surprised that this difference keeps changing, because you believe that this experience of surprise—of seeing a new type of equivalence—would precipitate meaningful conceptual development from additive to multiplicative reasoning.

Once satisfied with your educational design rationale, you set off to realize it in the form of some activity in which your students would engage. That is, you attempt to create for your students the interaction conditions that would give rise to an experience that you view as pivotal for learning the target content you have set for the lesson. Working either in a digital or material environment, you decide to create a scenario, materials, and task drawing on Aesop's parable of the Tortoise and the Hare. You will place images of the two protagonists at the starting point of a number-line racetrack, and you will guide students to advance the tortoise 2 units for every 3 units they advance the hare.

You try out this activity with young mathematics students. They find the activity engaging and eventually infer that the hare's lead over the tortoise keeps growing every go (1, 2, 3, 4, etc.). And yet the students are never surprised by this fact—it appears obvious to them due to their familiarity with the story. It would appear that the key learning experience you attempted to elicit was obviated by the scenario, materials, and task you had chosen.

You wish to improve the activity, and so you consider changing the appearance of the two objects from a tortoise and hare to some two other figures, such as nondescript circles. However in so doing, you realize, the very notion of two objects moving in parallel at different speeds toward a common target would potentially be lost, because the modified display would lack any familiar context that immediately prompts the desired scenario of a running competition between two agents of disparate athletic prowess.

You conclude, along with many other researchers in the past, that bringing familiar context into the process of learning mathematical concepts is not unproblematic; it is in fact riddled with tradeoffs (e.g., Uttal, Scudder, & DeLoache, 1997). A familiar scenario can instantly orient students toward relevant elements of an instructional activity as well as the elements' anticipated behaviors, but this very familiarity with the situated context might deprive the students of critical opportunities to struggle with inferring these behaviors and coordinating them with other knowledge they bring to the situation. On the other hand, one could begin with textbook definitions and solution procedures for proportional equivalence and only later apply these acontextual routines to everyday situations, and yet those routines would initially bear scant meaning for the students. It seems as though both approaches—from the situated to the symbolic, or from the symbolic to the situated—can be problematic. Is there a third option?

In this chapter, we will make the case for a third option. And yet this third option might appear quite different from the other two, because it highlights the educational role of the *physical actions* students perform as they manipulate objects in mathematics lessons. That is, teachers and researchers usually focus on how students select, arrange, and transform manipulatives in the working space (i.e., the *planning* and *product* of action) and what that could mean conceptually; few focus on how students coordinate their hands so as to move the manipulatives per the task objectives (i.e., the *process* of action; but see Abrahamson, 2004; de Freitas & Sinclair, 2012; Kim, Roth, & Thom, 2011; Nemirovsky, Kelton, & Rhodehamel, 2013).

In the activities we will discuss, students first learn to enact a new movement form and only later they ground that form in particular contexts as well as generalize it as mathematical rules. The students initially learn the movement form by way of solving an interactive manipulation problem involving two virtual objects, one per each hand. This instructional methodology draws on theories from the cognitive sciences that depict mathematical reasoning as the mental simulation of sensorimotor activity (Barsalou, 2010; Hutto, Kirchhoff, & Abrahamson, 2015; Landy & Goldstone, 2007; Vygotsky, 1997). Through their efforts to solve a new two-hand movement coordination, students may come to perceive the world in a new way. Educational designers can create conditions where this new moving/perceiving pattern is the meaning we experience and sustain for a particular mathematical concept that we are studying, even before we use formal symbolic notation (Abrahamson & Bakker, 2016).

The chapter will focus on comparing generic versus situated objects with respect to how students interact with the objects and what they infer from these interactions. By *generic* we mean that the objects deployed in these activities are not contextualized as representing or referring to anything outside of the activity. They may afford goal-oriented interaction as tools for accomplishing a task, but the designer does not intend for them to signify or symbolize for all students any particular meanings from some other domain, at least not initially. The term "generic" (non-specific) also alludes to its cognates "genus" (a class of things), "generative" (bearing potential for growth and application), and "generalize" (produce inference from a case), all perceived as potential attributes of these instructional materials. We think of generic objects as less likely than *situated* objects to evoke rich experiential contexts or narrative content—they are means of engaging in an activity without drawing on associations with what they resemble, denote, or connote. Clearly any choice of terminology comes with its ineluctable philosophical and theoretical baggage from the cognitive sciences literature, such as epistemological, ontological, and phenomenological assumptions about human perception and reasoning (e.g., Wilensky, 1991), so that perhaps an example will cut to the chase: With "*generic*" we are attempting to characterize the difference between a circle and a hot-air balloon. We wish to understand how this difference bears on the processes and consequences of learning.

Working with technological media, the objects employed in our pedagogical activities will be virtual. The situated objects will be iconic images of hot-air balloons, whereas the generic objects will be stark circles. The students will manipulate these virtual objects in their attempts to solve the interaction problem of making a screen green. As shall be reported, students respond to manipulation problems involving familiar objects by bringing to bear what they know about these objects, such as how hot-air balloons typically behave in particular contexts. The students thus perceive and manipulate the familiar objects to enact imaginary micro-scenarios that would be plausible with these objects in those contexts. For example, students may engage a virtual hot-air balloon by launching it vertically from the ground upward, but they are less likely to rotate it. By contrast, generic objects do not constrain the scope of potential perceptions and movements as much, because they conjure for the student less immediate sense of what might be plausible and implausible to do with them. For example, one would not be inhibited in rotating a simple circle as one would an icon of a hot-air balloon, and one would be less inclined to construe the circle as necessarily launching from the screen base as one would the hot-air balloon. We conjecture that students are likely to perceive and move stark objects in more ways than they would iconic objects and, consequently, potentially infer a greater range of mathematical rules.

Understanding the effects of objects on actions is important for teaching mathematics with manipulatives. If we hope to elicit from students particular ways of moving, because we see these ways of moving as critical for the learning process, then we should choose or create our manipulatives wisely with those movements in mind. This principle holds both for digital and material instructional resources (see Sarama & Clements, 2009, on "concrete" virtual manipulatives).

Below we present a technical section that will expand on the theories of learning that have motivated our research, focusing on literature that treats the relation between interactive objects and the forms of reasoning they enable (section "Theoretical Background"). We then detail the methods used in this study (section "Methods: Designing Constraints on Students' Sensorimotor Engagement of Manipulatable Elements in a Technological System"). Results and findings then follow (section "Results: Implicit Affordances of Manipulation Objects Mediate Student Strategies"), and we end with conclusions as well as implications for design and teaching (section "Conclusion").

# Theoretical Background

## *Framing the Debate*

Scholars of mathematics education tend to hold two diametrically opposed positions on best pedagogical practices for introducing new mathematical concepts (Abrahamson & Kapur, 2018; Nathan, 2012). The debate centers on the question of whether and when situated contexts should be employed in cultivating students' understanding of mathematical concepts. In particular, researchers debate on the optimal ontological nature of the objects that students are to consider as they solve instructional problems: Should these objects evoke specific, elaborate situations with rich contextual meanings, or should they be non-contextual "situation-agnostic" generic symbols and shapes? The progressive-formalization approach (see Fig. 1a; e.g., Goldstone, Landy, & Son, 2008; Gravemeijer, 1999; Noss & Hoyles, 1996; Ottmar & Landy, 2017) posits that students should begin from concrete situations and then progressively generalize, abstract, and formalize their understandings of the situations by creating, adopting, and using normative symbolical representations. In the course of adopting these mathematical visualizations and forms of discourse, cultural agents (such as designers and teachers) play key mediating roles in providing students with selected semiotic means of objectifying their emerging notions (Bartolini, Bussi, & Mariotti, 2008; Newman, Griffin, & Cole, 1989; Radford, 2003; Sfard, 2002). The formalisms-first approach (see Fig. 1b; e.g., Kaminski, Sloutsky, & Heckler, 2008; Sloutsky, Kaminski, & Heckler, 2005; Stokes, 1997), on the other hand, posits that students should first work with abstract representations, such as mathematical symbols and geometrical shapes, to enact and understand solution procedures; only then should they extend and practice these formal strategies by applying them to specific situated contexts. Each of these positions, we believe, holds merit, and yet each also suffers from the very shortcomings implicated by its critics. It could be that a third option exists that draws on the merits of each.

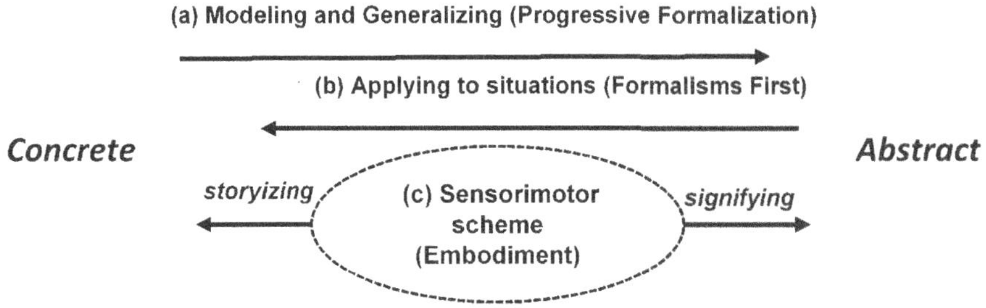


**Fig. 1** Positioning the **c** embodiment approach with respect to the **a** Progressive-Formalization approach and the **b** Formalisms-First approach

Inspired by the embodiment approach (Campbell, 2003; Chemero, 2009; Clark, 2013; Nemirovsky, 2003; Varela, Thompson, & Rosch, 1991), the educational approach portrayed in Fig. 1c positions sensorimotor schemes as the hypothetical epistemological core of mathematical learning and knowing. This approach responds also to calls (Allen & Bickhard, 2013; Arsalidou & Pascual-Leone, 2016; Varela, 1999) for renewed interest in Piaget's systemic theory of genetic epistemology (Piaget, 1968) as providing a viable alternative to the dominant paradigm of cognition as information processing. In line with our embodiment approach, we conjectured that students could encounter new mathematical concepts by first developing sensorimotor schemes and then both grounding these schemes in concrete situations (storyizing) and articulating the schemes in mathematical formalism (signifying; Howison, Trninic, Reinholz, & Abrahamson, 2011; see also Fuson & Abrahamson, 2005). Thus whereas we embrace the proposal to ground mathematical meaning in "our direct physical and perceptual experiences" (Nathan, 2012, p. 139), we decompose this idea by foregrounding and differentiating what we view as its two inherent phenomenological dimensions: sensorimotor schemes (goal-oriented movement) and situatedness (contextuality). We argue that these two dimensions have been conflated in historical debates (e.g., Barab et al., 2007; Bruner, 1986; Burton, 1999). That is, we maintain that learning activities can be created such that sensorimotor schemes are fostered either in contextual or acontextual situations, and we are interested in understanding the processes and consequences of these two instructional options.

To evaluate this embodiment approach to mathematics learning as it bears on pedagogical design, we began by formulating the hypothesis that different levels of contextuality have different effects on learning, and we assumed that sensorimotor schemes would mediate this effect. We believed more specifically that students would develop different sensorimotor schemes in low- versus high-context activities and that the low-context condition would prove advantageous.

To operationalize this hypothesis, we designed and implemented a learning activity complete with materials, tasks, and facilitation techniques based on the embodied-design framework (Abrahamson, 2006, 2009, 2014). In the empirical study reported in the later sections of this chapter, we varied the contextuality of a manipulation problem by either incorporating or not incorporating iconic information that would potentially cue particular narrative framings of the situation, and we measured for effects of this experimental variation on content-relevant qualities of students' behaviors as they engaged in solving the problem. Our study thus aimed to empirically evaluate the in-between embodiment position with respect to the ongoing debate of formalisms first versus progressive formalization, a debate which we now further detail.

## *Contrasting Approaches to Formalization*

Summarizing a rich research literature, Nathan (2012) has characterized two opposing approaches to mathematics education as follows:

- *Formalism first* proposes that students should encounter new concepts through abstract procedures and then map formalisms to concrete situations via application problems. For instance, a student might first learn the symbolic formula for adding fractions (finding a common denominator, etc.) and only later manipulate objects that serve to explain and demonstrate this algorithm; whereas,
- *Progressive formalization* proposes that students should encounter new concepts in the context of meaningful concrete situations and then abstract toward formal models of these situations by progressively adopting mathematical forms and nomenclature. In this case, per the prior example, a student would first manipulate objects to discover principles for adding fractions and only later learn the symbolic formula that represents this procedure.

Among the studies supporting the formalism-first approach, the work of Kaminski et al. (2008) and Sloutsky et al. (2005) are of particular relevance to this discussion. In their experiments, undergraduate students participated in pattern-learning mathematics activities, where the elements composing the patterns were either generic and acontextual (non-descript geometrical shapes) or situated and contextual (readily identifiable objects). In a subsequent transfer task in a novel yet structurally identical domain, the acontextual participants outperformed the contextual participants. Based on these results, the researchers concluded that generic instantiations of mathematical structures are pedagogically superior to their concrete correlates. Concreteness, they argue, necessarily bears irrelevant contextual features, and these negatively influence both learning and transfer. First, learners may miss cross-domain structural alignment as a result of perceptual dissimilarity between rich representations. For example, they would not see how both the situation of two interlocking gears and the situation of two buildings and their shadows instantiate the concept of proportionality. Second, irrelevant aspects of concrete representations are liable to draw the focus of learners' attention away from conceptually critical information. For example, a demonstration of proportionality with interlocking gears might orient students to note the circles' counter-rotation at the expense of noting the multiplicative relations between the circles' circumferences. The logic of this argument is that students learning from an example cannot yet know what this will be an example *of*, and so they cannot in principle separate the conceptual wheat from the contextual chaff. Finally, concrete objects are more likely to be interpreted as ontologically intact entities rather than as symbolizing something else and thus may have limited referential flexibility, which is vital for the transfer. For example, students who use a printed 10-by-10 grid as an organizational scheme to build an elaborate construction out of a set of 1-by-1-by-1 wooden cubes would be less likely to later use that same grid as a topographical map with numerical values in each cell standing in for the height of the column of cubes towering up in that cell. The concrete object (the grid) takes on functional fixedness as a thing onto its own rather than as a potential representation of something else, so that the students miss out completely on a key learning objective.

In contrast, the research of Goldstone and Son (2005) supports progressive formalization. In their experiments, undergraduates worked with computer simulations to learn about complex adaptive systems. The simulation featured visual elements of varying perceptual concreteness, for example foraging ants were represented either by dots or by iconic images of ants. Students' performance was compared in both the initial and transfer tasks. Students were divided into four groups: abstract then concrete; abstract then abstract; concrete then concrete; and concrete then abstract. The best performance on both the learning and transfer tasks was obtained in the concrete-then-abstract group (i.e., the progressive formalization approach). The authors interpreted their findings to suggest that progressive formalization helps learners by enabling them first to enter a specific domain with the aid of concrete cues and then abstract and generalize principles as this concreteness fades out (see Ottmar & Landy, 2017, for a mathematics example).

The embodiment approach put forth in this article: (a) borrows the Progressive-Formalization epistemological position that abstract notions are grounded in activity with asymbolic objects; yet (b) also partially subscribes to the Formalism-First ontological position that mathematical concepts should be grounded in acontextual entities. Thus on the one hand, as per Progressive Formalization, embodied-design learning materials are asymbolic. Yet on the other hand, per Formalisms First, embodied-design materials are acontextual (see Table 1). These asymbolic acontextual learning materials are thus designed so as to avoid evoking students' knowledge about a narrow set of situations, that is, to avoid cueing particular narratives that might circumscribe the range of meanings students bring to bear in solving our tasks. Similar to generic construction materials, such as sand, play dough, or building blocks, inherent qualities of these virtual resources are designed so as to enable specific interactions and combinations yet without pre-constraining what meanings students bring to bear as they use these resources. We explain what the objects can do but not what they are.

**Table 1** Comparison of three pedagogical frameworks according to symbolic and contextual attributes of learning resources

| | | Contextual | |
|---|---|---|---|
| | | No | Yes |
| Symbolic | No | Embodied design | Progressive formalization |
| | Yes | Formalism first | – |

## *Affordances and Constraints of Asymbolic versus Symbolic Manipulatives*

Pedagogical approaches inspired by constructivism and embodiment theory have highlighted the role of sensorimotor integration in students' cognition of mathematical concepts (Abrahamson, 2006; Gray & Tall, 1994; Nemirovsky, 2003; Steffe & Kieren, 1994; Thompson, 2013; von Glasersfeld, 1983). Our study considered from an embodiment perspective the effect of situatedness on the development of sensorimotor schemes prior to signifying the schemes in a discipline's semiotic register. We thus sought a theory of situated perception and action that would enable us to model, anticipate, and analyze for effects of experimentally varying an activity's situatedness.

Our focus on the relationship between the properties of objects that students manipulate and their actions on these objects led us to consider the theoretical notions of affordances and constraints as relevant to the goals of this study, bearing in mind the critical social role of cultural agents in creating and providing these objects and mediating their functions and forms of use. *Ecological psychology* (Gibson, 1977) theorizes an agent's potential actions on the environment as contingent on the agent–environment relations. An agent (e.g., a mathematics student) engaged in a particular task (e.g., solving an interaction problem) perceives opportunities for acting on objects in the environment (e.g., classroom manipulatives) in accord with these objects' subjective cues; the agent tacitly perceives the object as *affording* particular actions, that is, privileging certain forms of goal-oriented engagement (see Vérillon & Rabardel, 1995, for a complementary theorization of instrumental genesis). If you are attempting to exit a room, a door handle affords grabbing and rotating. Importing Gibson's interactionist views into educational research, Greeno (1994) modeled student learning as the process of attuning to constraints and affordances in recurring situations. Araújo and Davids (2004) further offer that an instructor can "channel" students' engagement in goal-oriented activity by controlling environmental constraints. That is, a teacher can organize a classroom space in which she steers students to engage manipulatives in particular ways she believes are conducive to learning targeted mathematical content (see Mariotti, 2009, for a complementary sociocultural view on semiotic mediation).

Still, to the extent that one subscribes to the constructivist thesis underlying this research, namely that sensorimotor learning grounds conceptual learning, *why might different degrees of the learning materials' contextuality afford different sensorimotor learning?* The answer, we believe, lies in the nature of these sensorimotor schemes vis-à-vis the particular features of the learning materials that the students mentally construct in the course of developing the materials' new perceived affordances. That is, a given situation may lend itself to different goal-oriented sensorimotor schemes. And whereas a variety of schemes may accomplish the prescribed task, some of these schemes may be more important than others for the pedagogical purposes of the activity. We hypothesize that *the situatedness (contextuality) of learning materials constrains which sensorimotor schemes the materials might come to afford*. Where particular contextual cues unwittingly preclude student development of pedagogically desirable affordances, the students' conceptual learning will thus be delimited.

In evaluating this hypothesis pertaining to the nature and quality of situated learning, we needed a theoretical construct that would both cohere with the embodiment perspective and enable us to implicate in our data which sensorimotor schemes students were developing. We realized we were searching for a means of determining how the students are mentally constructing the materials; what specifically they were looking at that mediated their successful manipulation. Such a theoretical construct already existed: an attentional anchor (see below).

An *attentional anchor* is a dynamical structure or pattern of real and/or projected features that an agent perceives in the environment as their means of facilitating the enactment of motor-action coordination (Hutto & Sánchez-García, 2015). Abrahamson and Sánchez-García (2016) demonstrated the utility of the construct, which originated in sports science, in the context of mathematics educational research. Abrahamson, Shayan, Bakker, and van der Schaaf (2016) studied the role that visual attention plays in the emergence of new sensorimotor schemes underlying the concept of proportion. They overlaid data of participants' eye-movement patterns onto concurrent data of their hand-movements. It was found that the participants' enactment of a new bimanual coordination coincided with a shift from unstructured gazing at salient figural contours to structured gazing at new *non*-salient figural features (even at blank screen locations that bore no contours at all). The participants' speech and gesture confirmed that they had just constructed a new attentional anchor as mediating their control of the environment (see also Duijzer, Shayan, Bakker, van der Schaaf, & Abrahamson, 2017).

For this study, we adopted the construct of an attentional anchor as a key component of our methods. We sought to characterize what attentional anchors students developed during their attempts to solve a motor-action manipulation task. By so doing we hoped to gauge for effects of varying the contextuality of learning materials (situated vs. generic) on student development of the sensorimotor scheme mediating an activity's learning goal. We hypothesized that the more situated manipulatives would constrain the scope of attentional anchors students develop, with the detrimental consequence of students missing out on interaction opportunities that the designer considered as pivotal for learning the target content.

## Methods: Designing Constraints on Students' Sensorimotor Engagement of Manipulatable Elements in a Technological System

The Mathematics Imagery Trainer for Proportion (MITp; see Fig. 2) sets the empirical context for this study. Students working with the MITp are asked to move two cursors so as to make the screen green and keep it green. Unknown to the students, the screen will become green only if the cursors' respective heights along the screen relate by a particular ratio. The color of the screen can change along a gradient from red through orange toward green, with the feedback for the correct ratio being a distinct base-color green. For instance, for a ratio of 1:2, the screen will be green only when one hand is twice as high along the monitor as the other hand. Students develop a variety of motor-action strategies to satisfy the task demand (Howison et al., 2011).

In the current study, images appear at students' fingertips when they touch the screen. These images are either generic crosshair targets (see Fig. 3a) or situated images (e.g., hot-air balloons; Fig. 3b).

We selected hot-air balloons as exemplars of situated images, because presumably they evoke a schematic spatial–temporal narrative—a script that includes normative (default) topological plotting on the immediately available frame of reference (begin from screen base), an orientation and destination (upward toward the top of the screen), and a displacement vector and schedule (steady pace of motion along a linear vertical trajectory). Moreover, different balloons could conceivably rise at different rates due to their idiosyncratic payload, fueling, and navigation, so that two balloons might rise side by side, in parallel, each at its own speed. When two hot-air balloons launch together from the same location, presumably also the script of competition is evoked, because the balloons' respective motion then expresses human presence and agenda—the balloon moving at a greater speed might distinguish its pilot as more skillful and victorious. Consequently, students' multimodal attention to the objects they are manipulating would manifest as a tacit contradistinction between two individual entities, each with its particular identity, animacy, and effort. As such, students might wield the bimanual operation by rapidly alternating their attention between the two objects, ensuring in turn that *each* is moving correctly, rather than perhaps seeking a sensorimotor means of integrating the movements as a *relation* between the objects, such as by focusing on the spatial interval *between* the objects as it changes. Consequently, we assumed that students working with the hot-air balloons, as compared to those working with the generic objects, would be less likely to select the spatial interval between the objects as an attention anchor facilitating their task-oriented manipulation.

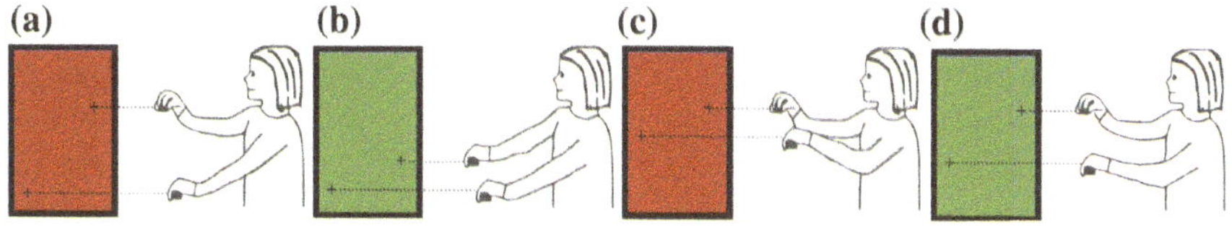


**Fig. 2** The mathematical imagery trainer for proportion (MITp) set at a 1:2 ratio. Compare (**b**) and (**d**) to note the different vertical intervals between the hands and, correspondingly, the different vertical (or diagonal) intervals between the virtual objects. Noticing this difference is presumed to be crucial for experiencing, then resolving a key cognitive conflict in expanding additive reasoning into multiplicative concepts

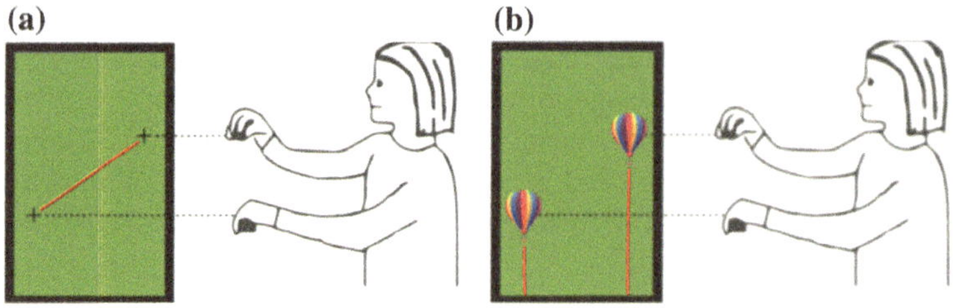


**Fig. 3** Experimental conditions and hypothesized attentional anchors: **a** generic crosshair targets cue the vertical or diagonal interval between the hands; and **b** situated images (hot-air balloons) cue the interval between each object and the bottom of the screen directly below it. In the actual experiments we used large touchscreens where the hands are on the interface

While hot-air balloon icons thus presumably constrain the range of potential interactions with the virtual objects (the "enactive landscape, Kirsh, 2013), students' tacit knowledge pertaining to how hot-air balloons behave also implicitly constrains them to manipulate the virtual objects along parameters relevant to the interaction. Namely, the software is programmed to respond only to the relative *vertical* location of these virtual objects (the $y$ axis), not their horizontal locations (the $x$ axis): The screen color is a function of the objects' relative distance from the bottom of the screen not its sides. The situated objects may thus better afford task-relevant manipulation as compared to the generic objects, thus minimizing exploration operations ("instrumenting," per Vérillon & Rabardel, 1995), just as a regular household wall-mounted light switch imposes vertical actions and precludes horizontal actions. Another set of situated objects designed for this activity were a pair of cars moving from the screen bottom to the top as per a birds-eye view of a racing track.

In both experimental conditions (generic and situated) students are led through a task-based semi-structured clinical interview. Following an unstructured orientation phase, in which the participants find several green locations, they are asked to maintain green while moving both hands from the bottom of the screen to the top. The interviewer then directly facilitates a coordination challenge, where the interviewer manipulates the left image and the student manipulates the right image. The student is asked to predict the green locations. This being a *semi*-structured interview, participants may experience additional opportunities to engage in tasks of finding green, maintaining green, and other unstructured exploration, either spontaneously or per the interviewer's suggestion. All along, the students are prompted to articulate rules for making the screen green. The interview was designed to last approximately 30 min, which included brief introductions and conclusions, with the core time equally divided between the two conditions, generic and situated.

We wished to investigate for attentional anchors that emerge during children's interactions with the technology. We reasoned that the attentional anchors would indicate what sensorimotor schemes the students developed. More specifically, we explored for an effect of experimental condition (generic vs. situated cursors) on the types of attentional anchors students construct and articulate (via speech and/or gesture). We also looked at the effect of condition sequence on the development of attentional anchors.

Twenty-five Grade 4–6 students participated individually in the interviews, 14 in the "generic-then-situated" condition and 11 in the "situated-then-generic" condition. In this study, we exclusively interviewed students around the numerical item of a 1:2 ratio, so as to minimize interview duration (see Abrahamson, Lee, Negrete, & Gutiérrez, 2014, for a study that explored other ratio items). These sessions were audio–video recorded for subsequent analysis. As our primary methodological approach, the laboratory researchers engaged in micro-analysis of selected episodes from the data corpus, focusing on the study participants' range of physical actions and multimodal utterance around the available media. Our working hypothesis, to iterate, was that the virtual objects' figural elements may cue (afford) particular sensorimotor orientations and thus "filter" the child's potential scope of interactions with the device. Namely, we analyzed for effects of the manipulatives' perceived affordances on participants' scope of interaction, bearing in mind that some interactions are more important than others for learning particular mathematical content.

## Results: Implicit Affordances of Manipulation Objects Mediate Student Strategies

A main effect was found. Below, we report our findings in each experimental condition by first describing participants' typical strategies and then illustrating these behaviors through brief vignettes. The section ends with comparing observed student strategies under the two conditions.

### *Generic Targets Afford the "Distance Between the Hands" Attentional Anchor*

In the trials where participants interacted with generic targets first, they began the activity by placing their left-hand- and right-hand fingertips on a blank touchscreen. Immediately they noticed crosshairs appear at the locations of their fingertips. In an attempt to make the screen green, the participants began moving their hands all over the screen with no apparent strategy, "freezing" their fingers as soon as the screen turned green. Eventually, participants oriented toward the spatial interval between their fingers, soon discovering that their fingers have to be a certain distance from each other at different heights along the screen. Finally they determined a dynamical covariation between the interval's size and height: the higher the hands, the bigger the interval must be (and vice versa). We turn to several vignettes (all names are pseudonyms). As we shall see, both participants will refer to an imaginary diagonal line connecting the cursors.

*Luke (age 10).* As he found various green-generating screen location, Luke commented about the space between his hands at these various locations: "It's the same angle. Well, I mean the line connecting them is the same direction" [4:53]. Later, he noted that the "[angle] is changing because my right hand is getting faster, so when this goes up that much (moves left hand approximately 2 in. on the screen) this one goes up at this much (moves right hand approximately 4 in. on the screen)" [11:10].

*Amy (age 9).* Amy reported her observation: "The diagonal [between the hands] at the top is different than [at] the bottom" [7:15]. Then later during the situated challenge, she said: "You have to make them different diagonally from each other to make it change color" [7:42].

Thus during the generic-target trials the participants not only noticed that the interval between their hands was changing in size, they came to see this interval as an imaginary line between their hands. In turn, this imaginary line—its size, angularity, and elevation along the screen—apparently served the participants in finding and keeping green, ultimately enabling them to articulate a strategy for doing so. This imaginary line along with attributed properties is an attentional anchor: It is crafted out of negative space to mediate the situated coordination of motor intentionality; subsequently this mentally constructed object serves to craft proto-proportional logico-mathematical propositions. This spontaneous appearance of a self-constraint that facilitated the enactment of a challenging motor-action coordination is in line with dynamical-systems theory (Kelso & Engstrøm, 2006).

Of the 14 students in this generic-then-situated experimental condition, 10 spoke about the interval between the hands still within the "generic" phase of their interview, and 8 of these 10 referred explicitly to its magnitude. Then during the "situated" phase of the interview, only 2 of these 10 students began to speak about the balloons as separate entities, focusing on the speed of each respective balloon, or reverting primarily to a focus on the color feedback of the screen to determine where to place the hands. The remaining 8 of these 10 students continued to use the interval line between their hands as a guide for making the screen green. These students' attention to the diagonal line was consistent, suggesting that this imaginary "steering wheel" had become perceptually stable in their sensorimotor engagement with this technological system.

## *Situated Images Afford the "Distance from the Bottom" Attentional Anchor*

Similar to the generic-then-situated condition, in the trials where students interacted with situated icons first, they began the activity by placing their left-hand- and right-hand fingers on a blank touchscreen. However in this condition they immediately saw hot-air balloons (not generic targets) appear on the screen. Thus, the virtual manipulatives in this condition are situational, even as the tasks are otherwise identical. Recall that these students worked first with the balloons and then with the crosshairs. As we will now explain, beginning with the balloons cued a narrative-based strategy, alluding to a frame of reference, that did not attend to the interval between the images but instead to each of these hot-air balloons' respective vertical distance above the "earth" (the bottom of the screen). This alternative sensorimotor orientation was so strong that it carried over to the crosshairs condition, so that by-and-large these participants were less likely to attend to the interval between the objects and thus were less likely to benefit from its potential contribution to their problem-solving strategy.

*Leah (age 11).* Having generated green for the first time, Leah noticed that when she moves one hand, the greenness dulls out toward red. Later, she described her strategy for making the screen green, referring gesturally to the hand's distance from the bottom of the screen: "I would say what I said before, where one hand chooses a place and the other hand chooses a color based on where the hand is, and you can adjust it to keep it green. Once you find that, you just need to keep it the same height [from the bottom]" [8:40]. Then in the next task, she maintained her strategy, saying: "When you move one hand up you need to move the other hand up so it's the same distance [from the bottom], but higher" [12:22].

*Jake (age 11).* Jake described his initial strategy in the form of a prescriptive rule, using the imperative grammatical mode, as though teaching another person how to accomplish the task:

> Try putting your hands together in the middle and then try moving one down or the other one up. One of the balloons should stay in the middle while the other moves [4:47].

Note how "middle" refers to that balloon's location along a vertical axis irrespective of the other balloon. Jake perseverated with this strategy throughout the set of challenges, moving his hands up along the screen sequentially rather than simultaneously. When later tasked to make the screen green with the stark targets, he appeared disoriented, noting, "This is harder because I don't have a starting point" [24:12]. Jake referred to the apparent absence of an "earth" as a grounding frame of reference for the cursors' vertical motion.

Of the 11 students who encountered the situated images first, 4 began to speak about the interval between the hands still during the situated condition, however these students did not elaborate about the line between the hands, and rather focused on each hand as a separate entity (e.g., stating that one hand controls color and the other controls brightness). During the second phase, in which they encountered the situated images, 2 of these 4 students as well as 3 of the 7 who had not attended to the interval demonstrated the emergence of this attentional anchor. The remaining students treated each of the two icons as separate entities throughout the entire interview, and hardly spoke about the interval between the hands. Collectively, these students were more inclined to treat the two objects on the screen as separate entities, focusing on the changing height of each object and the different speeds of the two objects as they move upward. Additional phenomena were encountered only in the iconic-then-stark condition. For example, one of the students (*Kate, age 11*), who spoke about the interval between the hands, used the length of iconic cursor itself to measure the interval. Kate explained her strategy for making green. It begins with placing the icons near each other at the bottom of the screen. Then, "in the middle there is one balloon between them, and at the top, two balloons between them. So it grows by one at a time" [06:45]. She accompanied this explanation with three quick demonstrations: at the bottom of the screen, in the middle, and on the top. When the icons were changed to the cars, Kate repeated her explanation:

> It's the same. They're right on top of each other at the bottom, and then in the middle it is like one car between them, and at the top—two cars. [08:32]

Later, Kate transferred this quantification approach to the generic condition, as follows:

> Um, let's say, move one of them [cursor], like, one length above the other, and then move the bottom one up until it's with another one, and then move the next like two lengths above, and then move the other one, and then here—four. [18:22]

Kate was well aware of the interval between her hands and in fact utilized it as an ad hoc unit of measurement so as to pace her bimanual ascent up along the screen (see Palatnik & Abrahamson, 2017, under review). Thus rather than negatively constrain her solution, as per our thesis, Kate's vignette provides a contrasting, if unique, non-protocol and idiosyncratic case of concreteness productively supporting progressive formalization.

## Summary

Participants who began the activity in the generic condition oriented toward the distance between their hands as their attentional anchor, whereas participants who began in the situated condition tended to treat the manipulatives as independent, untethered entities. It would appear that participants who began in the generic condition generated the interval as their attentional anchor because no other frame of reference was cued. Participants who began in the situated condition, on the other hand, followed the cued narrative implicit to the familiar images and therefore tended rather to visualize the two balloons as launching up from the ground.

It thus appears that objects bearing rich associative content introduce a new layer of baggage onto an interaction task, including forms, dynamics, hierarchies, and social conventions that guide the students' perception of the action space (on "framing," see Fillmore, 1968; Fillmore & Atkins, 1992). For instance, we typically think of hot-air balloons as "starting" at a point, such as the ground, at takeoff, and these evoked frames implicitly constrain the scope of possible attentional orientations to a situation, for example, by privileging the interval from each object down to the bottom of the screen at the expense of attending to the interval between the objects. In contrast, when manipulating stark cursors, there is no "starting point" as such, making it more likely that students attend to the interval between the hands. Presumably one could design icons that would draw students' attention explicitly to the relation between the two objects rather than viewing the objects as independent. Doing so, however, might come at the expense of two design goals: (1) enabling students to *discover* the target parameters (the behavior of a varying spatial interval would be evoked by the script rather than through exploration and would thus prevent eliciting students' inappropriate schemes, which in turn would prevent their experience of cognitive conflict that leads to reflection); and (2) opening up the scope of polysemous sensorimotor schemes (see Abrahamson et al., 2014).

Supporting our study's hypothesis, the results suggest an effect of situatedness on the construction of sensorimotor schemes. This finding is relevant to mathematics pedagogy, because sensorimotor schemes are theorized as mediating conceptual learning. It follows that *situatedness of instructional materials is liable to impede mathematical learning by precluding the emergence of sensorimotor schemes pertinent to a cognitive sequence toward the generalization of rules.* Whereas situatedness could, in turn, orient students precisely to the key parameters of the instructional design, doing so is liable on the other hand to narrow the manipulatives' enactive landscape and thus the scope of meanings that students bring to bear and develop through the interaction. Future iterations of this intervention would avail of eye-tracking (e.g., Abrahamson et al., 2016; Duijzer, Shayan, Bakker, van der Schaaf, & Abrahamson, 2017) and other multimodal learning analytics (Worsley et al., 2016) to corroborate students' oral and gestural report of attentional anchors and to expand our understanding of relations between situatedness and learning.

## Conclusion

Mathematics education researchers have long debated the question of pedagogical practices for introducing new mathematical concepts. The Formalism-First and Progressive-Formalization approaches offer diametrically contrasting positions on the question of whether concepts best develop from situated or generic learning materials. We tend to agree with the now-tempered view asserted by Day, Motz, and Goldstone (2015) that the "question of contextualization in instruction is neither simple nor settled" (p. 11; see also Goldstone & Sakamoto, 2003). Per their results, rich contextualization may encumber students' subsequent transfer of their understanding (see also McNeil, Uttal, Jarvin, & Sternberg, 2009). Yet we have contributed to the debate by offering that *the focus of situatedness research should be not on properties of the learning materials* per se *but on the sensorimotor schemes the materials may afford.* Thus, whereas Kaminsky et al. (2008) offer that "the difficulty of transferring knowledge acquired from concrete instantiations may stem from extraneous information diverting attention from the relevant mathematical structure" (p. 455), we refine that students' attention is diverted from the mathematically relevant *actions*. Richer materials, we have demonstrated, may unproductively constrain the scope of sensorimotor schemes students develop through engaging with the materials. In particular, richer materials may diminish opportunities for conceptual development, because they draw students' attention toward ways of thinking about the situations that, per the design, are less mathematically relevant. Students are liable thus to miss out on opportunities to think about the situation in ways that are critical for the educational success of an instructional sequence. On the other hand, where rich situated materials are designed so as to orient students explicitly on parameters that *are* relevant to the mathematical content, doing so would likely narrow the scope of meanings students bring to bear. For example, though we want students to attend primarily to the interval between the virtual objects, we wish for them to consider also the objects' relative speeds (see Abrahamson et al., 2014).

Students, that is to say children, are highly imaginative. They readily engage in pretense with generic objects, visualizing them one way and then another way. It is the *low* situativity of generic manipulatives that lends them to a greater variety of narratives and consequently a greater variety of sensorimotor orientations (see also Healy & Sinclair, 2007; Tahta, 1998). And so we agree with Uttal, Scudder, and DeLoache (1997) that sensory richness of manipulatives may derail certain forms of mathematics learning. But we stress that the issue here is not so much about sensory overload distracting from intended forms of engaging the objects. It is not about manipulatives but about *manipulating*—it is about task-oriented sensorimotor schemes students should develop in solving challenging bimanual motor-action problems. So the issue at hand is the hands' movements.

Goldstone and Son (2005) maintain that manipulatives combining concrete and abstract features facilitate students' learning and transfer better than those using uniform (e.g., only abstract) features. Similar, the tasks we used also combine elements of variable appearance. However, one might wish to bring into question the very dichotomy of concrete and abstract features. Per the constructivist approaches, concreteness is not an ontological trait but a phenomenological marker (q.v., Wilensky, 1991)—concreteness is the *result* of each student's inferential reflection on the movements of their own body, where action thus provides vital entry into the learning situation. We differentiate this sense of phenomenological concreteness from the concreteness of the icon per se, which in this case may constitute a source of superficial situatedness.

Our work bears implications for designing technologically enhanced embodied learning environments (see Lindgren & Johnson-Glenberg, 2013). Abrahamson and Lindgren (2014) called for further research to ascertain best principles governing designers' engineering of interactive materials, and in particular virtual manipulatives. The results of our study point to contextual advantages of generic manipulatives for the facilitation of anticipated learning outcomes toward conceptual understanding, at least in the realm of proportional thinking. Future work could examine how best to harness the affordances of situated manipulatives without interfering with the development of desired sensorimotor schemes. The field needs a deeper understanding also of cases where situatedness orients students toward *productive* engagement of instructional materials yet in so doing also narrows the scope of meanings students bring to bear (Abrahamson et al., 2014). Further research is necessary to understand how best to implement in classrooms technological media that enable students to enter conceptual domains by developing new coordinations toward new objects (e.g., see Negrete, Lee, & Abrahamson, 2013).

Learning is moving in new ways, and we should ensure that the tasks we create facilitate this moving. The perfunctory layering of contextual cues onto the objects learners are to manipulate might hit the 'engagement' goal yet in so doing quash the 'learning' goal (see also Abrahamson, 2015). In fact, sometimes the objects children manipulate might be so perceptually impoverished that there are no objects at all—just imagined objects. One might speak of mathematics students' right to bare arms.

**Acknowledgements** The research reported herein as well the writing of this chapter were supported by an REU (Rosen) under NSF IIS Cyberlearning EXP award 1321042.

## References

Abrahamson, D. (2004). Embodied spatial articulation: A gesture perspective on student negotiation between kinesthetic schemas and epistemic forms in learning mathematics. In D. E. McDougall & J. A. Ross (Eds.), *Proceedings of the Twenty Sixth Annual Meeting of the North American Chapter of the International Group for the Psychology of Mathematics Education* (Vol. 2, pp. 791–797). Toronto, Ontario: Preney.

Abrahamson, D. (2006). What's a situation in situated cognition? (Symposium). In S. Barab, K. Hay, & D. Hickey (Eds.), *Proceedings of the 7th International Conference of the Learning Sciences* (Vol. 2, pp. 1015–1021). Bloomington, IN: ICLS.

Abrahamson, D. (2009). Embodied design: Constructing means for constructing meaning. *Educational Studies in Mathematics, 70*(1), 27–47.

Abrahamson, D. (2014). Building educational activities for understanding: An elaboration on the embodied-design framework and its epistemic grounds. *International Journal of Child-Computer Interaction, 2*(1), 1–16.

Abrahamson, D. (2015). The monster in the machine, or why educational technology needs embodied design. In V. R. Lee (Ed.), *Learning technologies and the body: Integration and implementation* (pp. 21–38). New York: Routledge.

Abrahamson, D., & Bakker, A. (2016). Making sense of movement in embodied design for mathematics learning. In N. Newcombe & S. Weisberg (Eds.), Embodied cognition and STEM learning [Special issue]. *Cognitive Research: Principles and Implications, 1*(1), 1–13. https://doi.org/10.1186/s41235-016-0034-3.

Abrahamson, D., & Kapur, M. (2018). Reinventing discovery learning: A field-wide research program. In D. Abrahamson & M. Kapur (Eds.), Practicing discovery-based learning: Evaluating new horizons [Special issue]. *Instructional Science, 46*(1), 1–10.

Abrahamson, D., Lee, R. G., Negrete, A. G., & Gutiérrez, J. F. (2014). Coordinating visualizations of polysemous action: Values added for grounding proportion. *ZDM Mathematics Education, 46*(1), 79–93.

Abrahamson, D., & Lindgren, R. (2014). Embodiment and embodied design. In R. K. Sawyer (Ed.), *The Cambridge handbook of the learning sciences* (2nd ed.). Cambridge: Cambridge University Press.

Abrahamson, D., & Sánchez-García, R. (2016). Learning is moving in new ways: The ecological dynamics of mathematics education. *Journal of the Learning Sciences, 25*(2), 203–239.

Abrahamson, D., Shayan, S., Bakker, A., & van der Schaaf, M. (2016). Eye-tracking Piaget: Capturing the emergence of attentional anchors in the coordination of proportional motor action. *Human Development, 58*(4–5), 218–244.

Allen, J. W. P., & Bickhard, M. H. (2013). Stepping off the pendulum: Why only an action-based approach can transcend the nativist-empiricist debate. *Cognitive Development, 28*(2), 96–133.

Araújo, D., & Davids, K. (2004). Embodied cognition and emergent decision-making in dynamical movement systems. *Junctures: The Journal for Thematic Dialogue, 2,* 45–57.

Arsalidou, M., & Pascual-Leone, J. (2016). Constructivist developmental theory is needed in developmental neuroscience. *Npj Science of Learning, 1,* 16016.

Barab, S., Zuiker, S., Warren, S., Hickey, D., Ingram-Goble, A., Kwon, E. J., et al. (2007). Situationally embodied curriculum. *Science Education, 91,* 750–782.

Barsalou, L. W. (2010). Grounded cognition: Past, present, and future. *Topics in Cognitive Science, 2*(4), 716–724.

Bartolini Bussi, M. G., & Mariotti, M. A. (2008). Semiotic mediation in the mathematics classroom: Artefacts and signs after a Vygotskian perspective. In L. D. English, M. G. Bartolini Bussi, G. A. Jones, R. Lesh, & D. Tirosh (Eds.), *Handbook of international research in mathematics education, 2nd revised edition* (pp. 720–749). Mahwah, NJ: Lawrence Erlbaum Associates.

Bruner, J. (1986). *Actual minds, possible worlds.* Cambridge: Harvard University Press.

Burton, L. (1999). The implications of a narrative approach to the learning of mathematics. In L. Burton (Ed.), *Learning mathematics: From hierarchies to networks* (pp. 21–35). London: Falmer Press.

Campbell, S. R. (2003). Reconnecting mind and world: Enacting a (new) way of life. In S. J. Lamon, W. A. Parker, & S. K. Houston (Eds.), *Mathematical modeling: A way of life* (pp. 245–256). Chichester, UK: Horwood Publishing.

Chemero, A. (2009). *Radical embodied cognitive science.* Cambridge, MA: MIT Press.

Clark, A. (2013). Whatever next? Predictive brains, situated agents, and the future of cognitive science. *Behavioral and Brain Sciences, 36,* 181–253.

Day, S. B., Motz, B. A., & Goldstone, R. L. (2015). The cognitive costs of context: The effects of concreteness and immersiveness in instructional examples. *Frontiers in Psychology, 6.*

de Freitas, E., & Sinclair, N. (2012). Diagram, gesture, agency: Theorizing embodiment in the mathematics classroom. *Educational Studies in Mathematics, 80*(1–2), 133–152.

Duijzer, A. C. G., Shayan, S., Bakker, A., Van der Schaaf, M. F., & Abrahamson, D. (2017). Touchscreen tablets: Coordinating action and perception for mathematical cognition. *Frontiers in Psychology, 8*(144).

Fillmore, C. J. (1968). The case for case. In E. Bach & R. Harms (Eds.), *Universals in linguistic theory* (pp. 1–88). New York, NY: Holt Rinehart and Winston.

Fillmore, C. J., & Atkins, B. T. (1992). Toward a frame-based lexicon: The semantics of RISK and its neighbors. In A. Lehrer & E. Kittay (Eds.), *Frames, fields, and contrasts* (pp. 75–102). Hillsdale, NJ: LEA.

Fuson, K. C., & Abrahamson, D. (2005). Understanding ratio and proportion as an example of the apprehending zone and conceptual-phase problem-solving models. In J. I. D. Campbell (Ed.), *Handbook of mathematical cognition* (pp. 213–234). New York: Psychology Press.

Gibson, J. J. (1977). The theory of affordances. In R. Shaw & J. Bransford (Eds.), *Perceiving, acting and knowing: Toward an ecological psychology* (pp. 67–82). Hillsdale, NJ: Lawrence Erlbaum Associates.

Goldstone, R. L., Landy, D., & Son, J. Y. (2008). A well-grounded education. In M. DeVega, A. M. Glenberg, & A. C. Graesser (Eds.), *Symbols and embodiment* (pp. 327–355). Oxford, UK: Oxford University Press.

Goldstone, R. L., & Sakamoto, Y. (2003). The transfer of abstract principles governing complex adaptive systems. *Cognitive Psychology, 46,* 414–466.

Goldstone, R. L., & Son, J. Y. (2005). The transfer of scientific principles using concrete and idealized simulations. *Journal of the Learning Sciences, 14,* 69–110.

Gravemeijer, K. P. E. (1999). How emergent models may foster the constitution of formal mathematics. *Mathematical Thinking and Learning, 1*(2), 155–177.

Gray, E., & Tall, D. (1994). Duality, ambiguity, and flexibility: A "proceptual" view of simple arithmetic. *Journal for Research in Mathematics Education, 25*(2), 116–140.

Greeno, J. G. (1994). Gibson's affordances. *Psychological Review, 101*(2), 336–342.

Healy, L., & Sinclair, N. (2007). If this is our mathematics, what are our stories? *International Journal of Computers for Mathematical Learning, 12*(1), 3–21.

Howison, M., Trninic, D., Reinholz, D., & Abrahamson, D. (2011). The mathematical imagery trainer: From embodied interaction to conceptual learning. In G. Fitzpatrick, C. Gutwin, B. Begole, W. A. Kellogg, & D. Tan (Eds.), *Proceedings of the annual meeting of The Association for Computer Machinery Special Interest Group on Computer Human Interaction: "Human Factors in Computing Systems" (CHI 2011)* (Vol. "Full Papers," pp. 1989–1998). New York: ACM Press.

Hutto, D. D., Kirchhoff, M. D., & Abrahamson, D. (2015). The enactive roots of STEM: Rethinking educational design in mathematics. In P. Chandler & A. Tricot (Eds.), Human movement, physical and mental health, and learning [Special issue]. *Educational Psychology Review, 27*(3), 371–389.

Hutto, D. D., & Sánchez-García, R. (2015). Choking RECtified: Embodied expertise beyond Dreyfus. *Phenomenology and the Cognitive Sciences, 14*(2), 309–331.

Kaminski, J. A., Sloutsky, V. M., & Heckler, A. F. (2008). The advantage of abstract examples in learning math. *Science, 320,* 454–455.

Kelso, J. A. S., & Engstrøm, D. A. (2006). *The complementary nature.* Cambridge, MA: M.I.T. Press.

Kim, M., Roth, W.-M., & Thom, J. S. (2011). Children's gestures and the embodied knowledge of geometry. *International Journal of Science and Mathematics Education, 9*(1), 207–238.

Kirsh, D. (2013). Embodied cognition and the magical future of interaction design. In P. Marshall, A. N. Antle, E. V.D. Hoven, & Y. Rogers (Eds.), The theory and practice of embodied interaction in HCI and interaction design [Special issue]. *ACM Transactions on Human–Computer Interaction*, vol. 20, no. 1, 3, pp. 1–30.

Landy, D., & Goldstone, R. L. (2007). How abstract is symbolic thought? *Journal of Experimental Psychology. Learning, Memory, and Cognition, 33*(4), 720–733.

Lindgren, R., & Johnson-Glenberg, M. (2013). Emboldened by embodiment: Six precepts for research on embodied learning and mixed reality. *Educational Researcher, 42,* 445–452.

Mariotti, M. A. (2009). Artifacts and signs after a Vygotskian perspective: The role of the teacher. *ZDM—The international Journal on Mathematics Education, 41,* 427–440.

McNeil, N. M., Uttal, D. H., Jarvin, L., & Sternberg, R. J. (2009). Should you show me the money? Concrete objects both hurt and help performance on mathematics problems. *Learning and Instruction, 19,* 171–184.

Nathan, M. J. (2012). Rethinking formalisms in formal education. *Educational Psychologist, 47* (2), 125–148.

Negrete, A. G., Lee, R. G., & Abrahamson, D. (2013). Facilitating discovery learning in the tablet era: rethinking activity sequences vis-à-vis digital practices. In M. Martinez & A. Castro Superfine (Eds.), *"Broadening Perspectives on Mathematics Thinking and Learning"—Proceedings of the 35th Annual Meeting of the North-American Chapter of the International Group for the Psychology of Mathematics Education (PME-NA 35)* (Vol. 10: "Technology" p. 1205). Chicago, IL: University of Illinois at Chicago.

Nemirovsky, R. (2003). Three conjectures concerning the relationship between body activity and understanding mathematics. In N. A. Pateman, B. J. Dougherty, & J. T. Zilliox (Eds.), *Proceedings of the 27th Annual Meeting of the Int. Group for the Psychology of Mathematics Education* (Vol. 1, pp. 105–109). Columbus, OH: Eric Clearinghouse for Science, Mathematics, and Environmental Education.

Nemirovsky, R., Kelton, M. L., & Rhodehamel, B. (2013). Playing mathematical instruments: Emerging perceptuomotor integration with an interactive mathematics exhibit. *Journal for Research in Mathematics Education, 44*(2), 372–415.

Newman, D., Griffin, P., & Cole, M. (1989). *The construction zone: Working for cognitive change in school*. New York: Cambridge University Press.

Noss, R., & Hoyles, C. (1996). *Windows on mathematical meanings: Learning cultures and computers*. Dordrecht: Kluwer.

Ottmar, E., & Landy, D. (2017). Concreteness fading of algebraic instruction: Effects on learning. *Journal of the Learning Sciences, 26*(1), 51–78.

Palatnik, A., & Abrahamson, D. (2017). Taking measures to coordinate movements: Unitizing emerges as a method of building event structures for enacting proportion. In E. Galindo & J. Newton (Eds.), *"Synergy at the crossroads"—Proceedings of the 39th Annual Conference of the North-American Chapter of the International Group for the Psychology of Mathematics Education* (Vol. 13 [Theory and research methods], pp. 1439–1442). Indianapolis, IN: Hoosier Association of Mathematics Teacher Educators.

Palatnik, A., & Abrahamson, D. (under review). Rhythmic movement as a tacit enactment goal mobilizing the emergence of mathematical structures. *Educational Studies in Mathematics*.

Piaget, J. (1968). *Genetic epistemology* (E. Duckworth, Trans.). New York: Columbia University Press.

Radford, L. (2003). Gestures, speech, and the sprouting of signs: A semiotic-cultural approach to students' types of generalization. *Mathematical Thinking and Learning, 5*(1), 37–70.

Sarama, J., & Clements, D. H. (2009). "Concrete" computer manipulatives in mathematics education. *Child Development Perspectives, 3*(3), 145–150.

Sfard, A. (2002). The interplay of intimations and implementations: Generating new discourse with new symbolic tools. *Journal of the Learning Sciences, 11*(2, 3), 319–357.

Sloutsky, V. M., Kaminski, J. A., & Heckler, A. F. (2005). The advantage of simple symbols for learning and transfer. *Psychonomic Bulletin and Review, 12*(3), 508–513.

Steffe, L. P., & Kieren, T. (1994). Radical constructivism and mathematics education. *Journal for Research in Mathematics Education, 25*(6), 711–733.

Stokes, D. E. (1997). *Pasteur's quadrant: Basic science and technological innovation*. DC: Brookings.

Tahta, D. (1998). Counting counts. *Mathematics Teaching, 163,* 4–11.

Thompson, P. W. (2013). In the absence of meaning .... In K. Leatham (Ed.), *Vital directions for mathematics education research* (pp. 57–94). New York: Springer.

Uttal, D. H., Scudder, K. V., & DeLoache, J. S. (1997). Manipulatives as symbols: A new perspective on the use of concrete objects to teach mathematics. *Journal of Applied Developmental Psychology, 18,* 37–54.

Varela, F. J. (1999). *Ethical know-how: Action, wisdom, and cognition*. Stanford, CA: Stanford University Press.

Varela, F. J., Thompson, E., & Rosch, E. (1991). *The embodied mind*. Cambridge, MA: M.I.T. Press.

Vérillon, P., & Rabardel, P. (1995). Cognition and artifacts: A contribution to the study of thought in relation to instrumented activity. *European Journal of Psychology of Education, 10*(1), 77–101.

von Glasersfeld, E. (1983). Learning as constructive activity. In J. C. Bergeron & N. Herscovics (Eds.), *Proceedings of the 5th Annual Meeting of the North American Group for the Psychology of Mathematics Education* (Vol. 1, pp. 41–69). Montreal: PME-NA.

Vygotsky, L. S. (1997). *Educational psychology* (R. H. Silverman, Trans.). Boca Raton, FL: CRC Press LLC (Work originally published in 1926).

Wilensky, U. (1991). Abstract meditations on the concrete and concrete implications for mathematics education. In I. Harel & S. Papert (Eds.), *Constructionism* (pp. 193–204). Norwood, NJ: Ablex Publishing Corporation.

Worsley, M., Abrahamson, D., Blikstein, P., Bumbacher, E., Grover, S., Schneider, B., et al. (2016). Workshop: Situating multimodal learning analytics. In C.-K. Looi, J. L. Polman, U. Cress, & P. Reimann (Eds.), *"Transforming learning, empowering learners," Proceedings of the International Conference of the Learning Sciences (ICLS 2016)* (Vol. 2, pp. 1346–1349). Singapore: International Society of the Learning Sciences.

**Dana Rosen** (B.A. Interdisciplinary Studies, University of California Berkeley, 2016) is an honors graduate with a concentration on Technology in the learning sciences. Rosen completed her senior thesis exploring a model of comprehensive design for 21st-century education in the digital age. Her research interests include embodied design, narrative in mathematics education, and digital affordances of educational Technology. Publications include *Tradeoffs of situatedness: Iconicity constrains the development of content-oriented sensorimotor schemes*—which appeared in the proceedings of the annual conference of the North American chapter of the International Group for the Psychology of Mathematics Education (PME-NA 2016).

**Alik Palatnik** (Ph.D., Mathematics Education, The Technion- Israel Institute of Technology, 2016) is a researcher of student-centered and Technology-enhanced learning of mathematics. Palatnik has completed a post-doctoral fellowship at the Graduate School of Education, University of California, Berkeley and currently serves as a Lecturer at the Department of Mathematics Education, Shaanan Academic Religious Teachers' College, Israel. His research interests include project-based learning in mathematics, embodied design, and computer-supported collaborative learning of mathematics. Publications include "Sense making in the context of algebraic activities" in *Educational Studies in Mathematics,* 2017, *95* (3), 245–262 (with Boris Koichu).

**Dor Abrahamson** (Ph.D., Learning Sciences, Northwestern University, 2004) is a design-based researcher of mathematical cognition and instruction with an appointment as Associate Professor at the Graduate School of Education, University of California, Berkeley, where he runs the Embodied Design Research Laboratory. Abrahamson draws on constructivist–enactivist and sociocultural theories to inform the design of interactive educational materials. In turn, he analyzes empirical data collected in evaluation studies of these activities to develop theoretical models of mathematics learning as well as his pedagogical framework, embodied design. Abrahamson has published widely in the leading journals of his field and is recipient of grants from the Spencer Foundation and the National Science Foundation.

# Fingers-on Geometry: The Emergence of Symmetry in a Primary School Classroom with Multi-touch Dynamic Geometry

**Sean Chorney and Nathalie Sinclair**

**Abstract** In this chapter, we describe a research project with first grade children using a multi-touch dynamic geometry sketch. We approach our analysis through the lens of inclusive materialism (de Freitas & Sinclair, 2014), which considers the intra-actions involved in the child-device-geometry assemblages and thus to the way in which new mathematical ideas emerge in this assemblage. Drawing on the design experimentation methodology (de Freitas, 2016), we analyse the assemblage in order to study how concepts such as symmetry arise. We therefore seek to investigate the way digital technology can become a device for producing new concepts. We focus particularly on how the multi-touch environment, in which geometry objects can be continuously dragged with fingers, occasions new gestures and body motions that provide the basis for emerging geometrical ideas.

## Introduction

In this chapter, we experiment with the concept of symmetry in a grade one classroom where students interact with a dynamic geometry environment (DGE) on multi-touch tablets. In mathematics, a concept, in general, is seen as a robust, cohesive idea that represents all of its manifestations. For example, if the concept of symmetry is understood 'fully', a person should be able to apply, answer questions, and understand symmetry in all its instantiations. We believe this reductive approach to learning mathematical concepts relies too heavily on knowing as 'stored' mental knowledge. This perspective ignores the relevance that both tools and activities have on what it means to know a mathematical concept. Some

S. Chorney (✉) · N. Sinclair
Simon Fraser University, Burnaby, Canada
e-mail: sean_chorney@sfu.ca

N. Sinclair
e-mail: nathsinc@sfu.ca

 
N. Calder et al. (eds.), *Using Mobile Technologies in the Teaching and Learning of Mathematics*, Mathematics Education in the Digital Era 12,
https://doi.org/10.1007/978-3-319-90179-4_12

researchers have challenged the representationalist view of knowledge, arguing for a more performative practice of *knowing*, which happens in time, in context and through action (e.g., Pickering, 1995). That is, rather than interpreting experimental results in an abstract way, inductively drawing out a "schema" that a given child has constructed, they refuse the dichotomising of action and thought, focussing instead on the multiple ways in which bodies and materials engage in knowing. We challenge the idea that symmetry is a concept that is slowly developed and acquired, then schematised in a stable way in the brain. Instead we use the theoretical framework of inclusive materialism (de Freitas & Sinclair, 2014) which draws attention to the material nature of mathematical concepts, and thus to the concept as indeterminate, mobile and imbricated with the activities of children with tools. Following this set of assumptions, we also utilize a method of diffractive analysis (Barad, 2007, 2010, 2012; de Freitas, 2016), which will be described in more detail in a subsequent section of the chapter.

## The Importance of Symmetry in the Curriculum

Symmetry is an important mathematical topic that supports spatial reasoning and patterning. The topic appears in different forms in various curricula typically starting at a reasonably young age. In our jurisdiction, symmetry appears in the curriculum in grade 4 with line symmetry followed by two-dimensional shape transformations in grade 5, and then again, later in grade 9 with line and rotational symmetry. These stand-alone topics in grades 4, 5 and 9 are not the only time symmetry is addressed. Although symmetry is formalised in each of these grades, the word 'symmetry' is referenced in other mathematical areas such as problem solving and also in working with two- and three-dimensional objects. Other topics that draw on symmetry include analog clocks, direction, graphs, and working with parallel and perpendicular lines.

In primary school, children usually encounter symmetry as a property of shapes. They maybe asked, for example, whether a given shape (a butterfly, a square, etc.) is symmetric. Folding is frequently used as a means to determine whether a given shape is symmetry. In this study, in which we used a dynamic geometry environment, we used a more transformational approach involving motion, in which symmetry is the result of a reflection. The act of reflecting is an isometric transformation that 'reflects' a pre-image from one side of a line of symmetry to the other. In such a transformation, the image and the pre-image are equidistant to the line of reflection; and, the line joining the image and pre-image is perpendicular to the line of symmetry. These properties remain invariant as the pre-image is dragged on the screen.

Prior research has shown that children have a great deal of knowledge about symmetry long before they learn about it formally in geometry classes. For example, children spontaneously construct symmetrical figures during informal play at the pre-school age (Seo & Ginsburg, 2004). However, the importance of this implicit understanding has been under-utilized in mathematics education as well as under-represented in the research literature. Even when studied, the research has focused on "the development of children's ability to tell symmetrical figures apart, not to understand the relation between them" (Bryant, 2008, p. 34). For example, Bornstein and Stiles-Davis (1984) link the developmental progression of 4–6-year-olds with types of line symmetry. Namely, they found that 4-year-olds discriminated only vertical line symmetry, 5-year-olds, vertical and horizontal line symmetry, and 6-year-olds, vertical, horizontal and oblique symmetry. However, their study focussed exclusively on the visual identification of symmetry, rather than on the relationship between the various elements involved, such as the line of symmetry, the relationship of equidistance between the line of symmetry and both the pre-image and the image.

Based on their research of mathematics learning in the early years, Clements and Sarama (2004) propose that children should work with symmetry in the pre-K through to grade-2 years. They offer a developmental trajectory in which children begin at the pre-K level to create shapes that have line symmetry, then work in kindergarten and grade one to identify symmetry in 2-D objects. In grade two, children identify the lines of symmetry of various shapes. This trajectory also focusses more on identification than on properties of symmetry and relations between the pre-image and the image in reflectional symmetry, which, following Duval (2005), we see as significant parts of geometric thinking that can be engaged even at the early years.

In their study involving children in grades 2/3, Ng and Sinclair (2015) found that the use of a dynamic geometry environment developed dynamic and embodied ways of thinking about symmetry after engaging in teacher-guided explorations of a pre-constructed sketch called the "Symmetry Machine". In this sketch, which is also used in the present research, symmetry is preserved as different components of a diagram are moved, including the line of symmetry. While that research was conducted in a whole classroom setting using an interactive whiteboard, the present study also included as an addition the use of a classroom set of iPads, so that each student had the opportunity to directly manipulate the sketch.

As we will develop in the next section, we take symmetry to be a concept that cannot be separated from the tool, nor the user with which it is instantiated. In this study, when we speak of the concept of symmetry, we do not abstract it from the movement of fingers, eyes, bodies of students, nor the iPads, sketches and classroom dynamics. This choice is based on the idea of intra-actions (Barad, 2007) and the notion of assemblage.[1]

[1] *Assemblage* is a notion introduced by Gilles Deleuze and Félix Guattari, and later used both by Bruno Latour and Karen Barad. The article on Deleuze in the Stanford Library of Philosophy glosses it as follows: "'assemblages', that is to say, an emergent unity joining together heterogeneous bodies in a 'consistency'" (http://plato.stanford.edu/entries/deleuze/).

## Theoretical Framework

In this chapter, we integrate post-humanist and new materialist perspectives into both how we see and analyse mathematics teaching and learning. These perspectives, essentially, attempt to de-centre the human as the primary—or, indeed, only—agent in the learning process and to find ways of accounting for how matter matters in that process. These perspectives are rooted in broader philosophical developments associated with 'the ontological turn' that has emerged from feminist studies (see Barad, 2007; Braidotti, 2013; Haraway, 2008). Within the context of mathematics education, these perspectives have been adapted and refined to the context of educational research, in the form of *inclusive materialism* (de Freitas & Sinclair, 2014), which looks closely at the material specificities of mathematical experiences. This approach positions itself "within a tradition in which abstract thought and materiality are assumed to be entwined" (p. 3).

de Freitas and Sinclair draw primarily on the work of Barad and her concept of intra-action. Barad (2007) contrasts intra-action with interactions, where the latter assumes the coming together of entities that have pre-defined properties and characteristics. In intra-action, entities can be seen to be emerging from activity, that is, activity occurs first, and that activity creates and integrates the 'bounded' entities such as the iPad, the child and the mathematics. Combined with the post-humanist view, there is a shift away from individuating the student as an independent and well-defined body and how she is acquiring knowledge. Instead, the fixed boundaries of that body are disrupted in order to attend to the evolution of a tool, child and mathematics assemblage. Inclusive materialism, consequently, takes mathematical concepts to be material and emergent from particular intra-actions. Because the concept is material, and because—as Barad argues based on her analysis of experiments in physics, such as the two-slit experiment,[2] matter is indeterminate, concepts as well partake of the indeterminacy of matter. This challenges the traditional view of individuals abstracting conceptual knowledge from engaging with material objects. Rather than focussing on epistemological concerns, those related to what is learned by the student, inclusive materialism attends to ontological concerns, that is, what is the emergent material assemblage that gives rise to meanings.

In an inclusive materialist framework, mathematical concepts arise out of intra-actions between student and material and activity. In particular, we expand traditional approaches that see symmetry as a distinct idea or concept. We challenge the common pedagogical approach whereby different tools and different tasks will move students closer to the bigger picture of symmetry. Symmetry is not a *form* that

---

[2]Specifically, two slit experiments appeared to show that light was a particle or a wave, depending on the experimental apparatus that scientists used. Instead of seeing particles and waves as ontologically antithetical, as in classical (non-quantum) models of physics, Barad suggests that light, and therefore matter more generally, is *ontologically indeterminate*. It takes on a specific ontological form—it becomes determinate—through intra-actions with the measurement apparati.

is somehow buried *fait accompli* in matter and waiting to be conjured or evoked. We see symmetry as an emergence of different meanings from an entanglement of tool, task and student. Because of this view of symmetry, as a mathematical concept, the questions we can ask about teaching and learning are different. We avoid questions that assume strict boundaries between students, mathematics and tools, and that position mathematical concepts as fixed ideas waiting to be abstracted from experience. Our question is more ontological in nature since we will be concerned with what happens to an assemblage over time, how it changes, ruptures or renews.

## Methodological Framework

In this chapter we explore a diffractive methodology that draws on Barad's agential realism (2007, 2010, 2012). A diffractive analysis involves the reading of data through multiple theoretical insights in such a way to gain unpredictable and productive emergences; for example, Barad read Neils Bohr, the physicist, through Jacques Derrida, the philosopher. A diffractive analysis is thus less concerned with reflecting objectively a particular event and instead seeks to offer, in the words of Haraway (1992), "a mapping of interference" (p. 300). In the context of educational research, Lenz Taguchi (2012) uses the method of diffractive analysis to interpret data gathered from discussions she had with a boy who had made a bark boat. Her goal is to "make visible new kinds of material-discursive realities that can have transformative and political consequences" (p. 265).

What distinguishes Barad's diffractive analysis from that of Lenz Taguchi is that Barad's approach includes an experimental device, or an apparatus. Indeed, in her case, the apparatus is a machine used in physics laboratories that interferes with the environment (in the example she provides, light) and produces a new phenomenon (patterns on a screen). Experiments using this apparatus enable Barad to explore new ontologies, such as: What is light? What is matter? She sees the experimental interventions that she studies—with theoretical physics—as delving into the indeterminacy of matter, while also being 'the condition' of determinate meaning.

de Freitas (2016) has suggested that as educational researchers, we too, could conduct experiments that involve a *diffractive apparatus*. Such an experiment would be designed to explore new ontologies and to better understand the relations between matter and meaning that emerge in a particular classroom situation, for example. Imagine, as will be the case in this chapter, that the apparatus not only includes a particular educational digital tool, but also students' bodies and movements. A diffractive apparatus experiment would differ from methods based on theories of tool use in mathematics education research because of the way in which the apparatus is not simply taken as a mediator of learning (as in the theory of semiotic mediation elaborated by Bartolini Bussi and Mariotti (2008)), or a tool that students use in order to learn particular concepts (as in the theory of instrumental genesis (Artigue, 2002)). Instead, the tool is part of a diffractive apparatus that produce effects that help us see how meanings about the concept of symmetry and how it is entangled with the physical. This may sound surprising, given that symmetry is not exactly a new concept and that it is usually considered to be characterised by logical determination. And yes, what a diffractive analysis might show, in an experiment, is that the indeterminate nature of matter, which is a fundamental assumption of Barad's, entails indeterminacy about symmetry as well.

We thus use de Freitas's (2016) mobilisation of Barad's diffractive model in her elaboration of how an experimental device 'interferes' with the environment. Barad is exploring the indeterminacy of matter and de Freitas elaborates that the indeterminacy results partly from these devices that are part of experiments and consequent data collection. This diffractive methodology offers ways of exploring new ontologies and insight into the relationship between matter and meaning. In this chapter, our diffractive apparatus included DGE sketches that were designed in *Web Sketchpad* and used both in a whole classroom setting with a projector, and in pairs, with iPads. Given our theoretical framing, we assume that a concept is never a singular representation, nor is it an essence or form. Rather, in our diffractive apparatus, symmetry is indeterminate and will take on a specific ontological form in intra-action with the dynamic geometry apparatus (and other parts of the material surroundings). As such, our question becomes, what determinations of symmetry arise from our experimental setting? We invite the reader to consider the following as a thought experiment that tries to imagine what it would mean to adopt the theoretical perspectives we have outlined. We recognise that given the novelty of the methodological approach, there is bound to be some tension between our traditional focus on individual children and their actions with tools and on concepts, and our new attempt to focus on intra-actions. Nevertheless, we contend that a consideration of symmetry from this perspective will open opportunity for alternative, yet productive insights.

## Research Setting

At the start of 2016, from January to April, as part of a research project, we visited a grade one classroom in a public French-Immersion elementary public school in a North American west coast school. (Since this is the first year that the children are learning French, they often speak in English during class, and the teacher also sometimes addresses them in English.) We went every week and spent just over one hour working very closely with the regular teacher in organizing activities, discussing curriculum directions, and taking turns to teach. Nathalie (the second author) taught the class almost every time we visited. Sean also participated in teaching but was more often working with individual students during group work. Typically we went in the mornings before lunch. The classroom had a carpeted area in the front of the room where students often gathered as a group, there were also five tables set up around the room, to the sides and back of the room, where students could sit and work. Six students could sit at a table. Every session began in a whole classroom interaction, with the students at the carpet and an overhead projector connected to an iPad on which designed sketches were shown. This would be followed by pairwise explorations on the iPads, at the tables.

All classroom activities were videotaped, in the whole class gatherings, when students were sitting on the floor. The camera captured what was on the projected screen, all the children and the teacher. When students worked in pairs on the iPad around the room the video was focussed on one pair, and sometimes two, if the students were close enough to each other.

## The Apparatus and Sketch Design

The apparatus we look at in this chapter is a web-based variation of *The Geometer's Sketchpad* (Jackiw, 1991, 2001) that is used on the iPad. Different sketches with various functionalities were designed (available at www.sfu.ca/geometry4yl/). The sketches are open and exploratory in that there are no instructions explicitly given. In this chapter we work with the discrete Symmetry Machine sketch. In this sketch, there is a vertical line in the middle of the screen, which is the line of symmetry. On either side of the line are six coloured squares, two blue, two red, and two purple (Fig. 1a). When a coloured square is touched on the screen and moved, its image square moves so as to preserve the reflectional symmetry of the diagram as a whole. These squares move discretely on a square grid background. Dragging any square on one side of the line of symmetry will also move the corresponding square on the other side of the line of symmetry (see Fig. 1a, b). The discrete motion, as well as the use of the grid, was intended to help the children attend to the distance between a square and the line of symmetry. The line itself can be moved, right or left, which will move six of the squares in order to maintain symmetry. The line has a red point on it and when that point is dragged, the line can be rotated around so as to create diagonal or horizontal lines of symmetry (Fig. 1c).

## Outline of the Lesson (53 min)

The lesson we report on in this chapter was our first lesson using the Symmetry Machine. At the beginning of the lesson, the Symmetry Machine was projected on the front screen, the students were seated together on the floor in front of the screen and Nathalie was towards the back of the room with the iPad. Nathalie engaged the students in some questions relating to the sketches. The later part of the lesson had students working in pairs on the iPad at individual tables. Students were given set Symmetry Machine diagrams on paper and asked whether they could re-create the diagrams using the Symmetry Machine on the iPads. Not all of the diagrams were symmetric.

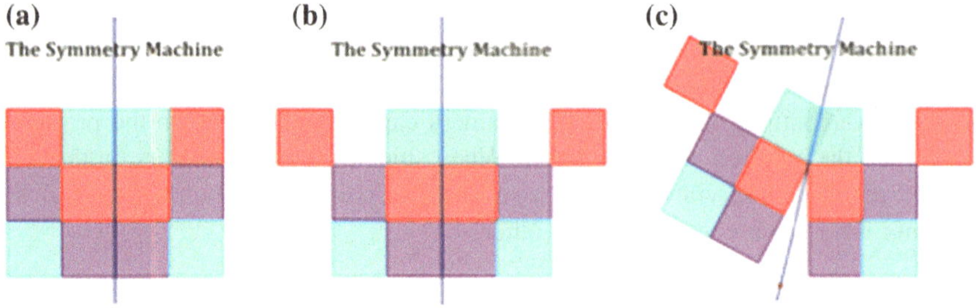


**Fig. 1** **a** The discrete symmetry machine; **b** after dragging one block away from the line; **c** after rotating the line of symmetry using the point visible near the bottom of the line of symmetry

## Becoming Symmetry

As per our stated methodology, our analysis of the video data follows the concept of symmetry as it becomes manifested through the experimental device of the iPad and the movement of the children. Below we present snapshots of that evolution. There are three main ones in the first ten minutes, while in the whole classroom configuration. Recall that this was the first time that the students were being formally introduced to the word 'symmetry' and were engaging in mathematical activity focussed on creating and manipulating symmetric shapes. Indeed, the first lesson was designed in order to introduce the students to symmetry by investigating its behaviour on the iPad, and not through a description or static example of it.

### *Symmetry as Twoness, Movement and Holes*

The first visible effects of the diffractive analysis are that 'symmetry' was seen as something that involves twoness.[3] In the first lesson, the students were gathered together as a group on the floor while Nathalie projected the image of the Symmetry Machine on a screen at the front of the classroom. There were six coloured blocks on either side of the vertical line. At first, many finger puppets were made on the

[3]Readers may find that their own conceptions of symmetry also involve some kind twoness as well. It is also implicit in more traditional ways of working with symmetry with young children where one evokes folding (so one side matches the other side, there being two sides) or mirroring (where what's in the mirror is the same as what is being mirrored, thereby also involving two things). We argue that in these situations, the emphasis is on sameness rather than on twoness. This is because attention is usually focussed on one side of the symmetry line, rather than on symmetry as a transformation of one shape to another. Since our goal is to study the determinacies of symmetry in this experimental setting, we examine the emergence of any and all such determinacies, whether they seem familiar, or not.

projected screen, but as soon as Nathalie moved the top, right-most square (which was red) to the right (which of course moved the corresponding red square on the other side of the line) the sound of student gasping was heard (see Fig. 2a).

4:20 Nathalie: What happened when I moved the red square?

4:21 Jonathan: Deux (*Two*) (two fingers up in the form of a peace sign (see Fig. 2b, left). Deux (*Two*).

4:22 Nathalie: Deux. Qu'est-ce que tu veux dire Jonathan? (*Two. What do you mean Jonathan?*)

4:27 Jonathan: Deux carrés bougent (*Two squares are moving*).

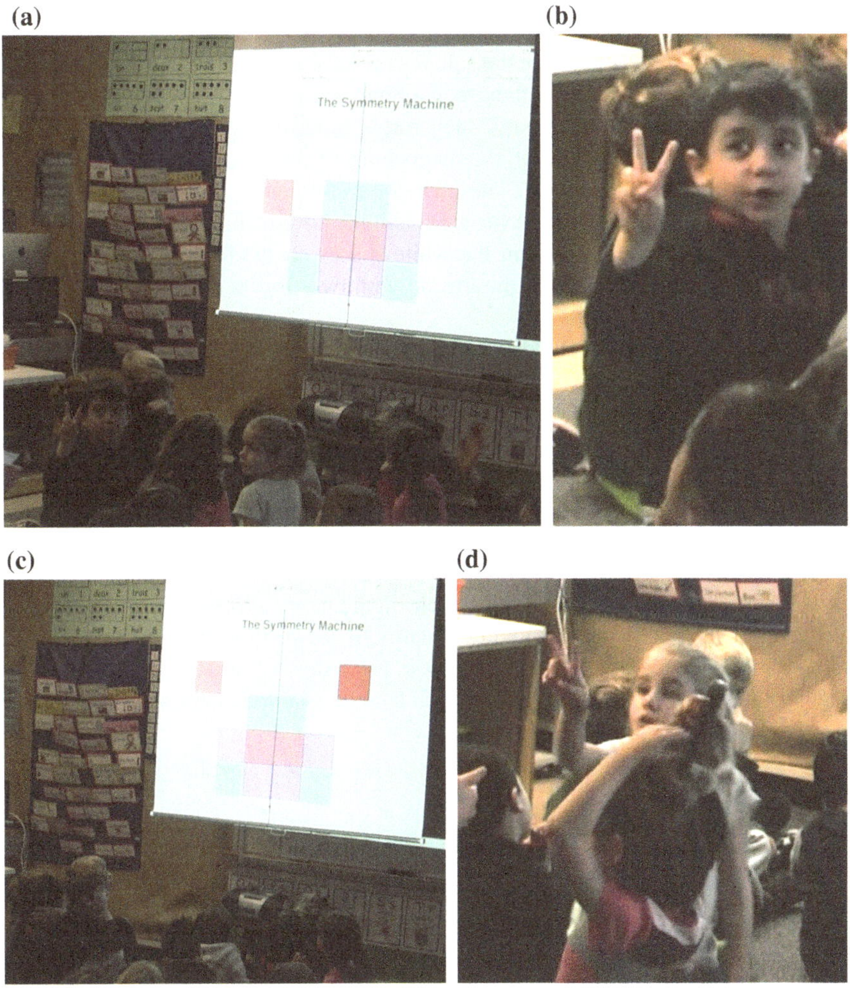


**Fig. 2** **a** Students gasp; **b** deux: two fingers up; **c** two red squares up; **d** peace: two fingers extended

4:34 Nathalie: Weston?

4:37 Weston: Deux carrés blancs. (*Two white squares.*)

4:40 Nathalie: Deux carrés blancs. Ah oui, il y a deux carrés blancs maintenant! (*Two white squares. Oh, yes, there are two white squares now!*)

Nathalie then moved the same red square up (see Fig. 2c) and several voices said "wow!" She asked what happened when the red square was moved towards the top. One voice said "deux carrés bancs", then several voices said "trois" (*three*) and then several other voices said "quatre" (*four*). When the same red square was moved towards the right, several "wow" exclamations were heard again as well as several numbers, including "four", "five" and "six".

Over the course of this period of time (1 min 22 s), two ideas emerged in relation to symmetry. The first is the notion of twoness, which occurs both in language and in gesture. Recall that the students were not looking at Nathalie's finger on the iPad, so they would just be able to see the squares move on the overhead screen. And that motion happened in pairs. Had they seen a finger move one square, as might be the case on an interactive whiteboard, they might have focussed less on the square being moved and more on what was happening to the image square. The idea of twoness emerges several minutes later, when the children begin working in pairs on the iPads (see Fig. 6a, where Ava is explaining how to move the squares). The students seemed to want to fill in the white spaces on the top row and Ava turns around, kneeling, and put her right arm up with two fingers extended (Fig. 2d), as in the peace gesture.

While the first comment "deux carrés bougent" addresses the motion of the squares, without saying anything specific about the way the motion happens (such as, moving away from the line), the next verbal comments focus less on what's moving than on what gets left behind. The two white squares are the empty squares that appear once the red square has moved. The movement of the red square leaves a kind of hole, which is like the negative space of the sketch. This hole is also characterized by its parity, first in the two white squares, then in the four white squares and finally in the six white squares. Interestingly, there are many more than two or even six white squares in the sketch, so the parity seems to focus specifically on the space created by the moving squares. It is worth remarking that had the squares not moved, holes could not have formed, so the iPad as an apparatus enables the discrete motion of squares to intervene in the concept of symmetry in a novel way.

## *Symmetry as Bringing Together*

In the initial activity, when the students were seated on the carpet as a group and Nathalie was moving the squares, she asked what would happen when she moved one of the purple squares (right-most square in the second row of Fig. 2a) upward. There was an approximate two-minute length of time during which the students

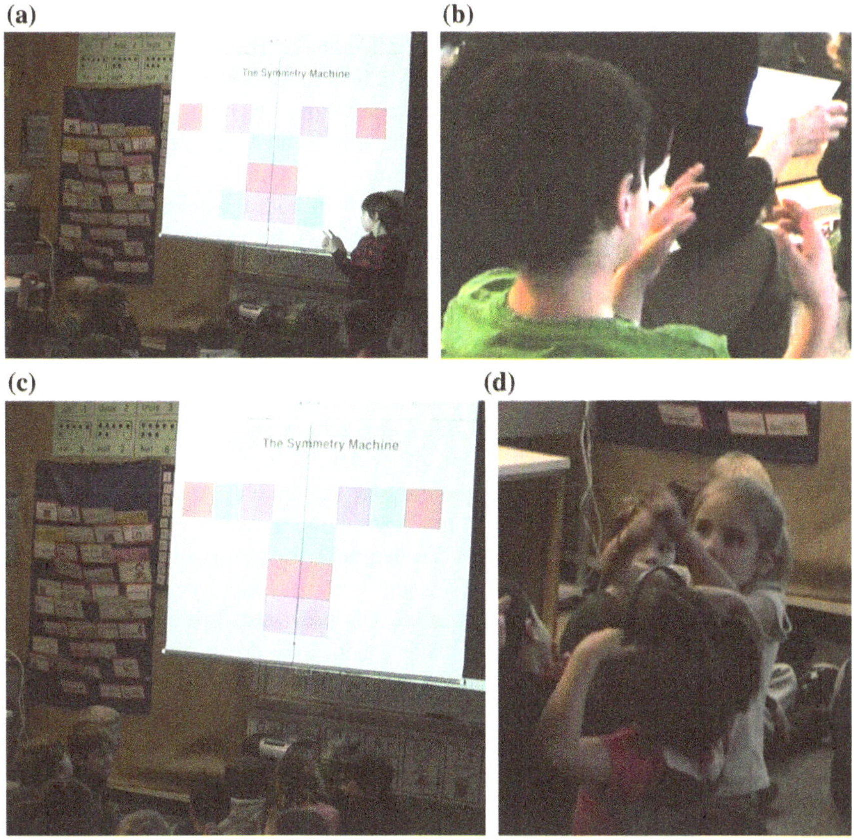


**Fig. 3** **a** What will happen; **b** thumb and finger gesture; **c** top row; **d** Ava's gesture

struggled to predict what would happen. Although there was a lot of mumbling, no one said anything that could be heard in the videotape recording. Nathalie moved the purple square (as shown in Fig. 3a) and then asked the students to predict what would happen if she moved the blue square (right-most square in the third row of Fig. 2a). One voice said, "it will make a white square". Michael's response was inaudible, but it was accompanied by a two-handed gesture in which his palms face each other and the thumb and fingers on each hand are a mirror image of each other (see Fig. 3b). After making this gesture, he began clapping his hands (and knees). He clapped once with his hands, once on his knees, four times on his hands, once on his knees, once on his hands, once on his knees, three times hands, one knees, he brings his hands together and rubs them, once on his knees and finally he brought his hands together and rubbed them.

Michael was sitting on a chair at the back of the room, while everyone else is on the floor, so the other students could not see what he was doing. The initial gesture, and then the more dynamic gesture of clapping (but without sound), both express the sense of twoness seen before, as well as the motion in the two hands coming together. Symmetry has moved from the projector screen to his own two hands. This shift of symmetry from the screen to Michael's own space, is a movement of symmetry, visual to physical, discrete to continuous, technology to body. The students then directed Nathalie to move the purple and blue squares up. Then, despite not being able to see Michael's gestures, Jessica made a clapping gesture as she explained that she wanted Nathalie to "put them both together", in order to fill in the white spaces on the top row of Fig. 3c. Several other children were asking for the same thing, but also mentioning the blue, purple and red squares. Ava made the same gesture shown in Fig. 3d and then brought her other arm up and moved her two hands together, as in a clapping gesture. She was speaking as she made the gestures, but her voice could not be heard above the other voices.

Ava's sequence of gestures combines the ideas of twoness and of bringing together (which includes the motion). But it also echoes the gestures of both Jessica and Michael, even though it is not at all evident that she had seen them. The bringing together does not just describe the way in which the coloured squares on either side of the line move, but in both Jessica and Ava's interventions, it also describes the filling of the white spaces. Furthermore, the point of contact of the two hands clapping can be seen as actualising the line of symmetry—that is, bringing forth an object (the line of symmetry) that was not previously present. Indeed, when the two squares touch, like when the hands touch, they do so right on the line of symmetry. Although that line is visible on the sketch, it has not been referred to yet.

## *Symmetry as Making Recognizable Shapes*

After a few squares had been moved on the screen, including the red one (twice) and the purple one (up) (see Fig. 4a), Nathalie again asked what had happened. Several children shouted out numbers, then Jonathan turned around and said "I" (identifying it as a recognizable letter). Jonathan continued by saying, "it's cutting the I" and lifted his hand and moved it down vertically. When asked what would happen if the purple square was moved up, another boy said, "it looks like a creeper". After moving more squares, as in the configuration shown in Fig. 4a, Nathalie moved the blue square on the bottom row to the right and someone said, "it's a T". Several children then began to ask Nathalie to move the blue square up so that it would reach the top row, eventually obtaining the diagram shown in Fig. 3c. At that point, several students said "whoa" and also shouted out "un T" (a T). Once there, as reported in the previous section, the students wanted Nathalie to move the squares so as to fill in the top row. Once the three squares had been moved into position, several students said, "that's a T" and one student said "awesome". Over the next few minutes, the children came up one by one to move other squares. Each time they did so, the other students commented on what "it looks like". For example, when the configuration in Fig. 4c was made, one child said, "it looks like a Chinese temple".

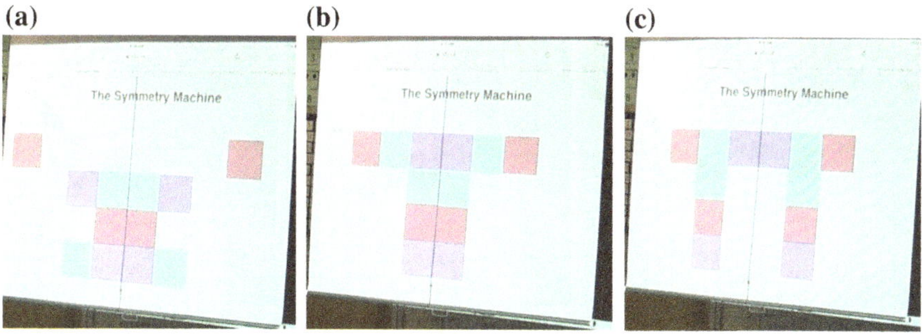


**Fig. 4** **a** Two reds and a purple moved; **b** a 'T'; **c** Chinese temple

From a focus on the local, that is, on individual squares and how and where they move, there's now a shift to seeing the collection of 12 squares as a whole, to a more global perception. The global perception of symmetry is typically the first one that students encounter, when they are asked to consider the symmetric nature of a heart, for example. In this case, it is only when the squares are moved into a certain configuration, that the children begin to talk about one whole shape, referring to it as a letter of the alphabet and a Chinese temple. This idea of the Symmetry Machine producing letters initiated the recognition of a T-like shape and the subsequent movement of the square to produce Fig. 4b. Thereafter, the talk was focussed on what the configuration looked like rather than on the number of white squares or the twoness.

Over the next ten minutes, the children worked on the task of trying to create a diagram that has been taped to the whiteboard using the Symmetry Machine. In turn, they explained where the squares should move. They described moving squares on the left as well as squares on the right. Throughout, they focussed on the overall, global configuration. They got several of the squares into place, but some children began to engage in other activities, so the classroom teacher decided that it was time to move to the pairwise activities with the iPads.

## *Symmetry as Joint Movement*

In the pairwise activity, the children were asked to reproduce a series of symmetric diagrams that were given to them on a piece of paper. Alik and Ava were given the diagram shown in Fig. 5a (bottom of the figure). Alik initially said that they could not make it and pointed to a square that is not symmetric with its corresponding square, saying "that square should be here" (pointing to the purple square to the right of the line of symmetry and then to the white square to its right). Nathalie urged them to try anyway. Ava put her finger on the purple square to the left of the line and moved it down, towards the line. Then Alik put his finger on the purple square to the right of the line and started moving it towards and away from the line

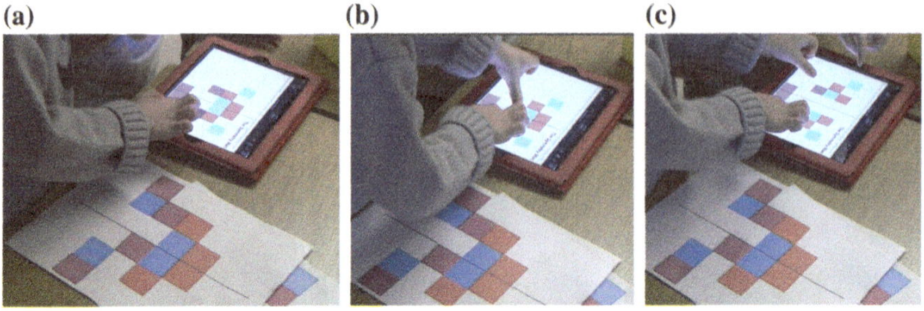

**Fig. 5** **a** Paper sketch; **b** thumb and index finger; **c** line of symmetry translated

(see Fig. 5a). He then put his thumb on the square (Fig. 5b) and moved both his thumb (left hand) and his index finger (right hand). He ended up moving the line of symmetry and translated all the squares on the left of the line one unit away from the line of symmetry (see Fig. 5c). Both Ava and Alik were surprised. He then used two index fingers to move the purple squares towards each other and Ava did the same with the two blue squares. They each moved the purple square several times, towards and away from the line and concur that they could not make the diagram. Alik said to Nathalie, "when we move one, it just ..." and then moves the purple square towards the line. He then pointed to the two purple squares on the piece of paper and explained that it's not symmetric because "that one (pointing to the purple square to the left of the line) is here" and "that one (pointing to the purple square to the right of the line) is on the line". Alik then went back to the Symmetry Machine and used one index finger, moving one purple square repeatedly back and forth. He then said, "there's supposed to be one there and one there" pointing to the right and to the left of the line.

## Symmetry as Lining up

Nathalie then offered a new diagram (Fig. 6a) and Alik and Ava each moved one square. They then paused and Alik said, "it's not symmetric". When asked why, Alik pointed to the purple square at the top left of the piece of paper. Ava pointed to that square too, then to the other purple square on the top right of the piece of paper (Fig. 6b). Nathalie asked "you can't make it?" and Ava shook her head and said, "because those two (placing one side of her hand on the page to form a diagonal line between the two purple squares) are supposed to be" (placing her two hands to form a line perpendicular to the line of symmetry (see Fig. 6c).

This sequence gives rise to yet new symmetry concepts. While it is tempting to see in the students' reasoning about the Symmetry Machine that they are showing awareness of the equidistant property and the perpendicularity property, it is evident from their actions, and especially their gestures, that the Symmetry Machine

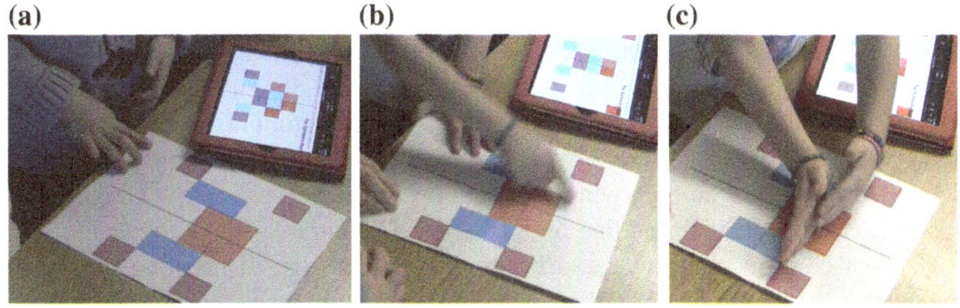

Fig. 6 a New diagram; b pointing; c perpendicular to line of symmetry

intervenes to give rise to new properties. In the first case, there is a shift from one-handed to two-handed dragging, then back again to one-handed dragging. In the first shift, Alik seems to want to force the squares to move in non-symmetric ways and when he sees that this is not possible (it just moves the line of symmetry), he goes back to the one-handed dragging, repeatedly making the squares come towards the line or move away. Therefore, it is less about the distance away from the line, which would emerge from a static configuration, than about the joint movement towards and away from the line. That the joint movement can be controlled by one square only emerges in the shift from the two-finger back to the one-finger dragging.

In terms of the perpendicularity, Alik and Ava barely move any of the squares from the Symmetry Machine before deciding that the diagram in Fig. 6a is not symmetric. Alik's continued pointing to the top left purple square suggests that he thinks it is out of place. The subsequent double pointing of Ava, which goes from one square to the other creates a virtual line that she then actualizes with her gesture. That line is not perpendicular to the line of symmetry (nor to the other 'lines' joining corresponding squares). With her second gesture, Ava shows what the correct line should look like, not necessarily in terms of the perpendicularity, but in reference to the pair of red squares that are already there.

## Discussion and Conclusion

Reiterating our objective of this chapter, we are not addressing epistemological issues of what was learned or how learning takes place, for this infers the concept either to be a priori, independent of context and tools or to be the result of the mediation of tools (which might subsequently become expunged from knowledge). Instead we focus on the *Web Sketchpad* as part of an experimental apparatus that can highlight the indeterminacy of material engagement.

We used a diffractive apparatus because it helps us see how meanings about symmetry are not only entangled with the physical, but can also be considered intrinsically indeterminate. As noted in the highlighted episodes, symmetry takes different actualisations and meanings at different times. It is the varying material effects that allow us to follow the concept and not individual understandings. For example, when Jonathan moved his hand up and down along the line of symmetry saying "it's cutting the I", this gesture initially emerges from his observation of the projected sketch. His gesture is also aligning with the line of symmetry on the screen so that when he is moving his hand up and down he is expressing, and in fact, materially actualising a line of symmetry. The intra-action of his gesture and the sketch confirm each other and become an assemblage of meaning making of symmetry. While the focus on the concept of symmetry enabled us to carry out the diffractive analysis, we found it more difficult to write about the assemblage. Indeed, our writing, following conventional style, evoked individual children doing individual actions (gesturing, dragging, speaking). More methodological innovation will be required in future work in order to adequately follow the entailments of our theoretical perspectives.

Nonetheless, through using *Web Sketchpad* as part of a diffractive apparatus, new meanings of symmetry emerged. The space provided by the Symmetry Machine created new ways of instantiating symmetry. The concept of symmetry was expressed in numerous ways, as reflected in the subsection titles of the previous section: as twoness, as making holes, as bringing together, as making recognisable shapes, as joint movement, as well as lining up. In each case, students were intricately tied to the Symmetry Machine and the activity by forming gestures and body motions (e.g., like clapping) expressing symmetry in both a material and indeterminate way. In the clapping gesture of Ava, Jessica and Michael, the line of symmetry is actualised in bringing their hands together. Symmetry is seen as instantiated in movement, in alignment and in bringing together. Our analysis, which focused on the shifting nature of the concept of symmetry, enabled us to attend more carefully to the gestures that emerged (from movements of objects on the screen, as well as from child to child) over the course of the lesson. We connect these gestures to Michel Serres' (2011) assertion that "there is nothing in knowledge which has not been first in the entire body, whose gestural metamorphoses, mobiles postures, very evolution imitate all that surrounds it" (p. 70). Serres is suggesting that the origin of knowledge is *not* understanding, which is about explanation and inference, but instead, is in the building of memory in the body, through gestures and movement. The mobility of the DGE can thus be seen as crucial to the changing ways in which the children moved and the continued new meanings for symmetry that emerged.

We are interested in challenging the a priori notion of symmetry and drawing attention, in particular, to the mobile device and how it influences and makes symmetry in different ways. We do not tie things up succinctly. Indeed, wrapping up this study with a cohesive conclusion is to contradict the very assumptions we began our study with. We merely tend to the assemblage of the diffractive apparatus and embrace the new and becoming of symmetry. By focussing on these new meanings, we did not track the ruptures and losses of meanings that resulted from the changing assemblage, such as the slipping away of the numerical value of two associated with the initial movements of the squares. In future work, more attention could be paid to this aspect of changing assemblages, to highlight the continuation of the mobility of symmetry.

## References


Artigue, M. (2002). Learning mathematics in CAS environment: The genesis of a reflection about instrumentation and the dialectics between technical and conceptual work. *International Journal of Computers for Mathematical Learning, 7*, 245–274.

Barad, K. (2007). *Meeting the universe halfway: Quantum physics and the entanglement of matter and meaning*. Durham, NC: Duke University Press.

Barad, K. (2010). Quantum entanglements and hauntological relations of inheritance: dis/continuities, spacetime enfoldings, and justice-to-come. *Derrida Today, 3*(2), 240–268.

Barad, K. (2012). On touching: The inhuman that therefore I am. *Differences: A Journal of Feminist Cultural Studies, 23*(3), 206–223.

Bartolini Bussi, M. G., & Mariotti, M. A. (2008). Semiotic mediation in the mathematics classroom: Artifacts and signs after a Vygotskian perspective. In L. English, M. Bartolini Bussi, G. Jones, R. Lesh, & D. Tirosh (Eds.), *Handbook of international research in mathematics education* (2nd revised ed., pp. 746–805). New York: Routledge.

Bornstein, M. H., & Stiles-Davis, J. (1984). Discrimination and memory for symmetry in young children. *Developmental Psychology, 20*(4), 637–649.

Braidotti, R. (2013). *The posthuman*. Cambridge, UK; Malden, MA: Polity Press.

Bryant, P. (2008). Paper 5: Understanding spaces and its representation in mathematics. In T. Nunez, P. Bryant, & A. Watson (Eds.), *Key understanding in mathematics learning: A report to the Nuffield Foundation*. Retrieved April 28, 2013, from http://www.nuffieldfoundation.org/sites/default/files/P5.pdf.

Clements, D., & Sarama, J. (2004). *Engaging young children in mathematics: Standards for early childhood mathematics education*. Mahwah, NJ: Erlbaum.

de Freitas, E. (2016). *Diffractive apparatus: Rethinking the design experiment in light of quantum ontology*. Washington, DC: Paper presented at the American Educational Research Association.

de Freitas, E., & Sinclair, N. (2014). *Mathematics and the body: Material entanglements in the classroom*. New York, NY: Cambridge University Press

Duval, R. (2005). Les conditions cognitives de l'apparentissage de la géométrie: Développement de la visualisation, differenciation des raisonnement et coordination de leurs fonctionnements. *Annales de didactique et sciences cognitives, 10*, 5–53.

Haraway, D. (1992). The promises of monsters. A regenerative politics for inappropriate/d others. In L. Grossberg, C. Nelson, & P. Treichler (Eds.), Cultural Studies. London and New York: Routledge.

Haraway, D. (2008). *When species meet*. Minneapolis, MN: University of Minnesota Press.

Jackiw, N. (1991, 2001). *The Geometer's Sketchpad* [Computer Program]. Emeryville, CA: Key Curriculum Press.

Lenz Taguchi, H. (2012). A diffractive and Deleuzian approach to analyzing interview data. *Feminist Theory, 13*(3), 265–281.

Ng, O., & Sinclair, N. (2015). Young children reasoning about symmetry in a dynamic geometry environment. *ZDM—The International Journal on Mathematics Education, 51*(3), 421–434.

Pickering, A. (1995). *The mangle of practice: Time, agency, and science*. Chicago: The university of Chicago Press.

Seo, K.-H., & Ginsburg, H. (2004) What is developmentally appropriate in early childhood mathematics education? In D. H. Clements, J. Sarama, & A.-M. Dibias (Eds.), *Engaging young children in mathematics: Standards for early childhood mathematics education*. (pp. 91–104). Mahwah, NJ: Erlbaum.

Serres, M. (2011). *Variations on the body*. [Translation: Randolph Burks]. Minneapolis, MN: University of Minnesota Press.

**Sean Chorney** is an Assistant Professor at Simon Fraser University in Burnaby, British Columbia. He is the editor of Vector, the British Columbia Association of Mathematics Teachers' journal. His research is focused on the material aspects of movement and gesture and the significance of such materiality in the construction of conceptual perspectives.

**Nathalie Sinclair** is a Professor and the Learning at Simon Fraser University. She is the author of several books, including co-author of *Mathematics and the Body: MaterialEntanglements in the Classroom*. She has also led the design of educational technologies, including the touchscreen app *TouchCounts*. She is the founding editor of the journal *Digital Experiences in Mathematics Education*.

# Touching Numbers and Feeling Quantities: Methodological Dimensions of Working with *TouchCounts*

**Francesca Ferrara and Ketty Savioli**

**Abstract** In this chapter, we discuss a classroom-based intervention that we carried out with a group of first grade children using a multi-touch iPad application called *TouchCounts*, as part of a research project aimed at developing children's number sense. We situate our work within an inclusive materialist approach to the study of mathematics education (de Freitas & Sinclair, 2014), which maps the various materialities at work in the mathematics classroom, as a way to attend to the relational entanglement of numbers, iPad and learners. Drawing on this vision, we analyse the methodological dimensions of using *TouchCounts*, conceptualising the application as an experimental means that allows us to better see methodological implications as they sustain mathematical encounters for learners. We investigate how event-like interventions emerge out of the material relations with the surrounding and engender new kinds of mathematical experiences with number. We focus particularly on a situation in which the number of children around a table became the variable under discussion providing the basis for emerging relational meanings of number.

## Introduction

In this chapter, we discuss an experience with a group of elementary school learners (six-year-old children) working with a multi-touch iPad application during their regular mathematics lessons. The experience is part of a wider research study aimed at improving mathematics teaching and learning in the primary grades, especially regarding the development of number sense. In addition, the study also aims to contribute to better knowledge of young learners' capacities for mathematical

F. Ferrara (✉) · K. Savioli
Università di Torino, Turin, Italy
e-mail: francesca.ferrara@unito.it

K. Savioli
e-mail: ketty.savioli@gmail.com

 
N. Calder et al. (eds.), *Using Mobile Technologies in the Teaching and Learning of Mathematics*, Mathematics Education in the Digital Era 12,
https://doi.org/10.1007/978-3-319-90179-4_13

understanding, in particular in the cultures of counting and early arithmetic. The novel iPad application used in the project is *TouchCounts*. Its multi-touch affordance is designed to encourage explorations with numbers in which children express themselves through fingers and gestures. Our interests in this chapter specifically are on the methodological dimensions of working with *TouchCounts* in the mathematics classroom. The rationale of our study originates from the direct experience of using the app in practice. This experience provoked questions about the kinds of tasks we asked the children to work with, the lines across which to develop classroom culture around number sense, and ways of inserting this activity into the mathematics curriculum. The methodological dimension has to do with these aspects emerging out of practice and shifts our attention to the challenge of reconsidering what constitutes mathematical activity more generally.

In the chapter, we focus on one specific aspect of the use of *TouchCounts* in the classroom, that is, what can be gained from the innovative touchscreen learning technology in terms of intervention and new possibilities for learning. The chapter, therefore, will address the following main issue: How does the use of *TouchCounts* change our interventions within the mathematics classroom, fostering new inventive ways of learning?

As mathematics education researchers, the main point we put forth is that a particular tool can be conceptualised as an experimental device with which we can better understand the methodological potential that sustains mathematical encounters for learners and engenders new kinds of mathematical experiences. Thus, we will eventually propose thinking of doing mathematics with *TouchCounts* as occasions for the children to change their habits of work within the classroom, to make sense of a new mathematics, to take unexpected directions in their mathematical activity.

## Theoretical Considerations

That the use of digital technologies in mathematics teaching and learning offers new ways of engaging with mathematical concepts and processes is not a novelty in mathematics education research. In fact, Calder and Campbell (2016), for example, underline that "the visual and dynamic elements of engaging mathematical thinking through digital technologies reposition both the types of knowledge and understanding required and the ways in which learning emerges, which simultaneously shape the learning experience in a range of interrelated ways" (p. 50). For the authors, the particular features of digital learning environments enable alternative ways to encounter, process, investigate and explain mathematical ideas. Fresh interest in the use of mobile technologies and apps in/for the teaching and learning of mathematics is becoming evident in the literature.

Calder (2015) asserts that learning through apps offers potential affordances for learning that are similar to those identified within other digital technologies. Calder and Campbell also claim that apps offer the opportunity for students to engage dynamically with mathematical concepts, while gaining immediate feedback to input. These technologies "activate the use of gestures, natural language, symbolic language, touching, tapping, dragging, artifacts etc. that enhance the student's sensory motor experience in space and time" (Santi & Baccaglini-Frank, 2015, pp. 226–227). Attard and Curry (2012) studied the integration of iPads into a grade 3 primary classroom and found that it influenced both teaching and learning practices and student engagement with mathematics. These authors additionally affirmed that "although it appeared all students were behaviourally and affectively engaged, not all were engaged on a cognitive level possibly due to a mismatch between their ability and the given task" (p. 80). Researchers also noted that children working with iPads seem to have high level of interest (e.g., Lange & Meaney, 2013).

While we share the positioning of these research studies about activity with mobile technologies as offering ways to afford new opportunities and spaces for teaching and learning mathematics, they tend to see the technology as that which causes learners to acquire existing meanings, therefore as subordinate to learners or practice. In contrast, we are more interested in a participationist vision of learning, which accounts for the relations between meaning and matter that emerge from learners' experiences with particular tools, and therefore treats knowledge as activity rather than representation. To work within this perspective, we adopt an inclusive materialist approach to the study of mathematics education (de Freitas & Sinclair, 2014). Inclusive materialism maps the various materialities at work in the mathematics classroom, with particular attention to the ways in which bodies are entangled with the material surrounding and the mathematics. It provides a way of attending to the relational entanglement of concepts, tools and learners, instead of seeing these as ontologically distinct entities. According to de Freitas and Sinclair, the human body does not have taken-for-granted borders. Instead, the boundaries between mind and matter, between the learners and the material surrounding, are mobile, constantly shifting, and reconfigure mathematical activity. For these researchers, concepts are also dynamic and indeterminate. Sinclair, Chorney, and Rodney (2016) take these assumptions to explore the social and affective dimensions of using *TouchCounts*. They avoid to see motivation "as something that begins activity so that cognition can kick in", or "joint activity as a means of sharing ideas developed individually" (p. 50). Complementary to the work of Sinclair et al. (2016), in this chapter we draw on an inclusive materialist vision of activity to focus on methodological dimensions of the use of *TouchCounts*. This gives us the opportunity to reconsider the research data that we have in our study, investigating the mathematical encounters of the students with *TouchCounts* as novel spaces for the emergence of unscripted possibilities in the intervention study.

This is also relevant in light of our commitment, as researchers, to better understanding and theorising the material dimension of intervention in the mathematics classroom. Stylianides and Stylianides (2013) argue that more attention is needed to research on classroom-based interventions in mathematics education. The term intervention refers here to the idea of "action taken to improve a situation" (Stevenson & Lindberg, 2012) and is used "in relation to the practice of teaching and learning mathematics in classrooms" (Stylianides & Stylianides, 2013, p. 334). According to this study, researchers tend to over simplify research in 'messy' environments such as actual classrooms if they feel this choice of context as possibly leading to their research being considered as methodologically poor. However, a high correlation seems to exist between classroom-based interventions and studies following a design experiment research methodology, due to the interventionist nature of design experiments (Cobb et al., 2003). Design experiments involve the design of an intervention and the study of its impact, for example by investigating possibilities for educational improvement, like when children engage in novel tasks. Interestingly, design experiments are also linked with the idea of close collaboration between researchers and teachers. This is relevant with respect to discourse we develop in this chapter, for which we refer to an intervention carried out together (both as educators, but one as researcher and the other as teacher).

We are not interested in shedding light on mechanisms of success, judging success or failure of activity, but precisely in the contingency of the event-like interventions as they are entangled with the specific app use. The inclusive materialist approach allows us to move towards this direction, helping us to study how material practice with mobile technology implicates unexpected lines of work within the classroom, which sustain speculative experiences and unscripted tasks that engender new kinds of mathematical encounters with the concept of number for learners.

## *TouchCounts* and the Enumerating World

*TouchCounts* is a multi-touch application for iPad designed by two Canadian researchers (Jackiw & Sinclair, 2014). It "is intended to offer an expressive environment in which learners could create and relate mathematical objects directly with their fingers and hands" (Sinclair & Heyd-Metzuyanim, 2014, p. 4). Essentially, it is a microworld for exploring number that offers two different worlds: an Enumerating world and an Operating world. They provide respectively ordinal and cardinal models of number. In this chapter, we draw attention only to the first world, which was used in our classroom-based activity. Every time one finger touches the iPad, a yellow disc is created on the screen, labelled with a numeral, and the numeral is said aloud. So, touch, sound and symbol come together. Each subsequent touch makes a new yellow disc appear with the next biggest numeral on it, which is also spoken aloud. The Enumerating world can be explored with or without the gravity mode on. When gravity is turned off, the discs stay where the touch was originally made, as shown in Fig. 1a.

As a multi-touch app, *TouchCounts* responds to the simultaneous touch of more than one finger. Therefore, in the case of a single five-finger touch, for example, five numbered discs appear simultaneously on the screen but aural feedback comes only for the last. Differently, five subsequent taps would create five numbered discs

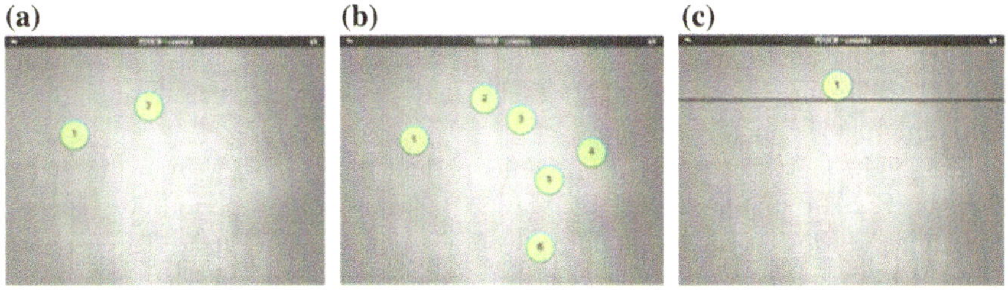

**Fig. 1** **a–b** The Enumerating world without gravity and **c** with shelf mode turned on

sequentially, counted one by one. This generally occurs when gravity is absent. Figure 1a, b present two frames of use of the Enumerating world. (From these images, however, it is impossible to tell whether the touches were made simultaneously or sequentially.).

When gravity is turned on, the discs fall down, then off the screen, unless they are 'caught' by the shelf (as in Fig. 1c)—which can be turned on or off. Tapping below the shelf makes discs fall away, while tapping above creates a disc that remains on the shelf, much in the same way that placing a book above a real shelf implies that it cannot fall. Thus, in order to have just a disc labelled 5 on the shelf, four taps below the shelf are first necessary—then a fifth tap, above the shelf. Doing this requires an awareness of the fact that 5 comes after 4. This highlights why, with or without gravity, in the Enumerating world, emphasis is put on ordinality, although the multi-touch, allowing for the recognition of both subsequent and simultaneous taps, also leaves room for cardinality.

## Context

The classroom-based activity, which is the focus of this chapter, involved a class of 24 children six-year-olds (10 boys and 14 girls), with low-medium socio-economic-cultural background, in a primary school in Northern Italy. At that time of the study, the children were attending grade 1. The activity was part of a wider research study aimed to provide directions for improving the teaching and learning of number sense, and the development of counting and early arithmetic (to date, the research engaged the children from grades 1–4). The study connects with the intended primary school mathematics curriculum that we find in textbooks. In the curriculum, number is first approached through numerosity, numeral recognition and writing, and counting of small quantities, while problematic situations about adding and subtracting are taken as a long-term aspect of learning number sense. However, since one goal of our research was to promote and provide opportunities for children to engage with new mathematical events and challenges, we chose to design our interventions—right from the beginning of primary school—in a way that challenged the sequential arrangement and alignment of the assumed curriculum or the progression of a hypothetical learning trajectory (e.g., Confrey et al., 2009). We much more thought of our experiments as occasioning stories (following Lockhart, 2009) to be told by learners and, through them, unfolding mathematical encounters.

The study took place over a 6-months period in 2014/2015, from the end of November 2014 to the middle of May 2015. During this period, we spent 9 three-hour sessions with the children working with *TouchCounts*. Three people were always present in the classroom: the regular mathematics teacher (the second author), one researcher (the first author) and one university student, who followed the activities in the context of her master degree work.

The activities were organised in various ways. There was group work, individual or pair written tasks, individual interviews for oral feedback and class discussion led by the researcher. In particular, for group work the class was divided into three groups of eight children each, all sitting around a table with a single iPad on it. This arrangement of the groups was determined by the fact that only three iPads were available in the classroom. The teacher worked with two groups, while the researcher worked with the remaining group. Otherwise, during class discussions, a unique iPad was available for use in the middle of the classroom, often connected to an interactive whiteboard allowing for screen projection in front of all the children. The individual interviews were conducted by the researcher in a silent corner of the school. Lastly, written tasks were mainly focussed on the diagrammatic activity of the children.

The university student used a camera to film all these phases of the study for the sake of analysis of classroom practice (in the case of group work, only one group was recorded). For us, this also had added value for design choices in the on-going development of the project, providing information about what was initiated by the children and did or did not emerge from previous activity.

In the first four meetings of the study, the children worked with the Enumerating world of *TouchCounts*, while in a second phase they started using the Operating world. The focus of this chapter is on the second day and, therefore, involved the use of the Enumerating world, with the gravity mode on and no shelf. More specifically, for this activity, the children were working in groups of eight around tables and were sharing an iPad, which was placed in the middle. The data presented in Sects. 5 and 6 (we will offer a translation of the dialogue into English) was selected because it revealed aspects of the intervention that are relevant in terms of entanglement with the physical environment (multi-touch app, arrangement of the children, single iPad use, etc.). Our goal in selecting the episode is to exemplify how interventions emerge out of the material relations with the surrounding and offer new opportunities for mathematical encounters. Through the analysis, we will also discuss what kind of mathematics arises from the assembling of children, numbers, researchers and *TouchCounts*.

## Ways of Making Five

In session one, the children played with the Enumerating world for the first time and explored it with the gravity mode turned on, experiencing the first relationships between their tappings and the counting of numbers appearing on the screen. In so doing, they essentially discovered that the app counts through engagement with the visible, the audible and the tangible. In fact, when the children produced numbers by finger tapping, they also heard the iPad speak numerals aloud and read their symbolic form on the yellow disc. Therefore, the production of numbers is entangled with the tap–*TouchCounts* voice–yellow disc. Then, the children began experiencing the making of small numbers, like five, nine and twenty-one, first sequentially and then all-at-once—making a number all-at-once is quicker than sequentially making a number. Note that, in the second case, making the number five can involve one single hand; however, making the number nine requires two hands; and, making the number twenty-one certainly requires the fingers of more than two children. These explorations brought forth discussions about different ways of making a given number, which was meant to encourage the use of multiple fingers at a time. In the end, the children were asked to individually produce drawings to show how they had made five.

During the second session, the first author was working with a group of eight children, again in the Enumerating world. The children had already used *TouchCounts* to make five all-at-once, each by tapping on the screen with the fingers of one hand, or using fingers of both their hands. After this activity, the researcher asked the children to work on a new task: the making of the number five with more than one child and constrained to using only fingers of one hand for each child. This task can be seen as contingently emerging out of the arrangement of the children and their entanglement with the multi-touch app. In fact, the number of children sitting around the table occasioned the inclusion of more than one child in the task. The simultaneity offered by the multi-touch affordance gave rise to new possibilities to generate the number five. These new possibilities in turn produced different potential encounters with five that required collaboration between the children for the creative making of the number with the app.

Passing from the two hands of one child to the right hands of two children was quite natural for the group, as if the right hand of the second child was borrowed from the left hand of one child who before had played alone. Even if this was the case, still new different ways of making five all-at-once with finger-tappings of the two hands arose from the physical use of the app and its intertwinement with the audible and the visible (a novel affordance of *TouchCounts*). These entanglements engendered again new possibilities of encounter with number five for pairs of learners. Number five started being experienced in a creative, mobile manner: instead of the full count or of the decomposition into 1, 1, 1, 1, and 1, opportunities for partitioning five into a couple of smaller numbers came into being—the partitions into 1 and 4 and into 2 and 3: (1, 4) and (2, 3), and vice versa (4, 1) and (3, 2), whether the order was relevant. This is very different from what is normally done in the classroom by counting and making five from a collection of five objects. Some children in the group began playing roles around the table, assembling in new different ways. Through the words "you make", the activity was reconfigured with two pairs of children entangled with different partitions of five (into couples of numbers) and with the audible feedback of *TouchCounts* repeating "cinque" ("five") for both finger-tapping combinations.

The presence of more than four learners around the table gave rise to a turn of attention to increasing the number of children involved in the task of producing five. The activity suddenly changed, entangled as it was with the multi-touch. New roles played around the table, new combinations of finger-tappings on the screen, new partitions of five emerged out of new collaborative attempts of making five. Unscripted bonds of fingers, quantities, table and app appeared, with the children first imagining and then creating new situations in which the fingers of three to four hands/children touch the screen to hear the iPad saying "cinque". The new partitions (2, 2, 1), (3, 1, 1) and (2, 1, 1, 1) came to be, for the group, new ways of seeing five, touching the screen with five taps and feeling the numerosity of five, implying the engagement of more children. It was at this time that the activity shifted once again, as we discuss in the next section.

## *Is There a Maximum Number of Children?*

Relating the increasing number of children to the corresponding number of fingers they might use to produce five is not a trivial task. When the coordinated bond of four children gave the partitioning of five into 2, 1, 1 and 1, the intervention of the researcher moved the children away from the physical use of the app, to focus on whether that was the only way of making five using the combined fingers of four children. Once more, the intervention arose from the exploratory assembling of children and the multi-touch app, which gave rise to the issue of order that the four children eventually followed in touching the screen. This implied a further shift of attention to the extreme situation in which order would never matter, that of a maximum number of children.

Researcher: Is there a maximum number of children with whom to make five, if we use only the fingers of one hand?

Pietro: We can do it so that one puts five (*turns to the iPad, which is in the middle of the table, and points to the screen with his left hand fingers open*) and stop! The others don't put anything.

Researcher: No no no, I want to know whether there is a (*emphasis*) maximum number of children.

Pietro: In which sense, maximum?

Researcher: Whether it is possible to use more children (*looks at the children*).

Pietro: No, there isn't (*convinced, looks away. Lucia, silent on the right side of the researcher, directs her gaze to the researcher and mimes a "no" shaking her left index finger in the air*).

Researcher: Cause, for now, we arrived at five with four children. Can we (*emphasis*) increase the number of children to make five?

Pietro: No, no.

Researcher: Can we have more than four children to make five?

Lucia, Pietro: No (*speak together*).

Alice: Yes! (*aloud. The other children look at her*).

Researcher: How?

Alice: Should we do it? (*points to the iPad in the centre, which was in stand-by mode. The researcher turns it on*)… one could do (*stands up*)… one child put one (*counts one with her right hand thumb up in front, looks at it; Fig. 2a*).

Pietro: No (*stands up*), she said that one has to put all the fingers of one hand (*opens his left hand fingers in the air in front, holds them with the other hand, looks at the researcher. Alice looks at him*).

Researcher: No, no, I said using only one hand.

Alice: One child (*aloud*) puts one.

Pietro: One child puts one (*looks at Alice. Alice turns towards him*).

Alice, Pietro: The other puts one, the other puts one, the other puts one, the other puts one (*standing up, keep the count together, taking each finger of one hand—right hand for Alice, left hand for Pietro. They look at each other and at each other's fingers; Fig. 2b, c*).

Researcher: How many children do we need?

Pietro: Five (*looks at the researcher, indicates five with his left hand fingers open up; Fig. 3a*).

Alice: No, let's make one, one, one, one, one (*beats on the table five times in five different positions; Fig. 3b*).

Pietro: One, one, one, one, one (*counts again each finger of his left hand, moves closer and closer to the centre of the table*). Five! (*emphasis*) I put one (*prepares his left index finger close to the screen of the iPad*).

**Fig. 2** **a** Alice: "One child puts one"; **b–c** Alice, Pietro: "The other puts one, the other puts one"

**Fig. 3** **a** Pietro: "Five"; **b** Alice: "let's make one, one, one, one, one"; **c** Fingers together

Lucia: One (*put her right index finger over the screen*).
Sofia: One (*adds her right index finger over the screen*).
Caterina: One (*adds her right index finger over the screen*).
Alice: One, two, three, four (*counts the fingers ready to touch*), five (*adds her right index finger; Fig. 3c*).
Pietro: Five. Go! (*all the five index fingers now touch the screen and the iPad says aloud "five"*).

There was a little pause, which provided some suspense before the researchers' next question.

Researcher: For you… six children?
Pietro: One puts one, one puts one, one puts one, one puts one and one puts one.
Alice: No, it's what we just made.
Researcher: Six children.
Pietro: It's not possible.
Alice: It's not possible.
Researcher: Why?
Pietro: Cause there are too many children… too many hands.
Alice: But one puts one more finger and it makes six.

We begin our analysis here with the material implication of the researcher's intervention in the explorative task of making five through the collaboration of many children. However, attention was drawn to the existence of a maximum number of children in a way that was unexpected for the children. Pietro first made present the other extreme situation, the easiest one of a maximum number of fingers constrained to the one hand of one child ("one puts five", "the others don't put anything"), and, therefore, of a minimum number of children, the one implied in the very first encounters with five and the iPad. The number of children is entangled with the production of five with the physical app ("We can do", and the eye and the hand both directed to the screen). Pietro then expressed surprise, asking to make "sense" of the maximum, as the researcher repeated her precise question. Therefore, the possibility of further increasing the number of children, with respect to the four who were implicated in the last experience of partitioning five, unexpectedly emerged from the researcher and her assembling with the children and *TouchCounts*. While Lucia and Pietro seemed convinced that this was an impossibility, Alice, silent so far, but following with her sight the different coordinated bonds of number five, suddenly spoke the possibility aloud ("Yes!"), summoning the attention of the others. We see at this point a change of perspective: the activity was reconfigured in a new activity of enumerating the fingers that each engaged child might use in order to accomplish the task. Alice stood up, occupying with her body a wider space than that around the iPad or over the table, so as to recall what the task was about: the partitioning of five through a collaborative tapping of more children. She saw the number of children in the number of fingers ("one child puts one"), but Pietro was confused about the one hand and all its fingers, and stood up as well to affirm his point of view.

The researcher's intervention emerged from the contingent doubt and helped quickly overcome the confusion. This prompted Alice to restart her counting of the children and Pietro to coordinate with her ("One child puts one", "The other puts one, the other puts one, the other puts one, the other puts one"). Their ways of enumerating together and moving and gazing towards each other were already a new way of partitioning five into ones and the collaborative single taps of five children to make five all-at-once ("let's make one, one, one, one, one" for Alice, "Five" and "One, one, one, one, one" for Pietro). The material entanglement of the children with the multi-touch world was evidently actualised: first Alice beat on the table in five different positions as to capture five different touches on the screen, then Pietro moved closer to the screen of the iPad, with his body and his left index finger, to be the one child putting one finger ("I put one"). Caterina, Lucia and Sofia partook in this relational movement that brought the children back to using the iPad —towards the centre of the table and closer to each other, preparing their right index fingers for the collaborative touch. Alice counted the ready suspended fingers ("One, two, three, four") and completed the set with the missing finger ("five"). The iPad was a site of engagement and coordination of more children. Therefore, the possibility of increasing the number of children to five emerged out of and in the ways of enumerating all the necessary fingers and finally checking the production of five ("Go!"). This moved the intervention of the researcher in the direction of investigating the impossibility of having a further increase of children ("For you… six children?"). Initially, Pietro extended the previous idea of using one finger for each child, but Alice pointed out that that was what they had just made, underlining this impossibility in talk. Alice and Pietro expressed the impossibility in two different ways: for the boy, it was the case of too many children, therefore too many hands; the girl explained it with respect to the previous encounter with five, which involved that putting one more finger the app would make six.

Five was now configured as a set of five ones and, contemporaneously, as the maximum number of smallest quantities into which five could be partitioned. This made apparent for the children first that a maximum number exists and, then, that it is the one that allows for the decomposing of five into the smallest quantities (impossibility of having six children tapping together with one finger). The children around the table became variable numbers: each of them could put from one to five fingers, acting in a specific moment as the specific number one to five. This was generative of unexplored possibilities and relational meanings of number arising from the activity. For example, if one child had acted as a three and another one had acted as a two, and they had switched their fingers/roles/order, the situation would have implied two different partitions of the same numbers, (3, 2) and (2, 3), and the privileged relation of commutativity. Changing to three children would have implied new configurations for number five, and the different privileged relation of associativity as emerging from the assemblages of fingers. We might have widened further the situation, for example exploring which number we would be able to partition into the smallest quantities whether we engaged all the children in the group. We might have worked with new bigger numbers, eventually eliminating the constraint of using only fingers of one hand for each child, and possibly thinking of the biggest number we might (de)compose using all the hands of eight children.

The number of children was a variable offered by the use of the multi-touch and the arrangement of the group, which implicated the researcher's subsequent interventions that changed this number keeping a given number as target of the activity with the Enumerating world (five, here). Even though the children started from numbers, we see in this episode how they worked on concepts such as variable and maximum, and how this work emerged from the contingency of their material entanglements.

## Conclusive Remarks

In this chapter, we attempt to investigate methodological aspects of working with *TouchCounts*. To this aim, we have discussed a classroom-based activity that involved groups of eight children in the same classroom working on the task of making five in the Enumerating world. We hope that our discussions shed light on the event-like nature of the researcher's interventions and on how they are contingent, entangled with the specific use of the multi-touch app and emerging out of the material relations with the surrounding. As such, these interventions engender new kinds of mathematical experiences for the children and occasion new encounters with the number five—a mathematical entity that normally does not receive such singular attention. The inclusive materialist perspective that we assume in our study allowed us to draw attention to the material entanglement of the researcher, the children and *TouchCounts*. It also helped us study how material practices with the mobile technology implicate new unscripted insights within the classroom, which sustain speculative engagement with the concept of number. It is surprising that while the first question and its follow-up would probably have been assumed to be easy for the children, they were not. The interesting thing is that this gives us a new way of understanding five because it gives us a new situation in which five is encountered.

In our activity, the multi-touch prompted the children and the researcher to attend to the number of children as a relevant variable in the activity of making five, focussing on the new activity of making five in different ways. Thoughtful investments of the children with the app were implicated in such a new activity, which became an unusual one of decomposition or partition of number five into smaller quantities. The partitioning is not unique, therefore implying that children might partition five in many different ways. In order to succeed at making different partitions of five, a focus on the sizes of the small quantities and the overall numerosity was needed. Relations between the number of children engaged and the number of fingers for each child had to be discovered. The various ways of seeing five emerged out of new bonds of fingers, tappings, numerals, quantities and counts, of new entanglements of five with the multi-touch and the children in the group. The activity was reconfigured through these new events, which originated new coordinated encounters with five for the children, as well as different partitions of five. Each new partition implied a new/different/wider set of children touching the screen with their fingers, in turn increasing engagement and collaboration within the group, and the app always pronouncing "cinque" as the invariant of the activity. Our children told stories of five that arose from the activity and the unexpected interventions, stories speaking about changing numbers of fingers/children in relation to changing partitions of given numbers. The metaphor of story captures the emergent unfolding of relational meanings of number sense, drawing attention to the material contingency of the activity with the iPad. As Lockhart (2009) claims:

> "Mathematical structures, useful or not, are invented and developed within a problem context, and derive their meaning from that context. Sometimes we want one plus one to equal zero (as in so-called 'mod 2' arithmetic) and on the surface of a sphere the angles of a triangle add up to more than 180°. There are no "facts" per se; everything is relative and relational. *It is the story that matters, not just the ending*." (p. 17, *emphasis in original*).

It is not that the task required the use of number five, as is usually the case in school mathematics. Instead, five derived a meaning from the particular situation and interventions (and not simply with respect to the activity of counting a collection of five objects). Therefore, the smaller quantities in the making of five had to be different rather than simple abstract things with no character, because they were actualised from different children. The five smallest ones became full of personality, so that there was Alice's one, Pietro's one, Caterina's one, etc. and what was added did come to matter for the children.

With the classroom-based activity with *TouchCounts*, our children began telling stories that involved numbers, fingers and multiple touches, but also changing numbers (variable and maximum were implicated). We like to see their mathematical doing not only in terms of creatively touching numbers but also in terms of imagining and feeling quantities, making new meanings about number sense emerge from the material engagement with the mobile technology and from the joint activity over time.

## References

Attard, C., & Curry, C. (2012). Exploring the use of iPads to engage young students with mathematics. In J. Dindyal, L. Cheng, & S. Ng (Eds.), *Proceedings of the 35th annual conference of the mathematics education research group of Australasia* (pp. 75–82). Singapore, SG: MERGA.

Calder, N. (2015). Apps: Appropriate, applicable and appealing? In T. Lowrie & R. Jorgensen (Eds.), *Digital games and mathematics learning: Potential, promises and pitfalls* (pp. 233–250). Dordrecht, The Netherlands: Springer.

Calder, N., & Campbell, P. (2016). Using mathematical apps with reluctant learners. *Digital Experiences in Mathematics Education, 2*(1), 50–69.

Cobb, P., Confrey, J., diSessa, A., Lehrer, R., & Schauble, L. (2003). Design experiments in educational research. *Educational Researcher, 32*(1), 9–13.

Confrey, J., Maloney, A., Nguyen, K., Mojica, G., & Myers, M. (2009). Equipartitioning/splitting as a foundation of rational number reasoning using learning trajectories. In M. Tzekaki, M. Kaldrimidou, & H. Sakonidis (Eds.), *Proceedings of the 33rd conference of the international group for the psychology of mathematics education* (Vol. 2, pp. 345–352). Thessaloniki, Greece: PME.

de Freitas, E., & Sinclair, N. (2014). *Mathematics and the body: Material entanglements in the classroom*. New York, NY: Cambridge University Press.

Jackiw, N., & Sinclair, N. (2014). *TouchCounts* [software application for the iPad]. https://itunes.apple.com/ca/app/touchcounts/id897302197?mt=8.

Lange, T., & Meaney, T. (2013). iPads and mathematical play: A new kind of sandpit for young children. In B. Ubuz, Ç. Haser, & M.A. Mariotti (Eds.), *Proceedings of the eight congress of the European society for research in mathematics education* (pp. 2138–2147). Ankara, Turkey: Middle East Technical University.

Lockhart, P. (2009). *A mathematician's lament: How school cheats us out of our most fascinating and imaginative art form*. New York, NY: Bellevue Literary Press.

Santi, G., & Baccaglini-Frank, A. (2015). Forms of generalization in students experiencing mathematical learning difficulties. *PNA, 9*(3), 217–243.

Sinclair, N., & Heyd-Metzuyanim, E. (2014). Learning number with TouchCounts: The role of emotions and the body in mathematical communication. *Technology, Knowledge and Learning, 19*(1), 81–99.

Sinclair, N., Chorney, S., & Rodney, S. (2016). Rhythm in number: Exploring the affective, social and mathematical dimensions of using *TouchCounts*. *Mathematics Education Research Journal, 28*(1), 31–51.

Stevenson, A., & Lindberg, C. A. (Eds.) (2012). *New oxford American dictionary* (3rd ed.). Oxford, UK: Oxford University Press. http://www.oxfordreference.com/view/10.1093/acref/9780195392883.001.0001/acref-9780195392883.

Stylianides, A. J., & Stylianides, G. J. (2013). Seeking research-grounded solutions to problems of practice: Classroom-based interventions in mathematics education. *ZDM: The International Journal on Mathematics Education, 45*(3), 333–341.

**Francesca Ferrara** is Associate Professor of Mathematics Education in the Department of Mathematics, at Università di Torino, Italy. Her research focuses on the role and uses of technology in mathematics teaching and learning; how theories of embodiment and the body have informed and changed our ways of doing research in the field; the role of visualization and imagination in mathematics thinking. Her recent work also examines how new perspectives offer reflections on mathematical practice, and gamification paradigms applied to mathematics education. She is one of the authors of the Italian translation of the book Lakoff, G. & Nunez, R. E (2000). *Where Mathematics Comes From: How The Embodied Mind Brings Mathematics*

*Into Being*. New York, NY: Basic Books, and editor of the volume *Emerging Perspectives on Gesture and Embodiment in Mathematics*. Charlotte, NC: Information Age Publishing. She is member of the scientific committee of the Italian Association for Research in Mathematics Education.

**Ketty Savioli** is a primary school teacher since 25 years, with a Master Degree in Mathematics from Università di Torino, Italy. She collaborates as a teacher-researcher with the mathematics education researchers at the university and she leads the Project AVIMES Piemonte for the assessment and improvement of the school system. She took part in many National projects about mathematics education and about learning and school assessment, funded by National agencies. Her research mainly focuses on the role and uses of technologies and laboratorial activities in the mathematics classroom and on the development of argumentative processes. She presented many research results in National and International conferences (like EARLI, PME, CIEAEM and ICTMT). In her recent work, she collaborates with the International TIMSS evaluation system for grade 4. She is member of the Italian Commission for the Teaching of Mathematics.

# Approaching Secondary School Geometry Through the Logic of Inquiry Within Technological Environments

Carlotta Soldano and Ferdinando Arzarello

**Abstract** The chapter illustrates a pedagogically innovative way of using mobile technology to support the transition from an empirical to a more theoretical and logical approach to geometry. We propose an approach that shows the possibility of discovering geometric theorem statements and appreciating their universal truth using a suitable pedagogical design that draws on the work of the Finnish logician J. Hintikka as well as on Dick and Zbiek's notions of pedagogical, mathematical and cognitive fidelities. We implement it through game-based activities, namely group activities in which, first, students play a game in a dynamic geometry environment (DGE) and then, guided by the questions contained in a worksheet task, investigate the geometric property on which the game is designed. In the worksheet task, students are asked to act as detectives using the game to investigate, formulate and check conjectures. In order to analyse the students' productions we use a cognitive model elaborated from Saada-Robert's psychological model, which properly describes the cognitive modalities and the empirical versus theoretical and logical approaches to geometry.

**Keywords** Game-based activity · Logic of inquiry · Game theoretical semantic Fidelities

## Introduction

In this chapter, we discuss an approach to elementary geometry in secondary school, which aims to support students' transition from an essentially empirical stance on geometric properties to a more theoretical and logical one. In fact, many researchers in the literature point to a gap between the two stances (Fischbein, 1982;

---

C. Soldano (✉) · F. Arzarello
University of Turin, Turin, Italy
e-mail: carlotta.soldano@gmail.com

F. Arzarello
e-mail: ferdinando.arzarello@unito.it

 
N. Calder et al. (eds.), *Using Mobile Technologies in the Teaching and Learning of Mathematics*, Mathematics Education in the Digital Era 12,
https://doi.org/10.1007/978-3-319-90179-4_14

Duval, 1991; Healy & Hoyles, 2000; Reiss, Klieme, & Heinze, 2001; Dvora, 2012; Thompson, Senk, & Johnson, 2012). All these studies, based on different arguments, emphasise the big distance between the usual forms of argumentation and the formal aspects of proof. (Toulmin's model 2003 is used for discussing this difference in depth.)

Our approach recognises the gap but is oriented towards the possibility of cognitive unity (Boero, Garuti, & Pedemonte, 1996, Arzarello, Bussi, Leung, Mariotti, & Stevenson, 2012) and draws heavily both on the work of Hintikka, the *Logic of inquiry* (LI) and the *Game Theoretical Semantic* (GTS). It also connects to the work of Dick (2008) and Zbieck et al. (2007), as elaborated by Larkin (2015), which concerns *Fidelities*. The former has allowed us to rethink an approach to the logical aspects of mathematics, which is more akin to students' ways of thinking since it is based on game-based concepts. The latter has allowed us to properly frame our approach within a technological environment, so as to evaluate the degree of fidelity with which students can learn geometry within a DGE.

In this chapter, we first provide a sketch of the Fidelity, LI and GTS frameworks, then we describe our methodology and research design. After this introductory part, we will illustrate our research using on an example activity that was used in a research experiment with 7th grade students: first, we will describe it using an a priori analysis (Artigue & Perrin-Glorian, 1991), then we will show some excerpts of the a posteriori analysis of one of the videotaped groups of students. Finally, we will discuss the main points and findings of the research.

## Theoretical Framework

We introduce here the two main theoretical frameworks upon which we based our research and the concrete way according to which they have been adapted to our design. First, we will discuss the LI framework and then the Fidelities one.

### *The Logic of Inquiry*

LI is a new form of logic developed by the Finnish logician Jaako Hintikka (1999). The idea at the base of this logic is not new: it consists in conceiving of rational knowledge-seeking as implicit or explicit questioning and of the production of logical inferences as the result of inquiry-interrogative processes.

LI is a form of reasoning that is often used in detective stories by investigators. Using an extract of a Sherlock Holmes episode, Hintikka illustrates the equivalence between a deductive argument and a chain of questions by transposing the former into the latter. The model of LI is elaborated in the form of a game, called an *interrogative game*, between an idealized *inquirer* and a source of answers, called *nature* or *oracle*. The oracle can be "the database stored in the memory of a computer, a diagnostic handbook, a witness in a court of law, or one's own tacit knowledge partly based on one's memory" (Hintikka, 1998, p. 32). The inquirer, starting from a given theoretical premise, should establish a given consequence. At each stage of the game, the inquirer can make a deductive move or can address a question to the oracle. If the oracle responds, the answer can be used as an additional premise; such a move is called an *interrogative move*. The answers that introduce new entities generally assume the form of an abduction (Peirce, 1960).

LI changes the usual way according to which formal logic is presented, but Hintikka shows that his approach is perfectly coherent with the standard one. LI is more akin to the "natural" ways according to which people (students included) reason in everyday situations. This is particularly evident in problems where people are asked to solve mathematical games. Hintikka systematises this showing that LI is particularly suitable to shape reasoning in game theory: he elaborates this issue theoretically, which led to his Game Theoretical Semantic. The GTS is a form of semantic whose truth definition relies on the concept of strategy in game theory. The truth of a sentence is established through *semantical games* (Hintikka, 1998) which are two-player games between a verifier and a falsifier. We illustrate their dynamics with the following formula.

$$\forall x \, \exists y S[x, y]$$

In the *semantical game* associated with this formula, the falsifier chooses a value $x_0$ for $x$ and the verifier tries to find a value $y_0$ for $y$ such that $S[x_0, y_0]$ is true. According to Hintikka, the finding of a suitable $y_0$ is a veritable test case of the truth of the sentence if "the value $x_0$ of $x$ is chosen in the most unfavourable way as far as the interests of the verifier are concerned" (Hintikka, 1998, p. 24). Hence if there exists a winning strategy for the verifier of the game then the formula is true; otherwise, it is false. The game theoretical definition of truth is very different from the usual Tarski recursive definition (1933, 1983). Tarski's definition follows a bottom-up model: starting from atomic formulas, it shows the truth of complex ones (for example, the truth of a formula like $A \wedge B$ depends on the truth of A and that of B). Hintikka's definition follows a top-down model: starting from a complex formula, it shows its truth moving towards the atomic ones, applying the principles and the rules of semantical games.

## *Fidelities Issues*

Larkin (2015) examined 54 apps aimed at developing geometrical conceptual understanding and showed that they are generally limited in supporting such a goal. He based his evaluation on the constructs of *pedagogical, mathematical* and *cognitive* fidelity (Dick, 2008), which are crucial components in students' learning of geometrical concepts.

*Pedagogical fidelity* is defined by Dick as "the degree to which a student can use a tool to further their learning" (p. 343). He suggests that "a pedagogically faithful tool will lend itself to describing moves in terms of interactions with the mathematics (e.g., "I graphed this function", "I created this triangle" or "I measured this area") rather than in term of interactions with the tool (e.g., "I went to this menu", "I change this mode", or "I set the preferences to" etc.)" (pp. 334–335). *Mathematical fidelity* is defined by Zbiek, Heid, Blume, and Dick (2007) as the "faithfulness of the tool in reflecting the mathematical properties, conventions, and behaviours (as would be understood or expected by the mathematical community)" (p. 1173). *Cognitive fidelity* refers to "the faithfulness of the tool in reflecting the learner's thought processes or strategic choices while engaged in mathematical activity" (Zbiek et al., 2007, p. 1173). Larkin remarks that "cognitive fidelity can be viewed largely in terms of the external representations provided by the tool" (p. 342).

The three fidelities are not independent of each other: in a certain sense, it is important to consider, so to speak, second-order fidelities, namely whether and how the three components are consonant and related to each other. This can be ascertained with an a priori analysis of their mutual relationships to elaborate a specific task, e.g. checking the consonance between the mathematical and the cognitive fidelity and consequently designing the pedagogical one. In the design of the pedagogical fidelity, the role of the teacher is also crucial: no tool is independent of the teacher's implementation and orchestration of the didactical situation (see the discussion of this point in Arzarello et al., 2012). The analysis allows a researcher to elaborate a pedagogical design, where the (first and second order) fidelities are considered and its validity can be checked based on the data from teaching experiments in classrooms, where it is concretely developed.

These considerations constitute a methodology for designing suitable tasks within technological environments, which satisfy the three "fidelity tests", as described in Larkin (2015). Our project, is based on this methodology and, as pointed out in the introduction, takes into considerations the difficulties between the usual mathematical approaches of arguing and proving in mathematics (mathematical component) and the behaviours of students (cognitive component) who are given proofs in the classroom.

Our study tries to bridge the gap pointed out in the research mentioned above between the cognitive and the mathematical fidelities. We have found a solution from the mathematical side in Hintikka's (1998, 1999) machinery of LI and of GTS, as sketched out above. In fact, Hintikka's framework allows for a perfect consonance between the standard frame of logical deductions in mathematics and that of games. This also permits a suitable approach to proof from the cognitive side.

Following the approach of the LI and the GTS we have developed game-activities implemented on a tablet or a computer using the DGE GeoGebra. Our game-based activities satisfy the pedagogical fidelity within a complete consonance frame with the other two fidelities. These activities invite secondary school students to play a game in the DGE and then to inquire about its geometrical meaning in order to discover the statement of the theorem on which the game is based.

In order to analyse students' cognitive processes while they are involved in mathematical activities, we partially modified Saada-Robert's (1989) cognitive model in order to adapt it to the analysis of game-based activities. This model has already been used by Arzarello et al. (2002) in order to analyse the cognitive modalities activated by students involved in DGE-based dragging activities. This tool allows us to analyse students' cognitive processes and draw some conclusions that corroborate our hypothesis that the game-based activities enable shifting from empirical to more theoretical and logical approaches to geometry.

## Methodology and Research Design

Seven game-based activities were developed and used in an experiment in three classrooms during 2015 and 2016. The design of the game-based activities was tested and reformulated through pilot studies developed in 2014. In each experiment, the first game-based activity was preceded by an introductory lesson in which an inquiry approach to mathematics was presented through the analysis of an excerpt of a Sherlock Holmes dialogue. The students involved were in grades 7, 9 and 10. Classrooms, schools and teachers were chosen according to the availability of tablets in the school. The teachers involved in the research suggested which students to videotape, choosing those who do not get disturbed by the camera. The students knew that the game was meant to guide them in the discovery of the statement of a geometric theorem and that they should behave as detective in the DGE environment.

A game-based activity consists in a game that students play in groups of three using the GeoGebra tablet App and a worksheet task. The teacher formed the groups by putting together students who have similar mathematical competences. Each group had at its disposal a tablet or a computer and the worksheet task. Before starting the activity, the teacher read the rules of the game to the students, ensuring that everyone understood.

The rules of the games are designed to trigger the dynamics of Hintikka's semantical games, as sketched in the theoretical framework. The worksheet task was designed to guide students in the geometrical interpretation of the game and in the discovery of the property on which the game is based.

In the game, two players, a verifier and a falsifier, play against each other making their moves in turn according to two opposite goals. Each one controls a particular dynamic object and has to reach a specific goal. When the verifier reaches his/her goal, he/she produces a configuration that shows a specific invariant,[1] for example, in the game that we will describe below ("The game of two circles"), he/she always creates "tangent circles". The falsifier, through his/her moves, tries to realise the

[1]By invariant, we do not mean a property preserved by dragging, but a property repeated each time a player reaches the goal.

opposite goal: destroying the invariant produced by the verifier and creating a new configuration without it: in the example, the falsifier creates "non-tangent circles". Starting from the new configuration, the verifier drags objects in order to produce a new configuration that shows the invariant property again. Theoretically, both the verifier and the falsifier can always reach their goals, hence the end of the game is established by an hourglass time limit. Note that the players do not know the geometric invariant (e.g. tangent and non-tangent circles) they create and destroy through their moves. For example, the verifier's goal is moving a slider so that two values shown in the GeoGebra window are equal, while the falsifier's goal is moving another slider so that the two values are different.

The design of the games required the adaptation of a logical frame for educational purposes. It was the most challenging and creative part of the study. The design of the games makes students perceive geometric properties and conditional links between them and experience the impossibility to find a counterexample to their existence. Games not only help students grasp in a natural way the logic of quantifiers but also make mathematics more interesting, engaging and motivating. The importance of using game for educational purposes has been deeply studied by many researchers (Prensky, 2001; Swan, 2012; De Freitas & de Freitas, 2013).

The geometric theorem on which the game is based is not part of the classroom knowledge; the students are supposed to discover it by acting as detectives in the DGE: they must investigate what properties their slider controls, what the values shown represent and use them to understand the geometric meaning of the game. The didactical purpose of the game-based activity is two-fold: on the one side, gaining geometrical knowledge, and on the other, appreciating the universal truth of the discovered geometrical theorem statement through the game dynamics. Each game-based activity concludes with a classroom discussion in which the discoveries of the groups are shared and commented on by the teacher.

In every game-based activity we videotaped two groups of students while they played the game and answered the questions contained in the worksheet task. The collected data were analysed using a cognitive model that characterises the sequence exploration-conjecture-checking according to six different modalities: *ascending, descending, neutral, detached, logical control* and *deductive modality*. The ascending and descending modalities are based on the psychological model developed by Saada-Robert (1989) and already used for the analysis of mathematical activities involving dragging in a DGE (Arzarello et al., 2002). In order to properly identify the different modalities, we considered a variety of simultaneous students' productions: statements (verbal and written), actions in the DGE and representations.

- The *ascending modality* characterises the cognitive processes during the exploration phase towards the formulation of a conjecture: the situation on the screen is explored using the dragging tool with the aim of finding interesting facts.
- The *descending modality* characterises the cognitive processes during the checking phase, it is different from the formulation of the conjecture since it

concerns the control on the conjecture: the situation on the screen is now explored using dragging with the aim of checking the conjecture.

- The *neutral modality* characterises the cognitive processes during the formulation of a conjecture [possibly as an abduction (Arzarello et al., 1998)]. It characterises the passage from the ascending to the descending modality, marking the evolution of the way the subject is looking at the situation, from a "discovering investigating" to a "checking investigation".

When students' cognitive processes are described through the ascending/neutral/descending modalities, the game plays the role of an *oracle* while the students are the inquirers that pose implicit or explicit questions in the form of a controlled experiment. The *logical control* is a form of descending modality in which the control on the conjecture is more logical and less empirical: it does not involve the concrete use of the game, but just a verbal elaboration of facts previously observed within the game. The *deductive modality* is another form of descending modality in which the control on the conjecture is made with the Euclidean theory and not with the empirical experience or the elaboration of empirical experience made within the game.

In order to distinguish the cognitive modality that characterises the production of a conjecture regarding facts that are observed on the screen (neutral modality) from the production of a conjecture as the result of a mental elaboration of facts observed at a different moment of time, we introduced the *detached modality*. For example, in "the game of the two circles" if, in order to answer a worksheet question, students explored the situation using the game and formulated the conjecture-answer based on what they observed on the screen, we associate to these productions an ascending-neutral cognitive modality. On the other hand, if students formulated the conjecture-answer just by rethinking and elaborating what they had observed in a previous moment, we associate to this production a detached cognitive modality.

When students' cognitive processes are described through the detached/logical control/deductive modality, students' knowledge and/or the conceptual elaboration of the game experience play the role of oracle. This lens helped us to investigate whether and how the game-based activities triggered an evolution from students' perception to their statement of abstract ideas.

## The Game of the Two Circles: A Priori Analysis

The game-based activity presented in this chapter was used in an experiment with a classroom of 7th grade Italian students. The geometric property on which the game-based activity is designed regards the relationship between the reciprocal positions of the two circles and the sum/difference between their radii. The GeoGebra window is divided into two parts (see Fig. 1): on the right, there is the numerical window with sliders and measurements; on the left, there is the graphic window with the representation of the geometric objects.

**Fig. 1** Game-based activity 1

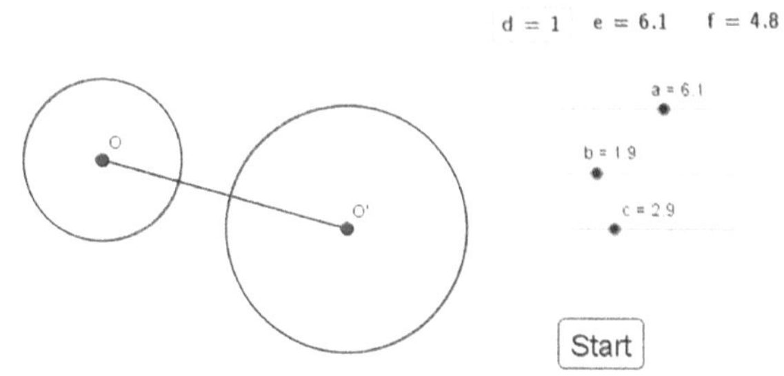


Sliders $a$, $b$ and $c$ control respectively the segment OO', the radius of the circle with centre $O$ and the radius of the circle with centre $O'$. The values $d$, $e$, $f$ are respectively the difference between the lengths of the radii, the distance between the centres and the sum of the lengths of the radii. When students drag sliders $b$ or $c$, they can observe the synchronic variation of one circle and of the values of $d$ and $f$.

Each group plays the game using one tablet; the game is opened in the GeoGebra App, while the rules of the game are contained in the worksheet whose translation is shown in Fig. 2.

Note that this game applies the formula "$\forall x \exists y | S[x, y]$" to the sliders $b$ and $c$ and to the values of $d$, $e$ and $f$ as follows: $x$ is a value of the slider $c$ and $y$ is a value of the slider $b$ so that the meaning of $S[x, y]$ is that "$e = d$ or $e = f$". The interpreted formula of the game is: 'For all values of the slider $c$ there exist a value of the slider $b$ such that $e = d$ or $e = f$'.

The students do not know a priori the meaning either of the sliders $a$, $b$ and $c$, or of the values $d$, $e$ and $f$: they can only see the effect of changing their values by looking at what happens in the left window (e.g. if $a$ diminishes, the length of OO' diminishes). Referring to Laborde's (2001) categorisation of the different types of

Player **B** controls slider *b*, player **C** controls slider *c* while player **A**, the referee, controls slider *a* and the hourglass. The goal of player B is to make $e = d$ or $e = f$, the goal of player C is to make $e \neq d$ and $e \neq f$.
At the beginning of each match, the referee chooses the value of *a* and turns the hourglass over. Each time a player reaches his goal, the referee turns the hourglass over and the turn moves to the opponent. If the player cannot reach the aim within the time on his/her hands, he/she loses. Take notes of the result of the matches in the following tables. Sign 1 point to the player who wins the match and 0 point to the one who loses. When you finish play sum up the partial score in order to establish the winner.

| | Score of the single matches | Total score |
|---|---|---|
| Player B | | |
| Player C | | |

After you play for some minutes, switch roles and play again.

**Fig. 2** Description of the game of the two circles

DGE tasks, we can identify it in the fourth category, since the game creates a sort of black box that could not be investigated in a non-digital environment.

During each match, the distance between the centres O and O' is kept fixed while the players vary the length of the radii. Player C is the falsifier of the game; s/he starts the match by dragging slider $c$ so that $e \neq d$ and $e \neq f$. In this way, the graphic window shows an example of non-tangent circles. Then player B, the verifier, makes a move by modifying the values of $d$ and $f$ through slider $b$ so that $e = d$ or $e = f$. When B accomplishes this move the graphic windows shows an example of tangent circles. In the graphic window, the circles look externally tangent when $e = f$ and internally tangent when $e = d$ (see the verifier standard example shown in Fig. 3). Conversely, each time player C reaches her/his goal s/he produces an example of (internally or externally) non-tangent circles. In the graphic window the circles can look secant, internal or external (see the falsifier standard example shown in Fig. 3). Since the interval of the sliders can take values from 0 to 10, players can also produce degenerate configurations. When player C moves slider $c$ to the value 0, the circle with centre O' degenerates to the point O' (see the falsifier non-standard example shown in Fig. 3). Player B, in order to win, moves slider $b$ so that the degenerate circle O' belongs to the circle with centre O (see the verifier non-standard example shown in Fig. 3).

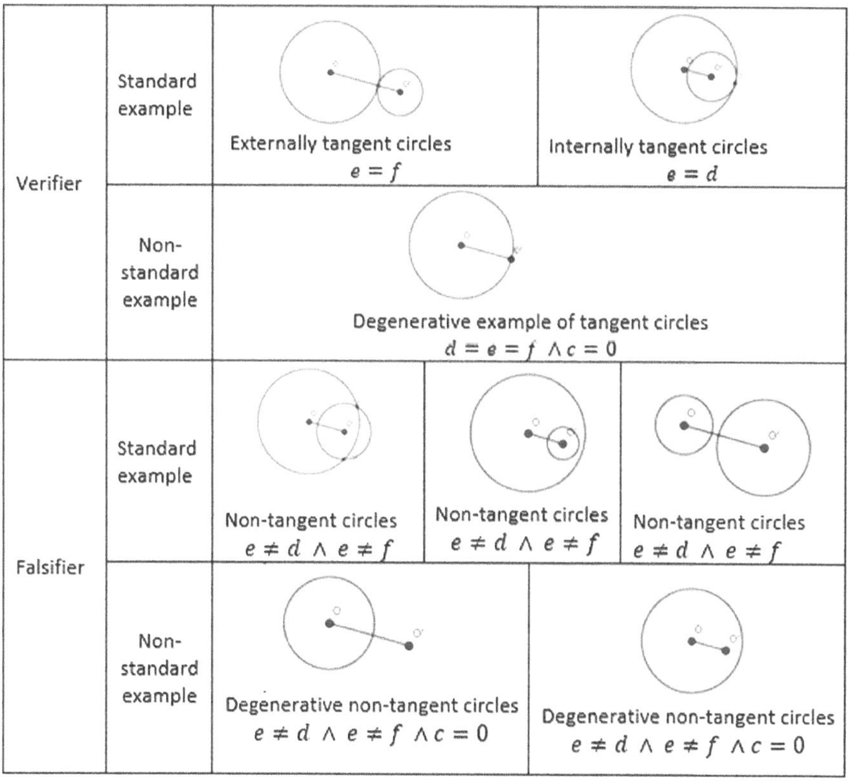


**Fig. 3** Example space of the game of the two circles

Answer these questions using geometric terms:

1. Which are the mutual positions between the two circles each time player B reaches his aim?
2. Which are the mutual positions between the two circles each time player C reaches his aim?
3. What do the sliders *a, b* and *c* represent?
4. What do the value of *d, e* and *f* represent?

**Fig. 4** Worksheet task of the game of the two circles

After playing the game the students are asked to answer the questions contained in the worksheet task (see the translation provided in Fig. 4).

The questions are intended to help students shift their frame of reference from the game to the geometric theory. The first two questions are meant to link the numerical window to the graphical one. In this way, students can discover the invariant configurations that characterise their moves. While playing, the focus of the students' attention is on the values *d*, *e* and *f*. Consequently, it can happen that the players do not notice the type of configuration produced in the graphical window.

Question number three is intended to link the values of the sliders to the measurements of precise geometric elements: the radii and the segment OO'. In order to answer this question, the students should observe the elements that change and the elements that remain invariant when they move the sliders.

Finally, question number four requires students to link the values of *f* and *d* to the sum and difference of the radii and the value of *e* to the distance between the centres. The answer to this question is more difficult to discover than the answers to the previous questions. In order to succeed, students must investigate the game situation in more depth.

The last question contained in Fig. 5 requires students to use discovered facts (answers to previous questions) and known facts (definitions of reciprocal position between circles) in order to discover the properties of sum/difference of radii and the distance between the circles' centres within the different circles' reciprocal positions.

## A Posteriori Analysis of the Activity

In order to investigate the role of the game-based activity in switching students' empirical approach to geometry to a more theoretical/logical one, we analyse some extracts of the actions and dialogues of the videotaped group made by An, Gi and Fe. According to our analysis, both groups were quite similar in terms of their productions, so we could have chosen either one to present as data here. The students had played the game for some minutes using the hourglass to establish the winner. After playing, they moved to the worksheet task. The following excerpt reports the translation of the dialogue of students while they answered the first question. We use parentheses to describe students' actions and square brackets to

Write on the same row of the following table mathematics terms and formula that are related to each other. *d(O,O')* stands for the distance between the centres of the circles, *R* stands for the greater radius and *r* for the lesser ones.

| A | B |
|---|---|
| Secant circles | $d(O,O') = R + r$ |
| Internally tangent circles | $d(O,O') > R + r$ |
| Externally tangent circles | $R - r < d(O,O') < R + r$ |
| Internal circles | $d(O,O') = R - r$ |
| External circles | $d(O,O') < R - r$ |

| A | B |
|---|---|
| | |
| | |
| | |
| | |
| | |

**Fig. 5** Last question of the game of the two circles

specify things that the students left implicit, but that can be useful to the reader for understanding the meaning of the sentence.

An B reaches the goal… *(creating a verifier's non-standard example, see Fig. 3)* It [the circle with centre O] could pass through O'

Gi No! Every time it [*d* or *f*] is equal to *a*, it [the circle controlled by B] has at least one point in common with the circle [controlled by C] *(looking at the figure produce by An)*
No it [the circle controlled by B] always has one point in common with the circle controlled by C *(moving slider b)*

An was in the *ascending modality*. He used the game as an *oracle* for exploring and formulating the conjecture. In this process, he created a degenerate configuration of tangent circles, which did not help him in the generalization of the answer. His approach to the game was empirical and not theoretical. Gi did not agree with An and formulated a new conjecture in *detached modality*: he recalled facts observed while playing the game. Then, using the game, he moved slider *b* in order to test the new conjecture and correct it: he was in *descending modality*. From this extract, we can notice that the game is used by the students as an *oracle* in the formulation and in the checking of conjectures, revealing students' cognitive processes and a more or less theoretical approach to it.

Similar dynamic processes were developed to answer the third question, hence we do not report it in the analysis, but we move directly to the analysis of the fourth question. While students were exploring the game while trying to discover the answer, the tablet switched off, hence the dialogue was carried on without moving the slider, but just by using the memory of the previous experiences, the students' cognitive modalities were forced to be *detached*.

Gi *d* controls this one *(pointing at the circle with centre O)*, *f* controls this one *(pointing at the circle with centre O')* and *e* controls…
An *e* is the distance [between O and O']
Fe But *d* and *f* are changing while here *it (pointing at the left part of the screen where the graphic window is supposed to be)* does not change.
Gi Wait a moment… answer 3 is wrong: *a* measures *e*, *b* measures *d* and *c* measures *f*.
Fe They [*a* and *e*, *b* and *d*, *c* and *f*] are not the same values!

Fe refuted Gi's conjectures using the *logical control*: he remembered that both *d* and *f* changed in the numerical window, while just one circle changed in the graphic window. The synchronic changing of two numerical values and only one circle contradicts the conjecture that links each value to one circle. In a natural way, the game supported students to develop logical argumentations: the experience lived during the game provided students with elements they could use to confute other points of view. This excerpt shows that the students were moving to a logical/theoretical level. When the tablet was again available, Fe repeated his reasoning supported by the game, moving again to the *descending modality*.

## Discussion

This chapter illustrates a teaching experiment that aimed to approach geometry in secondary school through games designed in the GeoGebra App. It is based on an elaboration of the "three fidelities" framework (Zbiek et al., 2007; Dick, 2008; Larkin, 2015) and has been extended to what we call its "second order" structure, where the integration of its components is considered in order to design the game. More precisely, the mathematical and the cognitive fidelities are integrated into the didactical game approach to geometrical properties using the LI frame.

The analysis tool examines the productions of students showing the complexity of the students' actions, which can lead to discovering and understanding of the mathematics underpinning the games. The game is the reference environment, on which the students' work is based and to which they refer while answering the worksheet task: the latter is cognitively marked by a strong modality change, namely from ascending/neutral/descending ones to detached/logical control ones. This change has also an epistemological connotation (within the mathematical fidelity frame), insofar as it marks students' transition from an empirical modality towards a logical/theoretical one.

The excerpts illustrate how "porous" the distinction between conceptual and procedural aspects of mathematics can be. Sfard and Linchevski (1994) haves nicely illustrated this point within the theory of reification: "there is an inherent process-object duality in the majority of mathematical concepts. It is the basic tenet of our theory [of reification] that the operational (process-oriented) conceptions emerges first and that the mathematical objects (structural conceptions) develop afterwards through reification of the processes" (p. 191). Our game-based approach can support the transition from the one to the other, which is generally not easy to observe in the field of geometry (Sfard refers only to the fields of number and algebra; see also Sinclair & Yurita, 2008): the worksheet task is the didactical tool (within the pedagogical fidelity frame) that triggers the transition. But this transition is possible insofar as there is a "natural" (cultural) attitude to pass from a played to a reflected-game in the game framework (Soldano & Arzarello, 2016): in any game it is natural to discuss it after it has been played in order to check why one has won or not. The methodology is within a Vygostskian frame. As pointed out in Arzarello et al. (2012):

> The teacher acts both at the cognitive and the metacognitive levels, by fostering the evolution of meanings and guiding the pupils to awareness of their mathematical status. […] From a sociocultural perspective, one may interpret these actions as the process of relating students' "personal senses" (Leontjev, 1964/1976, pp. 244 ff.) to mathematical meanings, or of relating "spontaneous" to "scientific" concepts (Vygotsky, 1978/1990, p. 286 ff.). (p. 108)

In our case, this role is played first by the worksheet task, which pushes the students to make the transition discussed above and, in a second moment, by the teacher during the following class discussion.

The major result of our research consists in the elaboration of tasks that satisfy both mathematical and cognitive fidelities while triggering and supporting a smooth transition from an empirical, DGE approach to a proof-oriented one. The research presented in the literature shows that it is problematic to have a common cognitive and mathematical (epistemic) fidelity. This difficulty creates serious obstacles in developing a coherent pedagogical fidelity. On the contrary, our approach has allowed us to produce an intertwined mathematical and cognitive fidelity through our pedagogical transposition of the GTS: in fact, our game-based activities are a viable way for students to enter in a natural way into the logical relationships between mathematical objects.

## References


Artigue, M., & Perrin-Glorian, M. J. (1991). Didactic engineering, research and development tool: some theoretical problems linked to this duality. *For the Learning of Mathematics, 11*(1), 13–18.

Arzarello, F., Andriano, V., Olivero, F., & Robutti, O. (1998). In: L. Magnani, N. J. Nersessian, & P. Thagard (Eds.), *Abduction and Scientific Discovery, Special Issue of Philosophica, 61*(1), 77–94.

Arzarello, F., Olivero, F., Paola, D., & Robutti, O. (2002). A cognitive analysis of dragging practises in Cabri environments. *ZDM: ZentralblattfürDitaktikderMathematik, 34*(3), 6–72.

Arzarello, F., Bussi, M. G. B., Leung, A. Y. L., Mariotti, M. A., & Stevenson, I. (2012). Experimental approaches to theoretical thinking: artefacts and proofs. In *Proof and proving in mathematics education* (pp. 97–143). The Netherlands: Springer.

Boero, P., Garuti, R., Lemut, E., & Mariotti, M. A. (1996). Challenging the traditional school approach to theorems: A hypothesis about the cognitive unity of theorems. In L. Puig, & A. Gutierrez (Eds.), *Proceedings of 20th PME Conference* (Vol. 2, pp. 113–120). Valencia, Spain.

Dick, T. P. (2008). Fidelity in technological tools for mathematics education. In G. Blume & M. Reid (Eds.), *Research on technology and the teaching and learning of mathematics: Cases and perspectives* (Vol. 2, pp. 333–339). Charlotte, NC: Information Age Publishing.

Duval, R. (1991). Structure du raisonnementdéductif et apprentissage de la démonstration. *Educational Studies in Mathematics, 22*(3), 233–261.

Dvora, T. (2012). *Unjustified assumptions in geometry made by high school students in Israel*. Ph. D. dissertation, Tel Aviv University—The Jaime and Joan Constantiner School of Education.

De Freitas, A. A., & de Freitas, M. M. (2013). Classroom Live: A software-assisted gamification tool. *Computer Science Education, 23*(2), 186–206.

Fischbein, E. (1982). Intuition and proof. *For the Learning of Mathematics, 3*(2), 8–24.

Healy, L., & Hoyles, C. (2000). A study of proof conceptions in algebra. *Journal for Research in Mathematics Education, 31*(4), 396–428.

Hintikka, J. (1998). *The principles of mathematics revisited*. Cambridge: Cambridge University Press.

Hintikka, J. (1999*). Inquiry as inquiry: A logic of scientific discovery*. Dordrecht: Springer Science + Business Media.

Laborde, C. (2001). Integration of technology in the design of geometry tasks with cabri-geometry. *International Journal of Computers for Mathematical Learning, 6*(3), 283–317.

Larkin, K. (2015). The search for fidelity in geometry apps: An exercise in futility. In M. Marshman, V. Geiger, & A. Bennison (Eds.), *Proceedings of the 38th Annual Conference of the Mathematics Education Research Group of Australasia*. Mathematics Education in the Margins (pp. 133–140). Sunshine Coast: MERGA.

Leontjev, A. N. (1976/1964). Problemi dello sviluppo psichico. In Riuniti & Mir (Eds.), *Problems of psychic development*.

Peirce, C. S. (1960). *Collected papers*. Cambridge, MA: Harvard University Press.

Prensky, M. (2001). Digital natives, digital immigrants Part 1. *On the Horizon, 9*(5), 1–6.

Reiss, K., Klieme, E., & Heinze, A. (2001). Prerequisites for the understanding of proofs in the geometry classroom. In M. van den Heuvel-Panhuizen (Ed.), *Proceedings of the 25th Conference of the International Group for the Psychology of Mathematics Education* (Vol. 4, pp. 97–104). Utrecht: PME.

Saada-Robert, M. (1989). La microgénèse de la rapresentation d'un problème. *Psychologie Française, 34*, 2–3.

Sinclair, N., & Yurita, V. (2008). To be or to become: How dynamic geometry changes discourse. *Research in Mathematics Education, 10*(2), 135–150.

Sfard, A., & Linchevski, L. (1994). The gains and the pitfalls of reification: The case of algebra. *Educational Studies in Mathematics, 26*(2/3), 191–228.

Soldano, C., & Arzarello, F. (2016). Learning with touchscreen devices: Game strategies to improve geometric thinking. *Mathematics Education Research Journal, 28*(1), 9–30.

Swan, C. (2012, May–June). Gamification: A new way to shape behavior. *Communication World, 29*, 13–14.

Tarski, A. (1933). "The concept of truth in the languages of the deductive sciences" (Polish), PraceTowarzystwaNaukowegoWarszawskiego, Wydzial III NaukMatematyczno-Fizycznych 34, Warsaw; expanded in English Translation in Tarski 1983. In J. Corcoran (Ed.), *Logic, semantics, metamathematics, papers from 1923 to 1938* (pp. 152–278) Indianapolis: Hackett Publishing Company.

Thompson, D. R., Senk, S. L., & Johnson, G. J. (2012). Opportunities to learn reasoning and proof in high school mathematics textbooks. *Journal for Research in Mathematics Education, 43*(3), 253–295.

Toulmin, S. E. (2003). *The uses of argument*. Cambridge: Cambridge University Press.

Vygotsky, L. S. (1978). *Mind in society: The development of higher psychological processes*. Cambridge: Harvard University Press.
Zbiek, R. M., Heid, M. K., Blume, G. W., & Dick, T. P. (2007). Research on technology in mathematics education: A perspective of constructs. In F. K. Lester (Ed.), *Second handbook of research on mathematics teaching and learning* (pp. 1169–1207). Charlotte, NC: Information Age.

**Carlotta Soldano** completed her Master's degree in Mathematics at the University of Torino with a thesis in mathematics education. She got the degree of Doctor of Philosophy (Ph.D.) in Mathematics on the 29th of June 2017. Title of the dissertation: "Learning with the logic of inquiry: a game approach within DGE to improve geometric thinking".

**Ferdinando Arzarello** is a Professor of mathematics at the University of Turin. His current major field of research is mathematics education. He studied the multimodal processes involved in the teaching and learning of mathematics, concentrating on the role of gestures and on the dragging practises within DGE. In his most recent research, he focused on the role of the logic of inquiry in the construction of new mathematical knowledge. He is author of more than 150 papers or chapters of books in important international journals or volumes and is a past president of ERME and of ICMI.

# Part IV
# Exploring New Forms of Communication to Make Mathematical Learning Visible

# Mathematics Screencasts for Teaching and Learning

**Linda Galligan and Carola Hobohm**

**Abstract** During our ongoing research into the use of mathematics screencasts at university, we have seen an increased utilization of mobile technologies both for teaching and learning. The ubiquity of mobile devices has allowed students and lecturers to create, curate and view screencasts far more easily than ever before. Whilst creating screencasts with such ease is deemed beneficial, one needs to caution that the quality of screencasts and inherent accuracy remains central to learning and teaching. As a result, our research has led us to the development of a tool for teachers and students to evaluate their own and others' screencasts. This chapter describes a case study of the use of mobile devices and screencasting in mathematics teaching, combined with the utilization of the evaluative tool in developing pre-service teachers' understanding of mathematics and how to teach it. It concludes with future directions in using mobile technologies to assist mathematical understanding and pedagogical content knowledge.

**Keywords** Screencasts · Tablet technology · Mobile technology
Understanding mathematics · Pre-service teachers · In-service teachers
PCK

For more than ten years, research has been undertaken on the use of various mobile technologies to support mathematics learning and teaching at the University of Southern Queensland (USQ). During this time mobile technologies have become cheaper, easier to use, and more accessible. At the same time various forms of video resources have been developed and utilized to support learning. According to Hartsell and Yuen (2006) online video-based instruction "brings courses alive by

L. Galligan (✉) · C. Hobohm
University of Southern Queensland, Toowoomba, QLD, Australia
e-mail: linda.galligan@usq.edu.au

C. Hobohm
e-mail: chobohm@usc.edu.au

L. Galligan · C. Hobohm
University of the Sunshine Coast, Sunshine Coast, QLD, Australia

 
N. Calder et al. (eds.), *Using Mobile Technologies in the Teaching and Learning of Mathematics*, Mathematics Education in the Digital Era 12,
https://doi.org/10.1007/978-3-319-90179-4_15

allowing online learners to use their visual and auditory senses to learn complex concepts and difficult procedures" (p. 31). In addition, we maintain that the kinesthetic sensory modality of students writing mathematics as well as the effort required to craft an explanation, incorporating gesturing and annotating, also increases students' learning and understanding of mathematics.

Increasingly mobile technologies enable users to migrate from traditional pen and paper to digital writing. This digital writing can easily be captured as a form of video instruction, called screencasts. Screencasts can also be augmented with text images or animations (Student Screencasting with the iPad, 2014). Sometimes screencasts are referred to as video podcasts (vodcasts) or podcasts. However, podcasts usually refer to audio only content. In our research, we consider screencasts as an extension of vodcasts in that they include freehand inking from mobile devices. In mathematics this allows for effortless writing and drawing. To produce these screencasts, authors use mobile tablet devices (i.e. with a touchscreen and stylus) and recording software. The screencasts are then uploaded to video libraries on the web (such as OneDrive, Evernote, or YouTube) to be viewed anywhere at any time. A screenshot of an example of a pre-service teacher explaining similar triangles is shown in Fig. 1.

Here the student, using a tablet, started with the typed question, triangle diagrams, and a ruler. He then recorded his screen writing and narrations with a mobile app. The resultant screencast (https://vimeo.com/134467682) was then uploaded onto Vimeo®.

Creating screencasts nowadays is a fairly easy process, however, ensuring good content is far more challenging. Sugar, Brown and Luterbach (2010) suggest that screencasts were originally developed to provide procedural information to students. While some screencasts today are being developed as a pedagogical tools (Heilesen, 2010), many merely capture classes held face to face. In addition, the focus of many of the mathematical recordings still appears to be more on procedural knowledge rather than any other form of mathematical knowledge. Yet, screencasts have the potential to do much more. In particular, the focus in our latest research is on student-produced mathematics screencasts as a tool for reflective learning and effective teaching.

Here we aimed to address the aspects of: what "understanding" mathematics means using taxonomies from Skemp (1976), Mason and Spence (1999), and Watson (2002); Pedagogical Content Knowledge (Chick, Baker, Pham, & Cheng, 2006; Shulman, 1987); and how to critique a screencast (Sugar et al., 2010) in terms

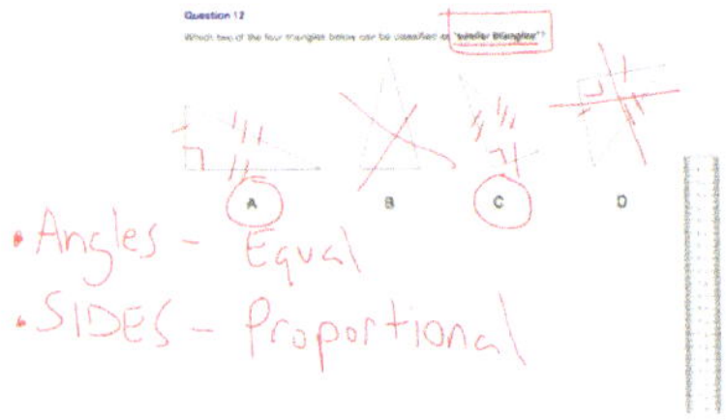


**Fig. 1** Example of an initial screencast by student in 2015

of structural elements (such as visual quality and delivery). This has allowed us to develop an evaluative tool, which guides students to critique screencasts and produce effective screencasts themselves (Galligan et al., 2017 (online)). This evaluative tool has four major components:

1. Purpose in terms of understanding mathematics
2. Pedagogical Content Knowledge (PCK)
3. Structural elements in terms of visual quality and delivery
4. Cohesion and Completeness in relation to a series of screencasts.

This chapter first summarizes our tablet-related research undertaken to date (Galligan & Hobohm, 2013; Galligan, Loch, McDonald, & Hobohm, 2015). It then reports on a case study of a USQ course where pre-service teachers used mobile technologies and associated software to produce mathematics screencasts with the guidance of the evaluative tool. The chapter then concludes with future explorations of combining the versatility of mobile technologies with screencasting to further enhance understanding and teaching of mathematics.

## Tablet-Related Research

In 2010 we highlighted the advantages of tablet technology in teaching one-to-many (the lecture), one-to-few (the tutorial) and one-to-one (individual consultations) (Galligan, Loch, McDonald, & Taylor, 2010). At that stage, most of the tablet-produced recordings were generated by lecturers. We trialled digital writing with on-campus students, noting the potential of the mobility of tablet technology to engage students and improve understanding. Since then, we have continued to use tablet technology to enhance teaching (Galligan & Hobohm, 2013; Galligan, Loch, McDonald, & Hobohm, 2015), similar to other universities (Loch, 2005; Loch & Donovan, 2006; Olivier, 2005; Al-Zoubi, Sammour, & Qasem, 2007; Anderson et al., 2004). Tablet technology was shown to enhance teaching and increase engagement in the classroom (Logan, Bailey, Franke, & Sanson, 2009), and for pre-service teachers, created a "truly transformational experience" (Kosheleva, Medina-Rusch, & Ioudina, 2007, p. 332). It also encouraged new approaches to teaching (Maclaren, 2014) including the creation of screencasts. There are time costs to the screencast producer (Corcoles, 2012), but if screencasts have an positive impact on a large number of students, then the time is well spent.

As mobile devices are becoming more ubiquitous, our focus has turned to the student, particularly pre-service and in-service teachers enrolled as university students. We have continued to refine our approach to support these students with creating screencasts, thus allowing us to assess their understanding of mathematics and how they teach it. Other research also focussed on student-produced screencasts. Croft, Duah, and Loch (2013) reported on an internship for undergraduate mathematics students to create screencasts for peers, finding that students who created the screencasts benefitted by gaining deeper understanding of mathematics concepts. Similarly, Wakefield et al. (2011) asked accounting students to produce screencasts for an assignment and found increased student engagement and performance. It has been documented in the past (e.g. Noss & Hoyles, 1996) that technology can be harnessed by teachers to become a window into student thinking. Now, with mobile technologies, teachers are in a better position to gain insight. In an elementary school setting, researchers have investigated student-generated screencasts using an iPad and Explain Everything® (Soto, 2014; Soto & Ambrose, 2016). Their studies found teachers were able to peer into students' mathematical thinking with screencasts, providing springboards for rich discussions. Soto (2014) concluded that screencasting "has the potential to transform the learning environment by allowing teachers to gain more insight into their students' mathematical thinking, encouraging students to reflect on their thinking and potentially influence the thinking of other students" (p. iii). Research by Richards (2012) at the middle school level, again using iPads and Explain Everything, found student-produced screencasts allowed them to document their own learning. In a study across three middle school classrooms in Germany, Ifenthaler and Schweinbenz (2016) found that while tablet PCs have the potential to "lead to new ways of designing personalized learning environments for the classroom" (p. 317), schools need teachers with good professional understanding, and knowledgeable technical staff.

Research has suggested that when students teach, they develop a deeper understanding of the material. However, Fiorella and Mayer (2013) argued that it had been unclear which features of teaching contributed to this learning. In their research with undergraduate students they found that "when students actually teach the content of a lesson, they develop a deeper and more persistent understanding of the material than from solely preparing to teach" (p. 281). They also found that learning gains even occurred with less than five-minute video-recorded lectures of the material, even if to an imaginary classroom. This suggestion of increased mathematical understanding has been found in other studies with mathematics students (Croft, Duah, & Loch, 2013). However, in these studies the nature of that understanding was not explored in any depth.

In 2012, when we first asked on-campus students to create screencasts, it was achieved relatively easily with university purchased tablets/iPads and face-to-face support. In order to provide on-line students with the same experience, it was only possible by mailing mobile devices to students in small numbers. However, by 2015, most students had access to iPads and other tablet devices, and recording programs such as Jing and Explain Everything were easy to use. In addition, cloud technologies streamlined the uploading and viewing of screencasts.

From our early research survey results, students indicated an increased understanding of mathematics as a result of creating and reviewing screencasts. However, students' responses in forum discussions did not focus on increased understanding (Galligan & Hobohm, 2013) or pedagogical content knowledge (PCK) when evaluating screencasts, instead focussing on procedural skills and structural elements. In our subsequent course development, we aimed to shift their focus to deeper understanding of mathematics. We first used Skemp's (1976) distinction between instructional and relational understanding. We further divided understanding mathematics using the Mason and Spence (1999) categorization of knowing-that, knowing-how, knowing-why, and knowing-to. Added to this "knowing" framework is knowing about usefulness in context (Watson, Geest, & Prestage, 2003). This latter "knowing" includes, for example, understanding ratios or decimals in the context of measurement. Pre-service teachers' PCK was also incorporated into our evaluative tool (Chick, Baker, Pham, & Cheng, 2006; Shulman 1987). The PCK elements included three categories:

- Clearly PCK (cognitive demands of the task, able to represent concepts and knowing target audience);
- PCK (content) Content knowledge in a pedagogical context (procedural knowledge, mathematical structure and connections and methods of solution); and
- PCK (context) Pedagogical knowledge in a content context (related to goals for learning).

Having identified the abovementioned components, we structured the evaluative tool into three categories: understanding mathematics; PCK; and structural elements of a screencast (Sugar, Brown, & Luterbach, 2010; Galligan et al. 2017 (online)). When creating a series of screencasts, we added an extra element of cohesion and completeness, encouraging students to create different screencasts based on the different "knowings". The aim was for students to use this tool to evaluate screencasts and produce effective screencasts themselves.

Our research asked two questions:

- What does an evaluative tool for mathematical screencasts look like?
- Does the use of the evaluative tool make a difference to the quality of the production and critiquing of student-produced mathematical screencasts?

## Case Study

While the trials were conducted over 4 years, this case study describes two courses offered in 2015: an undergraduate mathematics for teachers course; and a similar post-graduate course for in-service teachers with a 90% online enrolment. The total enrolment amounted to 50 students (35 undergraduate and 15 post-graduate). The courses shared many lectures, and the post-graduate students had access to the undergraduate Learning Management System (LMS). In the LMS site, pre- and in-service teachers (P/ISTs) shared the links of their own mathematics screencasts and peer critiqued others'.

Each course contained a number of elements where mobile technology and screencasts were used:

1. **Lecturer-produced screencasts**: After the first live lecture was delivered (and recorded) to on-campus students using a tablet device, all remaining lectures were pre-recorded to deliver content. On-campus students attended a two-hour workshop and online students a one-hour live virtual (Zoom®) session (also recorded). In the lectures, digital writing on tablets was actively used, particularly in the six weeks of mathematical content. The lectures also utilized the interactive quiz feature of Camtasia Studio®. In the live and virtual workshop sessions, tablet devices were used by the tutor and, at times, by the students, to explain mathematical concepts. Zoom allowed for screen-sharing and online annotation by both the tutor and the student.
2. **Assignment 1 where students created and linked their own screencast**: Using a mobile device, students had to record a screencast in which they explained a mathematics concept. Typically, students created screencasts on an iPad, a tablet PC, or a graphics tablet using ExplainEverything®, Jing or ShowMe®. All the recordings were predominately viewed via web linked cloud storage. This assessment instrument was used to identify common features of student-created screencasts, as well as gauge students' ability to create a screencast unguided. Students were encouraged to use their "warts and all" version regardless of imperfections such as errors and informal language often used in a classroom.
3. **Peer and Self Critique 1**: After students uploaded their screencast, they were asked to critique their own and others' first 'unpolished' screencasts via a dedicated online discussion forum. This instrument was used to identify students' ability to highlight features of a screencast without much initial guidance. The critiques were submitted as part of assignment 1. After students created and critiqued the first screencast, a discussion on the peer critiques was held in a lecture and subsequent workshop/Zoom sessions. It became evident that students were ill-equipped to critique screencasts, hence the evaluative tool was introduced to frame the discussion. We invited students to review their first screencasts in the forum, and this produced some discussion, albeit limited.
4. **Peer Critique 2**: Students were next asked to critique a set of mathematical screenscasts from a previous cohort with the help of the evaluative tool, and submit the critiques as part of assignment 2. The combination of the self-created and critiqued screencasts better prepared students to create more professionally produced and pedagogically aligned screencasts for the 2nd assignment.
5. **Assignment 2:** Students had to record a series of linked screencasts on how to teach a troublesome mathematics concept of their choosing that could be given to school students to aid in the understanding of the concept. This second set of screencasts was used to see if students could improve on initial screencasts with the use of the evaluative tool. Students were encouraged to restrict their screencasts to a maximum of five minutes. The screencasts were uploaded by students to their cloud-based video library for markers to access via web link.

In summary, the order in which screencasting tasks were introduced was deliberate and followed the development of the course over four years of trialling.

We introduced the creation of unpolished screencasts early in the semester to force students to engage with the technology and to promote active learning. At this stage examiner presence was highly supportive. The self and peer critiques were then introduced to demonstrate different ways of structuring and presenting the screencasts; and to showcase varied approaches for solving mathematics problems. Attention was given to foster a safe environment to encourage discussion of errors and engender a spirit of support. The screencast submissions and subsequent reflections were then followed by an introduction of the evaluative tool to prepare for the second assignment submission of sequenced screenscasts. The success of this approach, particularly for the online students, relied on easy access to mobile technologies, recording software and cloud storage.

## Method

In this research, (with ethics clearance) we used both a cooperative inquiry approach of iterated reflection and action (Reason, & Riley, 2008) to create the evaluative tool, along with a qualitative research approach of constant comparison (Glaser, 2008) to analyse findings. Participants in both courses were asked to create, self-evaluate and peer-critique screencasts that explained mathematical concepts as described above. The case study aimed to ensure that P/ISTs (a) learnt to produce quality screencasts using appropriate mobile technologies, (b) understood the mathematics more deeply, and (c) have a better understanding of how to teach mathematics concepts.

Apart from the assignments and the forum posts, data were collected from pre-and post-surveys to measure changes in attitudes and overall experiences in creating mathematical screencasts. The post-survey repeated similar questions to the pre-survey (about value, advantages and disadvantages of mathematics screencasts). In addition, students were asked about their attitude to screencasting (after having created their own), the mobile technologies they used, and their ratings of the importance of colour, legibility, clarity of voice, correct mathematics, completeness, clarity of explanation, comprehensiveness, and contextualising. We also asked specific questions about the use and value of the evaluative tool.

## Results

This chapter focusses on students' deeper understanding of mathematics and how to teach it more effectively with mobile technologies. Detailed results of this study, particularly around the development and effectiveness of the evaluative tool, can be found in Galligan, Hobohm and Peake (2017 online).

The following section incorporates results from peer and self-critique 1, and the post-survey, with a focus on three themes from the evaluative tool: understanding mathematics, PCK, and analysis of structural elements. Other results from the post-survey highlight the mobile technologies used and student opinion about the value of screencasting.

## *Understanding Mathematics*

Students' comments from peer-critique 1 often related to understanding mathematics, and they also related to PCK elements. The word "understand" was one of the top 20 words used in the comments (Galligan et al. 2017 (online)). The example below illustrates the "know why", and the structure of the screencast is typical of the cohort A student (IST15) critique suggested that the screencast could have included relational understanding (know why). Notice this student also commented on clarity. This was a common response from students as it is something that is immediately apparent when viewing a screencast (Fig. 2).

> Your screencast was very clear and succinct. If I were using your screencast for revision purposes or explanation purposes, I might have wanted to have asked the question, "**Why** do we have to reverse the inequality signs when we divide by a negative number …." (IST 15)

Students appreciated new ideas on approaches to teaching mathematics (i.e. Clearly PCK) and methods of solving as demonstrated by their peers (i.e. PCK Content). Even though most of the methods to solve problems were similar, students appreciated seeing their method used by others. At times, the approaches were quite different to those taken by the cohort. For example, one student mentioned the "cross-method" of factorising trinomials (know how). This was new to many students (four of whom explicitly commented on the discussion forum about this concept).

> I've never heard of factorising trinomials using the cross method, and now that I know how easy it is I might start using it to teach my students. I liked how you set out your page and used the cross in the middle to show which pair of factors is being multiplied by the other pair of factors. I also liked how you used trial and error to show students that working out answers is simply that, working out the right answer (IST 9)

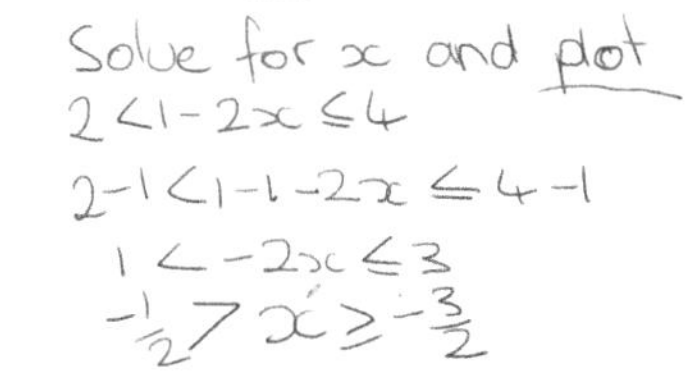


**Fig. 2** Screenshot of student solving an inequality (using an iPad, Jing and uploaded to screencast.com)

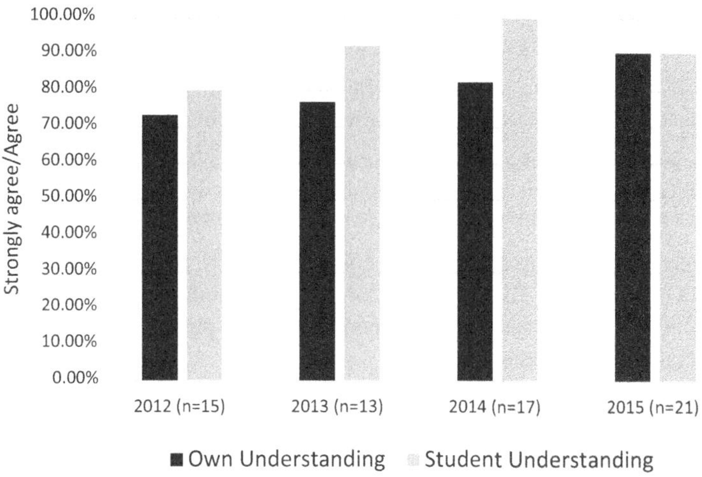


**Fig. 3** P/ISTs' opinion of the usefulness of screencasts to aid their students and their own understating of mathematics concepts (2012–2015)

In the post-survey, P/ISTs were asked to rate if their screencasts assisted their own and could assist their future students' understanding of mathematics concepts (Fig. 3). Over the four years of using the survey, 82% of P/ISTs strongly agreed/agreed, (with an additional 15% remaining neutral) that the process of creating screencasts assists *their own* understanding. Ninety percent of P/ISTs agreed, and the remainder were neutral that it could assist *student* understanding. While data are not directly comparable between cohorts of different years, it is interesting to note that agreement was slightly higher for both students and teachers for almost every year.

In an open-ended question on the process of peer reviewing, all students commented positively on the process, and resultant increased understanding. The following is a typical response from students:

> Providing reviews was extremely helpful in assisting my understanding of the topics. Receiving reviews was also excellent in pointing issues I may not have considered. (IST12)

Due to the nature of the course, we could not explore the full extent of students' understanding of particular mathematics topics, as highlighted by Fiorella and Mayer (2013). This is the focus of current screencasting research within an elementary mathematics course undertaken by many education students.

## *Pedagogical Content Knowledge (PCK)*

In the initial peer critiques, there were instances of students' thinking around PCK elements, but not as strong as the "understanding" theme. One "Clearly PCK" theme that emerged was knowing the target audience (examples IST6 and IST10) and knowing how to represent concepts (example IST6) in different ways:

> It's intentionally short (1 min) … At least it made me think about how succinctly we present information and different ways of saying things! (IST6)

> I liked that you explained that the method you were going to use by saying out loud… I did find it a bit confusing to follow only because I didn't have the original question to look at… it can be a good habit to encourage students to look back and re-read the question. (IST10)

However, as the course began to emphasize PCK, subsequent screencasts reflected PCK elements. For example, "original questions" were a feature of the start and end of screencasts (i.e. bumpers) created by students, along with more carefully crafted screencasts.

> Your audio comes across as very calm and well-delivered, with your text well organised. … I have seen some, even professionally-made ones, where the presenter constantly repeats him/herself, over-talking, and jumps around so much it can be confusing. This can leave the viewer … quite exhausted! (IST6)

In the post-survey analysis, open-ended questions were themed. One theme, about the use of screencasts in the classroom, focussed on the PCK element of repetition and efficiency, but also knowing their students.

> Honestly now I think it's one of the teaching tools I will definitely use in teaching. And that is like daily basis. …, students that are bit slow to get a concept can play and watch over and over again instead of asking the teacher ten times, which they wouldn't do anyway! (IST7)

> ….can link in more resources, can record your best version of teaching the material, can have writing pre-written so that it's legible, can have resources pre-pasted into it. (PST26)

> I like the fact that you can bring the outside world into the class room … that may help students stay attentive and help them think of mathematics as more than just a subject …. I also think there is room to use them for class and home help. Teams of teachers could share their work and have all classes in a school having similar lessons. (PST10)

In 2015, we asked students to rank how difficult it was to rate peer screencasts according to the evaluation tool. The results are tabled in Fig. 4, with ratings of extremely difficult/difficult, and extremely easy/easy combined. It became evident that structural elements (legibility, colour, and voice) were easier to rate than PCK elements (circled). The difficulty in rating abstract aspects is reflected in students' comments in the next section of this chapter.

We also asked P/ISTs in 2015 to rate their own screencasts using the evaluative tool (Fig. 5). Students were relatively critical of their attempts compared to markers' evaluation. We noted that PCK still appeared difficult to achieve, with half the students rating themselves average or below average. They also rated some aspects of the structural elements such as narration and mathematical language low, reflecting the difficulty in being able to articulate their thoughts appropriately and succinctly. This is evident in one student comment in the discussion forum: "Did anyone else find it hard to analyze themselves????".

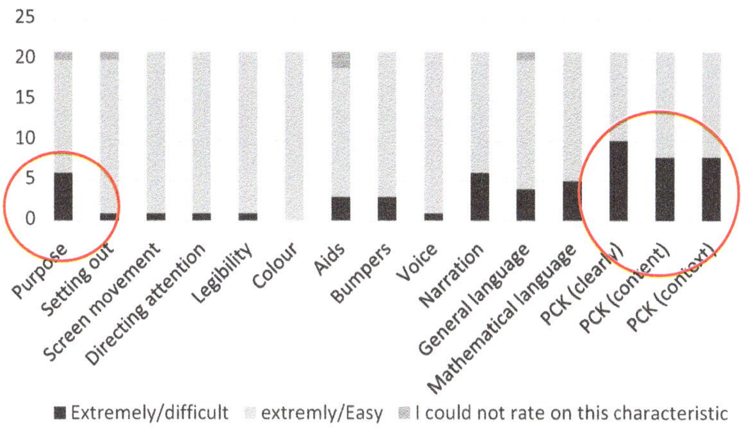


**Fig. 4** P/ISTs opinion of the difficulty ratings on peer screencast components/aspects (2015)

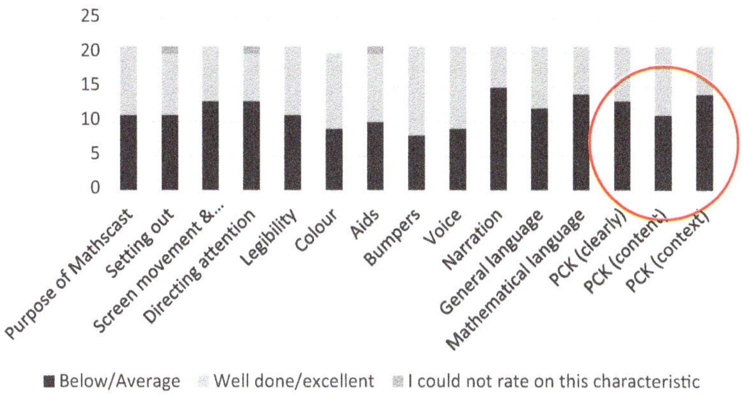


**Fig. 5** P/ISTs opinion of their own final screencasts (2015)

## *Analysing Screencasts Structural Elements*

Structural elements were divided into visual quality (setting out, screen movement, directing attention, legibility, colour and aids) and delivery (bumpers, voice, narration, general and mathematical language). Students in 2015 were invited to re-evaluate their first screencast with the evaluative tool. Self-critiques were non-compulsory and hence only seven P/ISTs volunteered to critique their own first screencasts. The students commented predominantly on visual elements (colour, setting out) and delivery (voice, general and mathematical language). A typical example can be seen in IST1 (Fig. 6):

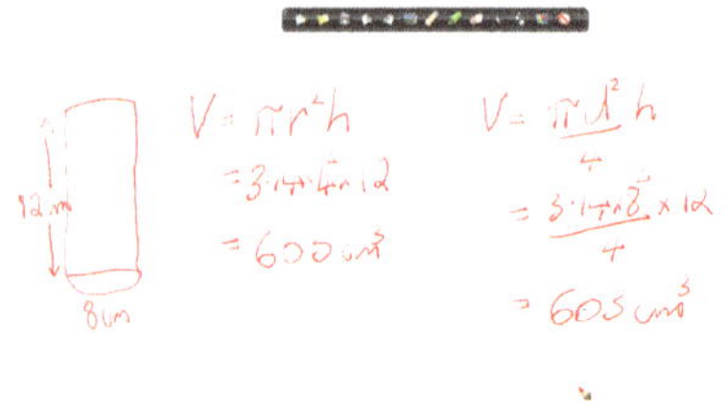

**Fig. 6** Screenshot of final moment in IST1's screencast (3.12 min)

> After viewing the reflective analytical tool, I can see that I need much improvement… but I could have used a different colour pen to direct the attention back to parts of the working out and diagram. However, it was a good idea that I used a diagram as an aid in the mathscast. My voice was very monotone. I found it challenging to talk and write at the same time. An idea might be to pause the video while I write and then talk. I believe that my general and mathematical language I used was suitable for teaching the concept. (IST 1)

Another example provided similar evidence of reflection on structural elements (i.e. setting out, legibility, and colour) and also included voice and language, which featured frequently across screencasts.

> After having a look at my Screenchomp screencast with the analytical tool, I think there is so much room for improvement. I could have used separate screens for the two different methods of finding the area. This would then have an effect on the legibility of my screencast. I am glad I used different colours, which did make it little bit more appealing and easy to follow. I also think I could improve on my delivery, maybe my tone could have [been] better, maybe I need to speak louder and at a slower pace. As for mathematical language part, I think I did alright, the vocabulary and terminology were appropriate. (IST 7)

Tools such as Explain Everything® have features that assist in the creation of well-structured screencasts. For example, it has a pause button to allow for uploading words or diagrams; options for pointers and highlighting to assist with directing attention; insertion of mathematics equations, audio, images, and editing capabilities. In addition, these tools are used in combination with the latest touchscreens and pens (such as Microsoft Surface Pro® and iPad Pro®) which allow for smooth, effortless writing.

## *Mobile Technologies Used*

At the start of the research project, mobile tablet devices were a novelty and not readily available to students. Tablets and their application in an educational setting were largely unexplored, but this changed substantially with the introduction of the iPad and other mobile tablet devices, along with access to cloud storage. This change is reflected in questions in the post-survey relating to what technology they used to create their screencasts.

The post-survey was used from 2012 to 2015. In 2012, of the 57 enrolled only 26% (15) completed the post-survey, whereas in 2015 of the 50 enrolled students, 21 completed the post-survey (38% completion rate). The most prominent change noted (see Fig. 7) was that by 2015, 67% of students were using an iPad to create their screencasts (up from less than 20% in 2012, when we first surveyed students), although a few students used their smartphones as seen in Fig. 8.

**Fig. 7** Devices used by students to create screencasts

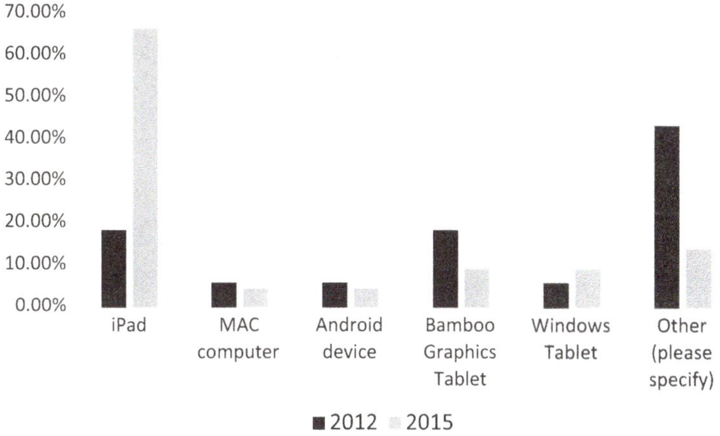


**Fig. 8** Student using a smartphone to record writing on paper

Survey questions over the four-year period indicated changes in the use of screencasting apps. In 2012, most students were using Jing® or PowerPoint® (Fig. 9), compared to 2015, when Screenchomp® and Explain Everything® became more popular along with easy access to cloud storage.

## *Value of Screencasting*

Due to the difficulty and subjective nature of rating screencasts and associated experiences, we provided various open-ended questions in the post-survey to identify additional aspects of the screencasting experience. Students were asked to comment on their own perception of advantages/disadvantages of screencasts and to describe any changes in emotions, feelings and attitudes towards screencasting. In 2015, five themes emerged around: improvement in screencast production; emotive attitude; teaching efficiency; changes in opinion, and disadvantages. Apart from statements on how students improved and understood the value of quality, engaging and accurate screencasts, students also welcomed the skill of producing such personalised educational artefacts through mobile technologies.

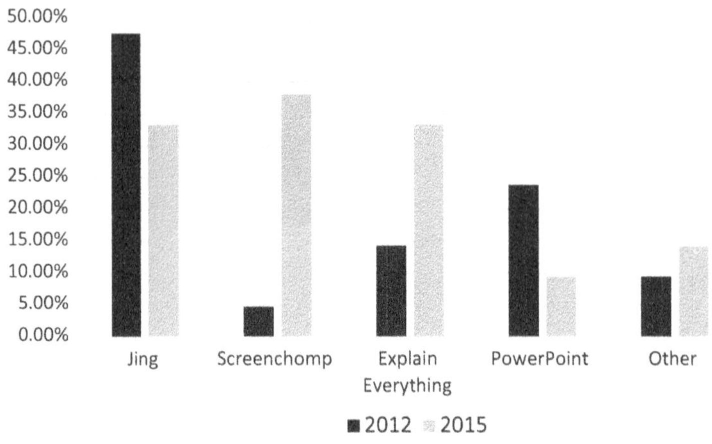


**Fig. 9** Proportion of students using programs to produce screencasts

Positive emotive language was used such as "so excited", "gained confidence", "mind blowing", "embraced the opportunity", I am really happy" and "desire to produce screencasts".

Another theme reflected students' shift in attitude positively from novice to professional screencaster. A typical example is illustrated from PST12 commenting on the emerging realization of the multimodal and social learning qualities of screencasting:

> To be honest, I thought they were a little redundant in a classroom situation and suited to long distance study only. My attitude has definitely changed regarding this, as screencasts allow a certain specificity that can be orchestrated which I imagine would be a lot harder in the ad hoc classroom environment. After watching some great screencasts online, I was much more relaxed this time around, enjoying the relaxed candor of presenters which veered me away from a rigid monologue I employed the first time. (PST12)

A final theme identified the disadvantages of screencasts, particularly on procedural vs relational understanding, and the potential distance between student and teacher. Because screencasts do not provide a live synchronous experience, other students commented on the inability for students to "interact with the teacher"; and the recordings added an "air of distance between the student and the teacher". This disconnect can also be seen in PST31's comment:

> If a teacher wanted to offer a (literally) tangible example of a maths problem, say, mixing up pancake batter in a classroom, a mathscast would not achieve this. Virtual reality is not tangible reality in this instance. (PST31)

The amount of effort to create a screencast (reflected in Corcoles (2012) research) is still real, and is reflected in this student's comment:

> I think that they are very valuable tools, yet it remains that initially there is a heavy impact upon time as I familiarise myself with the technologies, whilst also trying to familiarise myself with the demands of being a teacher. (PST anon)

This amount of effort has lessened because mobile and related cloud technologies are now much easier to use. Similarly, the culture of schools to embrace screencasting enabled through mobile technologies is changing. While one student related:

> I actually found myself at a screencast moment in my recent prac placement, but decided against it because they didn't 'do that' there so it seemed like too much of a stretch. (PST22)

Another commented:

> I completed my placement at the end of last term and was lucky enough to be placed with maths teacher who loves using technology in the classroom and teaches year 8 - 12

and went on to mention the use of Kahoot®, working with iPads and tablets, and Desmos®.

## Conclusion

Teaching and learning mathematics can be enhanced greatly by appropriate selection of mobile technologies. This chapter has outlined the use of screencasting and associated mobile technologies as an important approach to assist P/ISTs to understand mathematics and teach it. Our research undertaken to date has found the creation of carefully constructed quality screencasts evokes positive, even transformational, effects on in-service and pre-service teachers similar to that found by Kosheleva and colleagues (2007). Similarly, we have noted an increase in students' perception of improved mathematical understanding, as found in other studies (Croft et al. 2013; Kosheleva et al. 2007). However, there are caveats. Examiners have commented on the excessive time taken to mark these screencasts. Like all technology, screencasts are tools that should assist learning and teaching, but not at the exclusion of teacher intervention. In our research, we have found disadvantages such as reduced interactivity, and time consuming efforts to create well-crafted screencasts, as mentioned by Corcoles (2012). We wanted to use screencasts to peer into students' thinking in more depth, as Soto (2014) was able to do with school students, but students were reluctant to expose their errors in thinking. Our evaluation of the effects that student-created screencasts had on pre-service teachers' approach to broader pedagogy, is relatively exploratory to date. It needs to go beyond the confines of one course or one subject and into practice. As one student reflected:

> I am amazed at how many ways I can see uses for this every day now! I studied foreign languages for many years and I can see how this teaching area would also benefit by incorporating inking devices with narration. (PST20)

Meanwhile, our teaching has allowed us to identify and trial newer mobile technologies for online learning, and we are keen to explore the effects.

Can we see a future where students and teachers can capture artefacts, such as screencasts, anywhere, anytime, and with any device? We are seeing some of this now in blogs (e.g. Mayer, 2016), YouTube videos, or university-hosted centres (e.g. Mathematics and Statistics Help (MASH) Centre). It is even more imperative that teachers and students have a framework to critique and produce screencasts in order to ensure good pedagogical quality. Once the students know this process, our research can now focus on what triggers their understanding, and to what extent, so they develop a deeper and more persistent understanding of the material (Fiorella, & Mayer, 2013). Future research, within a mathematics content course, will explore the extent to which pre-service teachers' understanding of mathematical concepts improves due to their creation and peer critiquing of screencasts. In particular, it will probe students' own PCK related to what it means to fully understand and teach mathematics concepts. Such research will build on future developments of mobile technologies, their ease of use, and abilities to engage human presence.

## References

Al-Zoubi, A. Y., Sammour, G., & Qasem, M. A. (2007). Utilization of tablet PCs in electromagnetics education iJET. *International Journal of Emerging Technologies in Learning, 2*(2), 42–46.

Anderson, R., Anderson, R., Simon, B., Wolfman, S. A., VanDeGrift, T., & Yasuhara, K. (2004). Experiences with a tablet PC based lecture presentation system in computer science courses. *ACM SIGCSE Bulletin, 36*(1), 56–60.

Chick, H. L., Baker, M., Pham, T., & Cheng, H. (2006). Aspects of teachers' pedagogical content knowledge for decimals. In J. Novotna, H. Moraova, M. Kratka, & N. Stehlikova (Eds.), *Proceedings of the 30th conference of the International Group for the Psychology of Mathematics Education* (Vol. 2, pp. 297–304). Prague: PME.

Corcoles, C. (2012). The use of screencasts in mathematical and scientific education. *eLC Research Paper Series: Time factor in online teaching and learning on mathematics and physics*, (Vol. 4, pp. 6–10).

Croft, A., Duah, F., & Loch, B. (2013). I'm worried about the correctness: Undergraduate students as producers of mathematics screencasts for their peers—lecturer and student perceptions. *International Journal of Mathematical Education in Science and Technology, 44*(7), 1045–1055. https://doi.org/10.1080/0020739X.2013.823252.

Fiorella, L., & Mayer, R. E. (2013). The relative benefits of learning by teaching and teaching expectancy. *Contemporary Educational Psychology, 38*(4), 281–288. https://doi.org/10.1016/j.cedpsych.2013.06.001.

Galligan, L., & Hobohm, C. (2013). Students using digital technologies to produce screencasts that support learning in mathematics. In V. Steinle, L. Ball, & C. Bardini (Eds.), *Mathematics education: Yesterday, today and tomorrow. Proceedings of the 36th annual conference of the Mathematics Education Research Group of Australasia* (pp. 322–329). Melbourne: MERGA.

Galligan, L., Hobohm, C., & Peake, K. (2017 (online)). Using an evaluative tool to develop effective mathscasts. *Mathematics Education Research Journal*

Galligan, L., Loch, B., McDonald, C., & Hobohm, C. (2015). Conceptualising, Implementing and evaluating the use of digital technologies to enhance mathematical understanding: Reflections on an innovation-develpment cycle In P. Redmond, J. Lock, & P. Danaher (Eds.), *Educational Developments, Practices and Effectiveness: Global Perspectives and Contexts* (pp. 137–160). Hampshire, England: Palgrave Macmillian.

Galligan, L., Loch, B., McDonald, C., & Taylor, J. A. (2010). The use of tablet and related technologies in mathematics teaching. *Australian Senior Mathematics Journal, 24*(1), 38–51.

Hartsell, T., & Yuen, S. C.-Y. (2006). Video streaming in online learning. *AACE Journal, 14*(1), 31–43.

Heilesen, S. (2010). What is the academic efficacy of podcasting? *Computers & Education, 55,* 1063–1068.

Ifenthaler, D., Schweinbenz, V. (2016). Students' acceptance of tablet PCs in the classroom. *Journal of Research on Technology in Education, 48*(4):306–321.

Kosheleva, O., Medina-Rusch, A., & Ioudina, V. (2007). Pre-service teacher training in mathematics using tablet PC technology. *Informatics in Education-An International Journal, 6* (2), 321–334.

Loch, B. (2005). Tablet Technology in First Year Calculus and Linear Algebra Teaching. In M. Bulmer, H. MacGillivray, & C. Varsavsky (Eds.), *Proceedings of Kingfisher DELTA'05: Fifth southern hemisphere conference on undergraduate mathematics and statistics teaching and learning* (pp. 231–237). Brisbane: University of Queensland.

Loch, B., & Donovan, D. (2006). Progressive teaching of mathematics with tablet technology. *e-JIST, e-Journal of Instructional Science and Technology, 9*(2), 1–6.

Logan, M., Bailey, N., Franke, K., & Sanson, G. (2009). Patterns of tablet PC use across multiple learning domains: a comparison program. In D. Berque, L. Konkle, & R. Reed (Eds.), *The impact of Tablet PCs and pen-based technology on education: New Horizons* (pp. 83–92). La Fayette: Purdue University Press.

Maclaren, P. (2014). The new chalkboard: the role of digital pen technologies in tertiary mathematics teaching. *Teaching Mathematics and Its Applications, 33*(1), 16–26. https://doi.org/10.1093/teamat/hru001.

Mason, J., & Spence, M. (1999). Beyond mere knowledge of mathematics: The importance of knowing-to act in the moment. *Educational Studies in Mathematics, 28,* 135–161.

Mathematics and Statistics Help (MASH) Centre MathsCasts. Retrieved October 14, 2016 from http://www.swinburne.edu.au/student/study-help/mash/.

Mayer, D. (2016). dy/dan. http://blog.mrmeyer.com/.

Noss, R., & Hoyles, C. (1996). *Windows on mathematical meaning: Learning cultres and computers*. Dordrecht: Kluwer Academic Publishers.

Olivier, W. (2005). Teaching mathematics: Tablet PC technology adds a new dimension. In A. Rogerson (Ed.), *The mathematics education into the 21st century project: Proceedings of the eight international conference: Reform, revolution and paradigm shifts in mathematics education* (pp. 176–181). Malaysia: Universiti Teknologi Malaysia.

Reason, P., & Riley, S. (2008). Co-operative inquiry: An action research practice. In J. A. Smith (Ed.), *Qualitative psychology: A practical guide to research methods* (2nd ed.). London: Sage Publications.

Richards, R. (2012). Screencasting: exploring a middle school math teacher's beliefs and practices through the use of multimedia technology. *International Journal of Instructional Media, 39*(1), 55–68.

Shulman, L. S. (1987). Knowledge and teaching: Foundations of the new reform. *Educational Resaercher, 57*(1), 1–22.

Skemp, R. (1976). Relational understanding and instrumental understanding. *Mathematics Teaching, 77,* 20–26.

Soto, M. (2014). *Documenting students' mathematical thinking through explanations and screencasts*. Ph.D.: University of California, Davis.

Soto, M., & Ambrose, R. (2016). Screencasts: formative assessment for mathematical thinking. *Technology, Knowledge and Learning, 21*(2), 277–283. https://doi.org/10.1007/s10758-015-9272-6.

Student Screencasting with the iPad (2014). ETEC 510, 9 March. http://etec.ctlt.ubc.ca/510wiki/index.php?title=Student_Screencasting_with_the_iPad&oldid=59796.

Sugar, W., Brown, A., & Luterbach, K. (2010). Examining the anatomy of a screencast: Uncovering common elements and instructional strategies. *The International Review of Research in Open and Distance Learning, 11*(3), 1–20.

Wakefield, J. A., Frawley, J. K., Dyson, L. E., Tyler, J. V., & Litchfield, A. J. (2011). Increasing student engagement and performance in introductory accounting through student-generated screencasts. *AFAANZ Conference, Darwin, Austraila, July 2011 AFAANZ Conference Proceedings* (pp. 1–27). Melbourne.

Watson, A. (2002). Teaching for understanding. In L. Haggerty (Ed.), *Aspects of teaching secondary mathematics: Perspectives on practice* (pp. 153–163). London: RouteledgeFalmer.

Watson, A., De Geest, E., & Prestage, S. (2003). *Deep Progress in Mathematics: The Improving Attainment in Mathematics Project*. Oxford: University of Oxford Department of Educational Studies.

**Linda Galligan** is an Associate Professor and Coordinator of the Discipline of Mathematics and Statistics within the Faculty of Health Engineering and Sciences at the University of Southern Queensland. She teaches into first year mathematics, and has strong links with schools providing mathematics programs and activities for students and teachers. Her research includes language and mathematics; academic numeracy, and preparation for numeracy demands of university. More recently her focus has been on the use of Tablet technology to effectively teach and understand mathematics; and using modelling to improve pre-service teachers' deep understanding of mathematics.

**Carola Hobohm** is an Educational Designer with University of Sunshine Coast's Centre for Support and Advancement of Learning and Teaching. She holds a MBA in Knowledge Management. She has over 25 years' experience in ICT training and support in the private sector, and has been active in higher education learning and teaching for the past twelve years. Her research interests include effective and innovative uses of educational technologies in online and blended learning environments, with particular focus on tablet technologies and interactive media in academic disciplines of science, technology, engineering, and mathematics.

# The Use of Mobile Technologies in the Primary School Mathematics Classroom—Developing 'Create-Alouds'

**Anne Prescott and Damian Maher**

**Abstract** Traditionally, learning mathematics has often been limited to pen and paper and sometimes hands-on activities. Mobile technologies offer the opportunity to change practices within primary school mathematics classes. This chapter explores how Year 5 and 6 students worked collaboratively to solve a problem, explaining their mathematical thinking during that process. Their use of screen-casting apps such as *Explain Everything* and *Educreations* to produce 'create-alouds' helped them collaboratively understand and explain mathematical concepts. The apps also assisted teachers in being able to provide formative assessment and feedback to the students, while enhancing the 21st century skills of the students.

**Keywords** Tablets · Mathematics · Primary schools · Create-alouds
Assessment

## Introduction

"Students develop understanding and fluency in mathematics through inquiry, exploring and connecting mathematical concepts, choosing and applying problem-solving skills and mathematical techniques, communication and reasoning" [New South Wales Boards of Studies, n.d. (NSW BOS)].[1]

[1]NSW BOS: In NSW, Australia, the Board of Studies is the accrediting authority for schools. It is now known as NESA (New South Wales Educational Standards Authority).

A. Prescott (✉) · D. Maher
University of Technology Sydney, Ultimo, Australia
e-mail: anne.prescott@uts.edu.au

D. Maher
e-mail: damian.maher@uts.edu.au

 
N. Calder et al. (eds.), *Using Mobile Technologies in the Teaching and Learning of Mathematics*, Mathematics Education in the Digital Era 12,
https://doi.org/10.1007/978-3-319-90179-4_16

Thinking, reasoning or working mathematically is widely seen as providing an essential basis for future learning, for effectively participating in society, and for conducting personal activities. When thinking mathematically, a student uses previously acquired knowledge, skills and understanding to satisfy the demands of an unfamiliar situation. Many problem situations in everyday life involve the ability to reason, and the ability to approach problems in systematic ways. When primary school teachers are reticent about teaching mathematics and in particular, problem solving, their classes miss out on developing this essential skill.

> Students develop the ability to make choices, interpret, formulate, model and investigate problem situations, and communicate solutions effectively. They formulate and solve problems when they use mathematics to represent unfamiliar or meaningful situations, design investigations and plan their approaches, apply strategies to seek solutions, and verify that their answers are reasonable. (Australian Curriculum, n.d)

Vincent and Stacey's study (2008) showed that textbooks have "repetitive problems of low procedural complexity" (p. 82) and often problem solving has its own chapter with little connection to content. This reliance on textbooks often produces shallow teaching, which has been a feature of many Australian mathematics classrooms with considerable repetition while featuring an absence of deductive reasoning (Boaler, 2015; Hiebert et al., 2003).

Meaningful learning occurs when students build the knowledge and cognitive processes needed for successful problem solving (Mayer, 2002) and for this to occur, instruction must go beyond factual knowledge, and assessment must go beyond recalling or recognizing that knowledge. Students need to make sense of their experiences, pay attention to relevant incoming information, mentally organize that information, and integrate it with existing knowledge. These are all thinking mathematically processes, and therefore a universal theme in mathematics (Stacey, 2002).

While schools confront the serious challenge of disengaged students, many studies emphasise the need for more interesting, relevant classroom tasks to enhance engagement in learning (Russell, Mackay, & Jane, 2003). Engagement occurs when students are procedurally engaged within the classroom, participating in tasks and 'doing' the mathematics, and hold the view that learning mathematics is worthwhile, valuable and useful both within and beyond the classroom.

This project focused on the use of *Educreations* and *Explain Everything* apps on iPads as a means of enabling students to think mathematically by engaging with tasks in unfamiliar situations. *Explain Everything* and *Educreations* are screen-casting apps, allowing users to write and draw onto a space much like a traditional whiteboard. Different colours, pens and dusters can be used. Both apps allow the process to be recorded using voice over. *Educreations* is only available on iPads while *Explain Everything* can be used on both iPads and smart phones. Both apps were free at the time the research was undertaken but *Explain Everything* now has a cost. *Explain Everything* has more tools than *Educreations*, but essentially they are the same.

We sought to answer the following research questions:

How can the use of apps such as *Educreations* and *Explain Everything* facilitate students' understanding of mathematical concepts in the primary school mathematics classroom? How can these apps be used to support assessment and feedback?

## Literature Review

The focus of the review explores concepts associated with thinking, reasoning and working mathematically, using tablets and associated apps in the primary mathematics classroom, in particular, emphasising assessment and feedback practices.

### *Thinking, Reasoning and Working Mathematically*

Gordon Calvert (2001) discussed the move away from school mathematics lessons where students find the right answer by providing an agreed explanation and justification for a particular strategy and solution. This move coincided with the transition from mathematics being conceived as a solitary practice to one in which it becomes a group activity (Sullivan, 2011). Approaches fostering student thinking, reasoning and working mathematically are consistent with frameworks for quality teaching (Newmann, Marks, & Gamoran, 1996; NSW Department of Education and Training, 2003). Opportunities for thinking mathematically are at the core of many mathematics syllabuses (see for example the Australian Curriculum) because they:

- involve making decisions about what mathematical knowledge, procedures and strategies are to be used in particular situations—rather than simply having rules to follow—thereby allowing consideration of alternate solutions
- incorporate communication skills and ways of thinking that are mathematical in nature—giving students voice and developing critical thinking
- promote engagement in challenging mathematical investigations
- promote higher-order thinking
- develop deep knowledge and understanding
- develop students' confidence in their ability 'to do' mathematics
- connect learning to the students' real world (Queensland Curriculum and Assessment Authority, 2004).

## *Using Tablets and Create-Alouds*

In focusing on the concept of create–aloud, this paper draws on aspects of a think-aloud, which is a verbal technique traditionally used to support students in the development of their literacy (Wade, 1990). During the think-aloud, students can stop reading and verbalise their ideas with the teacher (Block & Israel, 2004), which helps bring the thinking out into the open so that it can be replicated in the future (Oster, 2001).

Teale and Martinez (1996) suggest that the most effective talk provides students with opportunities to reflect rather than expecting a quickly retrieved answer. Likewise, Dickinson and Smith (1994) found that talk was most beneficial when students reflected on story content or language.

### Ensuring Teaching Practice Is Current and Aligned with Learning Futures Expectations

In a create-aloud environment, students use apps such as *Educreation* and *Explain Everything* to create a movie as they engage in discussion individually or with each other (without input from the teacher) incorporating multimodal resources, which can support their mathematical development. It is through this discussion and development of a resource that students create and test a mathematical hypothesis. The focus is more on process than product. The teacher can then access the resource and assess students' understanding from their procedural explanations. The 'create-aloud' extends the idea of a verbal think-aloud to include multimodal resources as well as the opportunity to record the whole process they undertake in order to solve the problem.

A number of studies have been conducted focusing on the use of mobile devices and how they can support mathematical understanding for primary school students. Tablets allowed the teacher to introduce a wider range of teaching strategies (Attard & Curry, 2012) and enhance, augment and support deeper learning (Clark & Luckin, 2013).

Tablets allow multimodal interaction for input and output, including visual and audio interactions (Wang & Karlström, 2012). For example, in a study undertaken with Year 6 and 9 students in Greece, it was found that "tablets supported education by adding visual elements …" (Soykan, 2015, p. 240), which can be drawn upon to support mathematical meaning or thinking mathematically, as well as incorporating video, voice and written text.

There has been a recent focus by researchers on the use of screen-casting apps such as *Explain Everything* and *Educreations* to support mathematical understanding. These apps allow students to access content and, importantly, create content where they explain their mathematical understanding. According to Pelton and Francis Pelton (2013), "to explain their understanding of a concept is one of the best ways to both consolidate and assess [students'] understanding of the topic" (pp. 4843–4844).

## *Assessment and Feedback*

Another advantage of using tablets is they can effectively support assessment and feedback practices (Maher, 2013). Teacher assessment is an important part of the learning cycle because it enhances students' educational growth (Falchikov, 2005). Formative assessment is particularly effective as it enables teachers to understand student thinking, and then modify future lessons to ensure that content is scaffolded at the appropriate level to adapt teaching so as to close the gap between the student's current state of learning and the desired state (Heritage, 2007).

Tablets and apps like *Explain Everything* allow for richer narrative feedback by the teacher (Richards & Meier, 2016) because they allow students' narratives to be recorded and teachers can revisit the students' explanations. Using tablets also provides instant feedback (Attard, 2013; Ciampa & Gallagher, 2013), which has been shown to be more effective than delayed feedback (Mason & Bruning, 2001).

Teachers need to look for opportunities to give students effective and timely feedback that is not associated with grades (Wiliam, 2012). Traditionally, summative feedback has occurred through testing but opportunities for formative feedback can be seen to occur during collaborative work when teachers, their peers, and even themselves (also called self-assessment) reflect on their mathematics. In the whole class setting, both summative and formative feedback occurs when students share their process and product.

Peer feedback is a "communication process through which learners enter into dialogues related to performance and standards" (Liu & Carless, 2006, p. 280). It can be beneficial for learning and is a form of collaborative learning (Van Gennip, Segers, & Tillema, 2010). Importantly, peer feedback supports students in taking an active role in the management of their own learning. Peer feedback allows for more timely and frequent feedback to students compared to feedback only from the teacher. Peer feedback can enable students to better self-assess.

Self-assessment is also an important way to support mathematical understanding. Boud (1991) defines self-assessment as "the involvement of students in identifying standards and/or criteria to apply to their work and making judgments about the extent to which they have met these criteria and standards' (p. 4)." It is an important self-reflective activity. "Self-assessment can involve both description (i.e., these are the characteristics of my work) and evaluation (i.e., this is how good my work is and what it is worth)" (Brown, Andrade, & Chen, 2015, p. 444). Screen-casting apps like *Educreations* and *Explain Everything* are effective in supporting self-assessment as they allow students to revisit the process involved with their thinking.

While screen-casting apps are increasingly being used for mathematics in primary schools (Kearney & Maher, 2012), there has been limited research in this area. By listening in as students think aloud, teachers can diagnose students' strengths and weaknesses. "When teachers use assessment techniques such as observations, conversations and interviews with students, or interactive journals, students are likely to learn through the process of articulating their ideas and answering the teacher's questions" (National Council of Teachers of Mathematics, 2000 cited in Fouche, 2013). Tablets particularly lend themselves, therefore, to Assessment for Learning (formative assessment) as well as Assessment as Learning (self-assessment and peer-feedback).

## Theoretical Framework

The theoretical framework draws on aspects of Vygotsky's (1978) work, which viewed language as both a psychological and a cultural tool. He proposed that engagement in social interaction fosters the development of individual cognitive abilities. "Vygotsky (1986) showed that children need to rehearse language in different contexts and that through collaboration with and feedback from their peers and others they will learn to modify their ideas and refine their expression of them" (Monaghan, 2005, p. 84).

The place of talk is thus significant in helping students develop mathematical understanding and the notion of dialogic interactions are used to analyse the students talk, drawing on the work of researchers from the 'Thinking Together' project (Littleton et al., 2005; Mercer, 1996). The analytical framework also draws on the notion of multimodality, which examines how students draw upon a variety of resources, including written text, image, animation, sound (spoken words, sound effects and music) and colour to make meaning (Kress & Van Leeuwen, 2001; Jewitt 2006).

When students think out loud during a problem-solving process there is opportunity for peer- and self-assessment. They can reflect on the steps used to solve the problem in mathematics, write or record what they say, making thought processes as explicit as possible during task performance, hence giving observers insight into their cognitive processes (rather than only the final product). This allows teachers to observe which strategies students use, pinpointing the individual student's needs so there is the opportunity to provide appropriate instruction.

When students are given a task that enables them to think mathematically, using the 'create-aloud' environment, and are given feedback from themselves, peers and the teacher, they are enabled in their mathematics learning. Figure 1 shows how these three aspects of 21st century learning skills intersect in the central 'curved triangle' where 'create-alouds' enhance the possibilities for students thinking mathematically with formative assessment as a natural part of the process.

## Study Design

This project involved primary mathematics students and teachers using a number of apps on tablets enabling them to think mathematically in an inquiry-based approach in mathematics classes to enhance student learning.

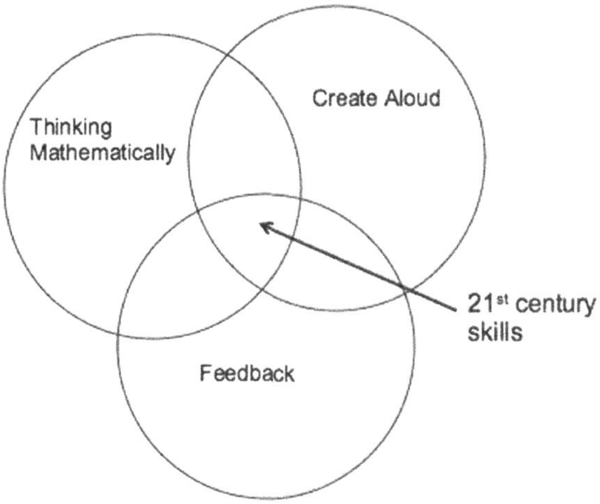


**Fig. 1** The thinking mathematically, create-alouds, feedback framework

The project draws on qualitative methodology (Lincoln & Guba, 1985), including aspects of case study methods (Miles & Huberman, 1994). The data collected included questionnaires and interviews completed by both Year 5 and 6 students and teachers, focus groups with teachers and students, and observations of lessons. Additionally, the 'create-alouds' created by students were also analysed.

## *Participants and Data Collected*

The project was conducted in a suburban NSW primary school with a population of approximately 840 students. Sixteen teachers from grade three to grade six participated in three professional learning sessions as a whole group over the four-week life of the study. The teachers' initial questionnaire asked them about apps they would like to use for the project. All sixteen teachers implemented the use of tablets into their teaching over the life of the project and twelve teachers completed a survey at the end.

Nine teachers in grades five and six participated in teaching that was documented in greater detail. In total, 200 students were observed engaging in one of the two lessons—How many people could fit on the basketball court? How many boxes could fit inside their classroom? These lessons were observed and video recorded. Either *Educreations* or *Explain Everything* apps were used by the students, depending on which apps they had on their tablets. Prior to the lesson each teacher and a researcher participated in a one-hour discussion of the lesson plan, its aims and objectives. A debriefing session followed each lesson. The teachers participated in a grade-level focus group. Twenty students were randomly selected from both grade levels to participate in one of four focus groups. Additionally, some of the 'create-alouds' were collected and analysed.

Table 1 describes how the students and teachers used the apps in their presentations, assessment and feedback. They had other apps on their tablets and used them during their investigations (Table 2).

**Table 1** *Explain Everything* and *Educreations* apps used by the students

| Apps | Thinking, reasoning and working mathematically | Assessment/feedback |
|---|---|---|
| *Explain Everything*<br>*Educreations* | • involve making decisions about what mathematical knowledge, procedures and strategies are to be used<br>• incorporate communication skills<br>• promote engagement in challenging mathematical investigations<br>• promote higher-order thinking<br>• develop deep knowledge and understanding<br>• develop students' confidence in their ability 'to do' mathematics<br>(Queensland Curriculum and Assessment Authority, 2004) | Teacher had information about all decisions because they were incorporated in the presentation<br>Assessment was ongoing because the teachers had access to the process and the product, not just the product<br>Two elements to the feedback. On the one hand students gave feedback to their peers and teachers gave feedback and on the other hand, there was the self-assessment process prior to the final submission |

**Table 2** Other apps used by the students

| App | Functionality |
|---|---|
| MagicPlan | MagicPlan creates a professional floor plan. It measures rooms and draws floor plans just by taking pictures |
| EasyMeasure | EasyMeasure shows the distance to objects seen through the camera lens of the device. The students aim their iPad to any object and it displays the distance towards that object on top of the camera image |
| RoomScan | RoomScan allows the user to draw floor plans by touching each wall with the device |

### *Data Analysis*

In analysing the data, thematic analysis was used to identify, analyse and report patterns within the data (Braun & Clarke, 2006, p. 79). This form of analysis is well suited when drawing on different sources, which in this study included spoken text, visual observations and the multimodal 'create-aloud' texts. All data were watched/ read and key themes identified. These themes were then cross-checked across each form of data.

## Results

### *Use of Multiple Modes of Representation*

One of the clear advantages noted in using the apps was that it provided students with an increased number of modes in which to capture and explain their mathematical thinking. For example, they had *Educreations* for their presentation and included information from a number of apps to get measurements, 3D diagrams, calculations, images, etc. The students explained in a focus group discussion that this made it easier for them to complete their work:

S: When we did our math activity we took a screenshot and we put it into the app.
R: Does that make it easier to have the visual image there when you want to work with it?
S: Yea.

Below is an example from one student's 'create-aloud'. The students used a measurement app to measure the room, importing the shape into the *Educreation* app to provide a visual representation (see MagicPlan in Table 2). Their student talk explained the mathematics in their diagram.

Student talk: To find the volume, we need to know the length, the width and the height. When we times all our measurements together it gives us volume (Fig. 2).

Students indicated in the questionnaires that they valued the use of multiple modes for recording their work:

S: The *Explain Everything* app helped me record work because it had a variation of different tools to use.
S: It helped us complete the activity and present it in a neat and comprehensive way by enabling us to type, draw, insert diagrams and record our voice.

Teachers in the focus group also expressed this idea in terms of supporting students' understanding:

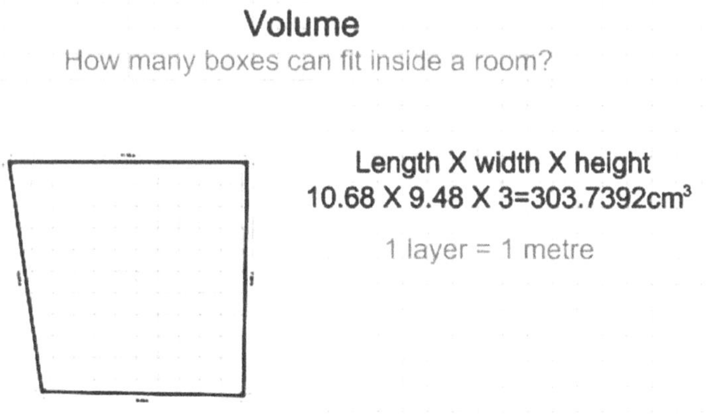


**Fig. 2** Written text used by a student to explain their calculations

R: Do you think the students having a record of what they're doing makes it easier for them to identify the process?
T: Yes, because they are drawing on both the verbal and visual.
R: Do you think students using the different modes helps them to learn in a better way or different way?
T: Different, I think it opens it up because it's not just two dimensional.

Drawing on the example above, students also believed that their ability to explain the mathematical concepts to their audience improved by using the *Explain Everything* app and that the audience would gain understanding from viewing their presentation. In the example above the students were able to verbally explain how they calculated volume whilst illustrating the different dimensions.

Drawing on an expanded number of modes assisted the students, as they discussed in a focus group session:

R: Do you think using *Explain Everything* is better than using pen and paper for Maths?
T: Yes, it helps you explain better.
R: How does it help you to explain better?
S: Because you can talk and you can write and you can draw.
S: *Explain Everything* let me draw and write to explain my answer, and also recording was one thing our whole group contributed to that helps the audience to know clearly how we got the answer in our own words.

## *Collaboration*

In the teacher focus group, the teachers stated that one of the important uses of the apps was that it allowed students to collaborate with each other. Some of the examples provided by the teachers were ways in which students decided how they would manage the investigation, what apps they would use and how they would explain their understanding. In particular, using *Explain Everything* or *Educreations* allowed them to critique their explanation and then easily make adjustments so their thinking was clear.

The use of the apps changed the way the teachers used collaboration in mathematics.

T: That's where group work comes into it—working together to discuss a problem. That's where collaborative group work is really important.
R: Do you think when the students use *Educreations* in groups that it gives you a different way of understanding their thinking?
T: I have changed the way I do maths because of this thing, the way that you can get kids to collaborate with each other.

And the student voice:

S: I enjoy collaborating with other students and using that app.

One of the benefits of technology is that it provided opportunities for the development of 21st Century learning skills such as collaboration, critical thinking, creative thinking, communicating, and digital literacy. The use of the apps facilitated such opportunities and, therefore, aided the development of those skills as illustrated by teachers in focus group discussions:

T: When you think about it, when these kids have to go into the workforce is not too many jobs now where people don't have to work collaboratively.
T: *Educreations* promoted 21st century learning by students communicating, collaborating.

## *Assessment*

One of the key themes that emerged from discussion with the teachers was the way in which the 'create-alouds' facilitated their assessment of the students. The teachers could follow the narrative of the students' thinking rather than just accessing the end product as illustrated in the following extract.

R: How do you find listening to what they have to say assist your thinking? Does it help you differently than if you just read something on a piece of paper?

Teacher: It's like 'do you understand the process?' because if you ask a child what is 2 × 3 they say 6. But using *Educreations* they explain the process like 'I have two groups with 3 things in each group therefore it is 6'. It gives them a much better opportunity to explain the process of multiplication than it does if you just give them the questions and they answer them.

The ability of the 'create-alouds' to capture students' understanding in a different way provided teachers with more information to support their assessment practices in both the spoken and written language that students used. During a focus group discussion, the teachers were asked how the use of the apps could support students:

T: I think it also exposes them to using mathematical language. This is an area of weakness for many children. The app allows them to verbalize that understanding using the correct language. In their books you can't get an actual vision of their understanding.

In the example below, (Fig. 3) the student explained her work by both writing out the process as well as talking through the process. This allowed the teacher to understand the type of both written and verbal language the student was drawing on.

Student talk: Fifty two by zero point seven which will gives us 36 cubic centimetres [sic].

In the extract above, the teacher can observe that the student understands the concept of volume and is using correct written terminology for cubic centimetres as well as saying it the correct way, even though the student arrived at the wrong answer. To assist students to be able to use mathematical concepts they learn in their everyday lives, it is important to assess and provide feedback on both their written and spoken mathematical language.

The app was important when the teacher sat down with the student to provide feedback. The ability to be able to go over the recording was valued by this teacher:

T: Coupled with this, get them to sit down and talk through it. The difference is when they recorded it you can listen to it over and over.

Not only was the teacher's ability to assess and provide feedback enhanced through the use of the 'create-alouds', so too was the opportunity for students to

**Fig. 3** Student drawing

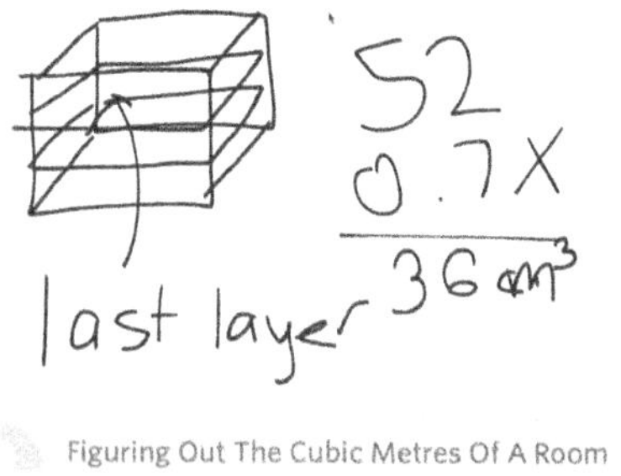

self-assess. The ability for students to set out their thinking and revisit it was fundamental in assisting them to critically reflect on their mathematics, especially as it was in their own voice with their own understandings, as explained by a teacher in the questionnaire:

T: Student listening, going back to their own explanations, is really quite powerful.

The students also commented on the ability of the app to allow them to set out their thinking in a procedural way as indicated by one student in a questionnaire response when asked how the use of the app helped them to learn mathematical content:

S: I used *Educreations* and it's good because if you need to show your class something you can record it then show it step by step.

During the construction of their 'create-alouds', we also observed students providing each other with feedback. The students focused on the mathematical content such as the process they might use to come to a solution with multiplication, as well as providing feedback about the layout of their 'create-alouds' and the types of media they wanted to use. Throughout these discussions there was evidence of listening, responding, questioning and debate allowing the students to shape and refine their mathematical thinking.

## Discussion

### *Multiple Modes of Representation*

The apps allowed students to engage in a think-aloud protocol drawing on multiple resources where they could explain the processes and steps involved in mathematical problem solving. The modes that they could draw upon included text, images, colour, audio and video. Traditionally, students have not been able to draw on all of these modes, but the use of audio and video added functionality to the explanation process. Significantly, the 'create-alouds' allowed students to show their thinking in ways that are not possible using pen and paper.

Both the teachers and the students saw the use of multiple modes of representation and the setting out of their work as a narrative as a way of supporting the learning process. The teachers felt that the students learned in different ways. The students were able to develop their written skills as well as their spoken vocabulary. The speaking-out-aloud practice is not normally an aspect of students' learning when using pen and paper. The verbalisation process enabled students to better develop their thinking, reasoning and working mathematically skills. When students checked the resource they had recorded as a 'create-aloud' movie, they engaged in a more critical way with their ideas than they would with pen and paper, where often revision is not part of the process. The students were able to externalise and make explicit their own thinking to themselves (Monaghan, 2005) and others. Through the revision of their final products an important aspect of self-assessment came into play because the students were conscious of producing a 'create-aloud' that would be accurate and able to be understood by others.

As well as learning and demonstrating mathematical skills and knowledge in using the apps, both the teachers and students believed that they were also developing one of the 21st century learning skills—collaboration.

Collaboration is an important mechanism in assisting students' understanding of mathematical concepts. Billett and Choy (2013) found that collaboration supported learners in extending their knowledge in ways that they could not achieve independently, especially in terms of peer feedback. "Vygotsky (1986) showed that children need to rehearse language in different contexts and that through collaboration with their peers they will learn to modify their ideas and refine their expression of them" (Monaghan, 2005, p. 3). Henderson and Yeow (2012) found that the use of mobile devices facilitated collaboration between children and also allowed them to more easily engage with content. The creation of their solution to the problem enabled the students to engage with the mathematics in ways they found exciting. Importantly, the use of the tablet facilitated the process of collaboration and peer review in new ways where students could better review what they had done (Heinrich, 2012).

Collaboration is not a new concept and many contemporary teachers use group work as a way of supporting mathematical learning for their students. However, the accessibility of the tablet facilitated collaboration between students in new ways. As stated by Fisher, Lucas, and Galstyan (2013), the "size, portability, versatility and tactile nature of the iPad are four of the main factors that contribute to its accessibility" (p. 176).

## *Assessment and Feedback*

Traditionally, these teachers mainly used pen and paper tests or projects, but during the create alouds they were able to assess students in new ways. The formative assessment gained while the students were working on their problems gave the teacher a better understanding of each student's knowledge of mathematical content and ability to reason during problem solving. While the former could have been gained from a pen and paper test or a report in a project, understanding the student's ability to reason during problem solving is difficult in that situation.

In using the 'create-alouds', the teachers' practice changed, allowing them to follow both the process of student understanding as well as the final outcome. Importantly, 'create-alouds' became artifacts that teachers could pause and rewind as well as discuss with students as part of their feedback and assessment practices. Assessing the 'create-alouds' meant teachers had time to critically reflect on each student's mathematical work, which they could then discuss with the student the following day. This allowed students to link back to their thinking rather than having to rely on their memory, which can sometimes inhibit the learning process for young people.

Providing such detailed information and assessment allows teachers to purposefully plan for future lessons as illustrated by Williamson-Leadley and Ingram (2013): "Educreations together enabled the teachers to gather more detailed assessment data about their students' mathematical learning and assist them in deciding their next steps of teaching" (p. 133). The students shared their creations with the class so all could learn from each other—having seen the problem, and struggled with it, seeing someone else's results was a powerful learning experience.

One drawback though, as pointed out by Soto and Ambrose (2016), was that it was very time-consuming to assess every student's work constantly (which links back to the narrative nature of resources produced via the apps). Instead, teachers could focus on a specific number of students each day or week. The students shared their completed 'create-alouds' to the teacher's computer using *Instashare*, allowing the teacher to assess as many 'create-alouds' as desired at a time that suited them.

## Conclusions

This chapter has outlined the use of screen-casting apps used in mathematics, which provided a range of different resources such as text, audio, images and video to support students as they constructed 'create-alouds'.

The multiple modes of the apps on the tablets encouraged the students to be creative and facilitated their understanding by allowing them to engage with the mathematics. The video and audio aspects of the students' creations enabled teachers to discern their students' ability to think mathematically as well as to understand the content. The 'create-alouds' could be assessed as stand-alone products or in a conference between the teachers and students. They also provided opportunities for students to critically evaluate their own thinking and pinpoint potential problems.

Collaboration is an essential part of learning in the 21st century with use of the tablets enhancing other skills such as digital literacy and creative thinking. The use of apps enhanced student engagement and provided opportunities for the teachers to explore collaborative tasks and formative assessments within the mathematics syllabus.

## References


Attard, C. (2013). Introducing iPads into primary mathematics pedagogies: An exploration of two teachers' experiences. In *Proceedings of the 36th Annual Conference of the Mathematics Education Research Group of Australasia* (pp. 58–65).

Attard, C., & Curry, C. (2012). Exploring the use of iPads to engage young students with mathematics. In *Mathematics Education: Expanding Horizons: Proceedings of the 35th Annual Conference of the Mathematics Education Research Group of Australasia*. Singapore: MERGA. Retrieved from http://www.merga.net.au/documents/Attard_&_Curry_2012_MERGA_35.pdf.

Australian Curriculum. (n.d). *Problem solving portfolio summary*.

Billett, S., & Choy, S. (2013). Learning through work: Emerging perspectives and new challenges. *Journal of Workplace Learning, 25*(4), 264–276.

Block, C. C., & Israel, S. E. (2004). The ABCs of performing highly effective think alouds. *The Reading Teacher, 58*(2), 154–167.

Boaler, J. (2015). *What's math got to do with it?*. New York: Penguin.

Boud, D. (1991). *HERDSA green guide, no.5: Implementing student self-assessment* (2nd edn). Sydney: Higher Education Research and Development Society of Australasia.

Braun, V., & Clarke, V. (2006). Using thematic analysis in psychology. *Qualitative Research in Psychology, 3*(2), 77–101.

Brown, G. T., Andrade, H. L., & Chen, F. (2015). Accuracy in student self-assessment: Directions and cautions for research. *Assessment in Education: Principles, Policy & Practice, 22*(4), 444–457.

Ciampa, K., & Gallagher, T. L. (2013). Professional learning to support elementary teachers' use of the iPod Touch in the classroom. *Professional Development in Education, 39*(2), 201–221.

Clark, W., & Luckin, R. (2013). *What the research says: iPads in the classroom*. London Knowledge Lab, Institute of Education, University of London.

Dickinson, D. K., & Smith, M. W. (1994). Long-term effects of preschool teachers' book readings on low-income children's vocabulary and story comprehension. *Reading Research Quarterly, 29*, 104–122.

Falchikov, N. (2005). *Improving assessment through student involvement: Practical solutions for aiding learning in higher and further education*. Routledge.

Fisher, B., Lucas, T., & Galstyan, A. (2013). The role of iPads in constructing collaborative learning spaces. *Technology, Knowledge and Learning, 18*(3), 165–178.

Fouche, J. (2013). *The effect of self-regulatory and metacognitive strategy instruction on impoverished students' assessment achievement in physics*. Doctoral dissertation, Liberty University. Retrieved from http://digitalcommons.liberty.edu/cgi/viewcontent.cgi?article=1711&context=doctoral.

Gordon Calvert, L. M. (2001). *Mathematical conversations within the practice of mathematics*. New York, NY: Peter Lang.

Heinrich, P. (2012). *The iPad as a tool for education*. NAACE and 9ine Consulting. Retrieved from http://www.maltonschool.org/files/ipad-research-papers/The-iPad-as-a-Tool-for-Education-study-NAACE-9ine-Consulting.pdf.

Henderson, S. & J. Yeow. (2012). iPad in Education—A Case Study of iPad Adoption and Use in a Primary School. In *HICSS'12 Proceedings of the 2012 45th Hawaii International Conference on System Sciences* (pp. 78–87).

Heritage, M. (2007). Formative assessment: What do teachers need to know and do? *Phi Delta Kappan, 89*(2), 140.

Hiebert, J., Gallimore, R., Garnier, H., Givvin, K. B., Hollingsworth, H., Jacobs, J., et al. (2003). *Teaching mathematics in seven countries: Results from the TIMSS 1999 video study*. Washington, DC: National Centre for Education Statistics, U.S. Department of Education.

Jewitt, C. (2006). *Technology, literacy and learning: A multimodal approach*. London: Routledge.

Kearney, M., & Maher, D. (2012). Mobile learning in maths teacher education: Driving pre-service teachers' professional development. *Australian Educational Computing, 27*(3), 78–86.
Kress, G., & Van Leeuwen, T. (2001). *Multimodal discourse: The modes and media of contemporary communication.* Oxford UK: Oxford University Press.
Lincoln, Y., & Guba, E. (1985). *Naturalistic inquiry.* Newbury Park: Sage Publications.
Littleton, K., Mercer, N., Dawes, L., Wegerif, R., Rowe, D., & Sams, C. (2005). Talking and thinking together at key stage 1. *Early Years: An International Journal of Research and Development, 25*(2), 167–182.
Liu, N. F., & Carless, D. (2006). Peer feedback: The learning element of peer assessment. *Teaching in Higher Education, 11*(3), 279–290.
Maher, D. (2013). Pre-service primary teachers' use of iPads to support teaching: Implications for teacher education. *Educational Research for Social Change, 1*(2), 48–63.
Mayer, R. E. (2002). Rote vs meaningful learning. *Theory into Practice, 41*(4), 226–232.
Mason, B. J., & Bruning, R. (2001). *Providing feedback in computer-based instruction: What the research tells us.* Retrieved February 15, 2007.
Mercer, N. (1996). The quality of talk in children's collaborative activity in the classroom. *Learning and Instruction, 6*(4), 359–377.
Miles, M., & Huberman, A. (1994). *Qualitative data analysis: An expanded sourcebook* (2nd ed.). Thousand Oaks: Sage.
Monaghan, F. (2005). Don't think in your head, think aloud': ICT and exploratory talk in the primary school mathematics classroom. *Research in Mathematics Education, 7*(1), 83–100.
Newman, F. M., Marks, H. M., & Gamoran, A. (1996). Authentic pedagogy and student performance. *American Journal of Education, 104,* 28–312.
NSW Department of Education and Training. (2003). *Quality teaching in NSW public schools.* Sydney: Professional Support and Curriculum Directorate. NSW BOS Syllabus. (2012). Working Mathematically. https://syllabus.bostes.nsw.edu.au/mathematics/mathematics-k10/working-mathematically-outcomes/.
New South Wales Board of Studies. (n.d). *Working mathematically outcomes.* Retrieved from http://syllabus.nesa.nsw.edu.au/mathematics/mathematics-k10/working-mathematically-outcomes/.
Oster, L. (2001). Using the think-aloud for reading instruction. *The Reading Teacher, 55*(1), 65–69.
Pelton, T., & Francis Pelton, L. (2013). Using an iPad to Explain Everything by creating interactive activities and vignettes. In *Proceedings of Society for Information Technology & Teacher Education International Conference* (pp. 4843–4847).
Queensland Curriculum and Assessment Authority. (2004). *Mathematics KLS Year 1 to 10. Thinking, reasoning and working mathematically.* Retrieved from https://www.qcaa.qld.edu.au/downloads/p_10/kla_maths_pd_trw.ppt.
Richards, R., & Meier, E. B. (2016). Leveraging mobile devices for qualitative formative assessment. In *Handbook of Research on Mobile Learning in Contemporary Classrooms* (pp. 94–115). IGI Global.
Russell, V. J., Mackay, T., & Jane, G. (2003). *Messages from MYRAD (Middle Years Research and Development): Improving the middle years of schooling.* Melbourne: Independent Association of Registered Teachers of Victoria.
Soto, M. M., & Ambrose, R. (2016). Making students' mathematical explanations accessible to teachers through the use of digital recorders and iPads. *Learning, Media and Technology, 41* (2), 213–232.
Soykan, E. (2015). Views of students', teachers' and parents' on the tablet computer usage in education. *Cypriot Journal of Educational Sciences, 10*(3), 228–244.
Stacey, K. (2002). Problem solving: A universal theme. In L. Grimison, & J. Pegg (Eds.) *Teaching secondary school mathematics* (pp. 208–229). Melbourne: Thomson Learning.
Sullivan, P. (2011). *Teaching mathematics: Using research-informed strategies.* Melbourne: Australian Council for Educational Research.

Teale, W. H., & Martinez, M. G. (1996). Reading aloud to young children: Teachers' reading styles and kindergartners' text comprehension. In C. Pontecorvo, M. Orsolini, B. Burge, & L. B. Resnick (Eds.), *Children's early text construction* (pp. 321–344). Mahwah, NJ: Erlbaum.
Van Gennip, N. A. E., Segers, M. S. R., & Tillema, H. H. (2010). Peer assessment as a collaborative learning activity: The role of interpersonal variables and conceptions. *Learning and Instruction, 20*(4), 280–290.
Vincent, J., & Stacey, K. (2008). Do mathematics textbooks cultivate shallow teaching? Applying the TIMSS video study criteria to Australian eighth-grade mathematics textbooks. *Mathematics Education Research Journal, 20*(1), 82–107.
Vygotsky, L. S. (1978). *Mind in society: The development of higher psychological processes.* Cambridge, MA: Harvard University Press.
Vygotsky, L. S. (1986). *Thought and language.* Cambridge, MA: MIT Press.
Wade, S. E. (1990). Using think alouds to assess comprehension. *The Reading Teacher, 43*(7), 442–451.
Wang, J., & Karlström, P. (2012). Mobility and multi-modality an exploratory study of tablet use in interaction design learning. In *mLearn* (pp. 276–279).
Wiliam, D. (2012). Feedback: Part of the system. *Feedback for Learning, 70*(1), 30–34.
Williamson-Leadley, S., & Ingram, N. (2013). Show and tell: Using iPads for assessment in mathematics. *Computers in New Zealand Schools: Learning, Teaching, Technology, 25*(1–3), 117–137.

**Anne Prescott** Associate Professor Anne Prescott works with mathematics teachers and teacher education students focusing on improving practice in the classroom. She started her career as a secondary school teacher of mathematics and is currently the coordinator of the primary and secondary mathematics teacher education as well as the Master of Teaching (secondary). Anne is the co-Chief investigator of a large Australian Mathematics and Science Partnership Program Grant with Dr Mary Coupland. The aim of the project is to develop resources to demonstrate to students and their teachers the importance of mathematics and statistics in our lives. She was also a researcher in the project: Investigating technology-mediated, project-based learning: Case studies of four schools. For a number of years, Anne has been working with teachers in a remote area of Nepal, improving the pedagogy within the classroom.

**Dr Damian Maher** is a senior lecturer in teacher education at the University of Technology Sydney in the School of Education. He focuses on the use of educational technology both in his teaching and research. His research spans primary schools through to universities. Some of the learning technologies he has investigated include the interactive whiteboard and mobile devices. He has investigated technology use across a number of subject areas including mathematics.

# Using Show and Tell Apps to Engage Students in Problem-Solving in the Mathematics Classroom

**Naomi Ingram, Keryn Pratt and Sandra Williamson-Leadley**

**Abstract** Show and Tell apps, which record students as they speak and write on a tablet, have a number of affordances for student learning in mathematics. One of these affordances is their utility in engaging students in problem-solving processes. Three iterations of research into Show and Tell apps present evidence that using Show and Tell apps for problem-solving can lead to improvements in the level and quality of student engagement. Students are encouraged to socially negotiate their understandings and Show and Tell apps can make student thinking more visible during this process. The apps also scaffold students to reflect on the processes they used for problem-solving.

**Keywords** Mathematics · Technology · Problem-solving · Engagement
Tablet · Show and tell apps · Group work · SAMR · TPACK

## Problem-Solving in Mathematics

It is generally accepted that learning mathematics involves more than mastery of facts and procedures (Schoenfeld, 1992). Students need to be actively involved in solving problems (Holton, Neyland, Neyland, & Thomas, 1999). They need to learn how to reduce a problem to a mathematical form and to make sense of it by using the tools of abstraction, symbolic representation, and symbolic manipulation (Schoenfeld, 1992). They need to "wonder why things are, to inquire, to search for solutions, and to resolve incongruities" (Hiebert et al., 1996, p. 12). Problem-solving is therefore an important classroom practice as reflected in

N. Ingram (✉) · K. Pratt · S. Williamson-Leadley
University of Otago, Dunedin, New Zealand
e-mail: naomi.ingram@otago.ac.nz

K. Pratt
e-mail: keryn.pratt@otago.ac.nz

S. Williamson-Leadley
e-mail: sandra.williamson-leadley@otago.ac.nz

 
N. Calder et al. (eds.), *Using Mobile Technologies in the Teaching and Learning of Mathematics*, Mathematics Education in the Digital Era 12,
https://doi.org/10.1007/978-3-319-90179-4_17

mathematics curricula worldwide. For example, problem-solving is at the heart of the Singaporean curriculum (Ministry of Education, 2012) and is one of the four proficiencies in the Australian curriculum: Mathematics (Australian Curriculum, Assessment and Reporting Authority, 2016).

Problem-solving has multiple meanings across the field of mathematics education (see Bransford & Stein, 1984). These meanings often depend on the individual's beliefs about mathematics and range from working on rote exercises to doing mathematics as a professional (Schoenfeld, 1992). In this chapter, a mathematics problem is defined as a question or situation where the method of solution is not immediately obvious (Holton et al., 1999). As such, whether something is a problem depends on a range of factors, including students' knowledge and experience in solving problems of that type.

Doing mathematics is "an inherently social activity" (Schoenfeld, 1992, p. 335) where students learn mathematics by socially negotiating meaning (Jonassen, Carr, & Yueh, 1998). By interacting with other students and teachers, students can be exposed to new concepts in a more sophisticated manner than if they were exploring them individually (Cavanagh, 2016). When students share their problem-solving with others, they need to explain and justify their methods and these are reinforced or improved as they adjust their thinking, using others' ideas and the results of their investigations (Hiebert et al., 1996).

Mathematics teachers can support students in problem-solving in a number of ways. They can develop a classroom culture that supports the social negotiation of problem-solving (Pennant, 2013). They can also ensure a wide range of materials are available (Jones, 2013) and they can work on specific aspects of the problem-solving process with students, such as using a mnemonic device to reduce a word problem to a mathematical form (Bureau of Exceptional Education and Student Services, 2010). In addition, they can use technology in ways that support problem-solving.

This chapter will explore how technology, in the form of Show and Tell apps, can be used to engage students in problem-solving in the mathematics classroom through recording students working collaboratively as well as using 'Think Aloud" protocols to record their solving of the problem. We will first explore how technology can and has been used in mathematics classrooms, and how it can enhance engagement, before considering how technology can be used to support students' problem-solving in mathematics. Finally, we will report on a body of work that has explored the use of Show and Tell apps in mathematics classrooms, focusing on how their use has enhanced students' engagement in problem-solving.

## Using Technology in Mathematics Classroom

Technology allows teachers to provide a wider range of opportunities and better cater for student needs (Conole, 2012; Hammond, 2010). For example, through access to encyclopaedic referencing sites (e.g., Wolfram Alpha) and Internet search engines, technology can easily supply knowledge when the need for it is identified during the problem-solving process. Technology can also supply solutions and commentary about well-known mathematical problems (e.g., the four colour theorem). It can organise and help to record solutions and strategies (e.g., Excel, Notability), or provide access to virtual manipulatives to support problem-solving (see https://nzmaths.co.nz/virtual-manipulatives). Technology in mathematics is often used to instruct the learners (Jonassen et al., 1998) or to judge the learner's response and provide feedback about the correctness of the response (Kim & Hannafin, 2011).

When teachers decide to use technology in their classroom practices, they need to consider that "teaching with technology is a difficult thing to do well" (Koehler & Mishra, 2009, p. 67), and depends on factors such as teachers' level of professional development, their experience in the classroom, and their technological, pedagogical, and content knowledge (TPACK) (Koehler & Mishra, 2009). One way of categorising how teachers use technology is the Substitution, Augmentation, Modification Redefinition (SAMR) model (Cavanaugh, Hargis, Kamali, & Soto, 2013; Puentedura, 2009). This model describes technology use on a continuum where technology can enhance teaching and learning in the form of Substitution or Augmentation, or can progress into transforming it, via Modification or Redefinition (Puentedura, 2009). As Puentedura explained, the lowest level of use involves using technology to complete a task that was previously possible, such as using a word processor rather than typing or handwriting a document (Substitution). Moving along the continuum, Augmentation involves some improvement in functionality, due to the affordances of the technology, such as using features like cut and paste in a word processed document, or online dictionaries. Within the transformation half of the continuum, Modification occurs when tasks can be redesigned because of the technology, such as using graphing packages or allowing for collaborative writing. The final level in the continuum is Redefinition and this occurs when teaching and learning tasks that would not have been possible without technology are implemented. An example of a task at the Redefinition level would be using virtual manipulatives where objects expand to show their nets.

Teachers also need to consider both the affordances and constraints of a particular technology before deciding whether or not the particular technology will enhance teaching and learning (Koehler & Mishra, 2008), and whether or not its use will be transformative or merely a substitute for what is already possible. Researchers have found that one of the affordances of technology is enhanced student engagement (Hammond, 2010), as well as benefits in terms of accessibility, diversity, communication, and collaboration (Conole, 2012; Hammond, 2010). There are also specific affordances associated with different technology devices. For example, tablets are portable, easy to use, and promote social interactivity (Blackwell, 2014; Ng, 2015). Koehler and Mishra (2009), Sinclair, Chorney and Rodney (2016) and Ladel and Kortenkamp (2012) are further examples of single use tablet technology apps for to illustrate the concept of affordances and constraints. If we consider this concept in regards to supporting students' learning in mathematics, a single use application, such as *Chicken Coop Fractions*, "affords students to practice their [fractional knowledge] but the constraint is that the teacher is not able to make changes to make it specific to an individual student's needs or context" (Koehler & Mishra, 2009, p. 6).

## Engaging Students in the Mathematics Classroom with Technology

Enhancing students' engagement is a particularly important affordance of technology as low levels of student engagement in mathematics is viewed as detrimental to student learning (Sullivan, McDonough, & Harrison, 2004). The construct of engagement has its roots in the broader literature regarding affect and is related in a complex way to elements of students' relationships with mathematics, including views of mathematics, feelings about the subject, and identities (Ingram, 2011; McLeod, 1992). Students' engagement in mathematics is deemed to be vital to their acquisition of knowledge and strategies (Sullivan et al., 2004). Student engagement has been associated with a variety of academic, social, and emotional outcomes (Christenson, Reschly, & Wylie, 2012; Reeve, Jang, Carrell, Soohyn, & Barch, 2004), including finding that engagement is positively related to achievement (Dotterer & Lowe, 2011), motivation (Attard, 2012), and emotional wellbeing (Reschly, Huebner, Appleton, & Antaramian, 2008). Although engagement can be interpreted broadly as a student's participation in school (for example, Dotterer & Lowe, 2011), here engagement refers to the "behavioural intensity and emotional quality of a person's active involvement during a task" (Reeve et al., 2004, p. 147). As such, in this setting, engagement is considered to be students' involvement in the mathematical activity of the classroom and their commitment to learning the mathematical content.

Students' engagement in mathematics can be described by both its level and quality. The level is related to the strength of the engagement. The quality is related to the student's unique engagement skills. These engagement skills, as described by Ingram (2011), include: perseverance (continuing to do a mathematical task despite experiencing difficulty); integrity (searching for understanding as well as the correct answer); intimacy (emotional engagement in mathematics); independence (solving problems autonomously); concentration (the skill of remaining focused on the mathematics); utilisation of feelings (using negative affect as a signal to persevere or change strategy); cooperation (discussing mathematics with others); and reflection (being self-aware of problem-solving processes). Various aspects of the mathematics classroom have been found to have an impact on students' engagement, including classroom culture (Sullivan et al., 2004), the use of games (Bragg, 2012), a high level of student autonomy and involvement in decisions (Calder, 2013; Skilling, 2014), and the use of relevant contexts and student interests (Skilling, 2014).

There is a common understanding that students find using technology naturally engaging (Kuiper & de Pater-Sneep, 2014). Indeed, a number of studies have shown that using technology can be engaging for a wide variety of students, in a range of learning areas, and in a variety of contexts (O'Rourke, Main, & Ellis, 2013; Williamson-Leadley, 2016). However, this is not always the case (Kuiper & de Pater-Sneep, 2014; Selwyn, Potter, & Cranmer, 2009). The degree of engagement when students use technology depends on a range of factors, including what technology is being used, how it is being used and to what purpose, the wider school context, and other contextual factors (Kuiper & de Pater-Sneep, 2014; Selwyn et al., 2009).

Technology has also been found to have an impact on engagement in the mathematics classroom (Attard, 2014). A number of studies identified an increase in engagement amongst students learning mathematics when they used various forms of technology (see Attard & Curry, 2012; Chen, Liao, Cheng, Yeh, & Chan, 2012; O'Rourke et al, 2013). In contrast, Kuiper and de Pater-Sneep (2014) found that the students in their study preferred to use books rather than drill-and-practice software packages. From the research that has been conducted, it appears that in mathematics, as in other learning areas, how engaging technology is depends both on what technology is being used, and how.

## Using Technology to Explore Problem-Solving in Mathematics

In order to explore the potential of technology to enhance engagement in problem-solving in mathematics, a form of technology with the appropriate affordances needed to be chosen. After consideration of a number of apps, a set of apps, described as Show and Tell apps (Williamson-Leadley & Ingram, 2013), were identified as having this potential. Show and Tell apps, such as *Educreations*, *Show Me*, and *Explain Everything*, are tablet apps designed to record the screen interactions of people writing on the tablet and talking in real time. *Educreations* was originally designed for teachers to record a mini-lesson, to "teach what they know and learn what they don't" (http://www.educreations.com). This involved a teacher using a tablet in much the same way as they would use a whiteboard; explaining the concept as they write notes or draw diagrams on the tablet. The key advantage of a Show and Tell app is that it records the audio and anything written or drawn on the tablet, and so could be later replayed. The affordances of Show and Tell apps, such as *Educreations*, for problem-solving include the real-time capture of students' discussion and engagement with the problem, while collaborating and/or thinking aloud, and also the functionality for them and others to review what has been recorded (Ingram, Williamson-Leadley & Pratt, 2016). Although these affordances can provide an insight into students' thinking, it must be acknowledged that it is not possible to fully access the internal thinking processes of students. However, when using these apps, students are encouraged to verbalise their thinking as they work through the problems, allowing deeper insight into their thinking processes than is possible by simply viewing their solution and/or working. As such, these apps have the potential to be used in transformational ways, as defined by the SAMR framework. That is, depending on their use, they can either modify or redefine teaching and learning as they provide information for teachers that is not possible to access in pen and paper-based solutions.

Ingram, Williamson-Leadley and colleagues (Ingram, Williamson-Leadley, Bedford, & Parker, 2015; Ingram et al., 2016; Williamson-Leadley & Ingram, 2013) have explored a number of different ways in which Show and Tell apps could be used in the mathematics classroom. Williamson-Leadley and Ingram (2013) explored how *Educreations* could be used for the assessment of primary students' numeracy through working with three primary teachers and then extended this work to investigate how Educreations could be used in primary and secondary mathematics classrooms (Ingram et al., 2015). This exploration was further expanded in 2016 to include how eleven teachers used a range of Show and Tell apps for tablets (Ingram et al., 2016) in their teaching. In the latter two iterations, after professional development, teachers were encouraged to explore the use of Show and Tell apps within their mathematics programmes over a period of two weeks, recording their written reflections in journals, collecting examples of students' work from the iPad apps, and also student written reflections on their experiences. The teachers had a range of teaching experience and taught students between the ages of 5 and 14.

This body of research indicated that students working with a Show and Tell app were perceived by teachers to have higher levels of engagement in mathematics, with this engagement being of a higher quality than was likely to occur without the apps. It was also apparent that Show and Tell apps had affordances for open-ended problem-solving. An example of an affordance was evident when Oscar, a Year 10 student, stated that the Show and Tell app "gives the ability to showcase your thought process and be able to review how you approach a problem (Ingram et al., 2015, p. 29). Karen, a primary teacher, found that having students use a Show and Tell app to record their thinking when solving open-ended problems and then sharing their work with each other generated discussion after they followed the way another student has solved the problem using a different strategy (Ingram et al., 2015). This chapter extends our research agenda by explicitly focusing on the level and quality of student engagement when using Show and Tell apps for problem-solving.

## Methodology

To explore how Show and Tell apps enhanced the engagement of the students when problem-solving, the data sets across all three iterations of the Show and Tell project were re-analysed for examples of students using the apps to support problem-solving. The main data set included the reflective journals and transcribed interviews of 15 primary and one secondary teacher after two weeks of using Show and Tell apps in the classroom. Examples of student work were collected by these teachers and reflective data collected from 15 Year 10 students.

The qualitative coding was guided by our research into engagement and problem-solving and our previous research, which indicated that using the Show and Tell apps affected the level and quality of students' engagement. The first step was to identify any data that related to the use of Show and Tell apps for problem-solving. The first iteration of coding then saw this data separated into being related to either the level or quality of engagement. The level of engagement was not separated into discrete levels. Rather, any reference referring to the amount of engagement was sought, for example, "it helps with getting them more engaged in the mathematics." The data was then coded for quality using the engagement skills (described above). For example, when a teacher described focus, this was related to concentration. Three themes emerged related to the data on quality engagement on problem-solving: socially negotiating the mathematics, visible thinking, and reflection, and these coding categories structured the findings.

The following sections present these findings. With the exception of Hannah and Karen, teachers who were researcher/participants in the second iteration, pseudonyms have been used to protect the identity of the teachers and students. The study was conducted after the University of Otago Human Ethics Committee granted ethical clearance and informed consent was given by the teachers, students and their parents.

## Findings

Show and Tell apps were used for problem-solving in a similar way across all classrooms. Depending on the age of the students, the problems ranged from one step addition, subtraction, and multiplicative problems to multi-step, open-ended word problems (e.g., Fig. 1). The solution or method was not immediately obvious to the student and they had the opportunity to "get stuck" (Hannah). Most teachers chose the problems themselves, although some teachers had a range of challenging problems available for students to choose from.

Most students worked in groups of up to four students around one tablet, although occasionally individual students worked on the same problem on separate

Four people can paint a fence with 30 panels in 3 hours. How long would it take five people to paint a fence with 25 panels?

**Fig. 1** An example of a problem given to 14 year olds

tablets. The teacher or a student put a written, typed, or photographed problem on a blank page within the Show and Tell app. The students then worked to solve the problems, recording their drawings, jottings, and solutions, using the same or subsequent pages. The students were explicitly encouraged to think aloud (that is, verbalise their thinking) during the process of problem-solving. This enabled both their work on the tablet and their thinking to be recorded simultaneously. The students often played back the recording to themselves, other students, the teacher, or the whole class.

## *Level of Engagement*

According to the teachers, students engaged highly in problem-solving when using Show and Tell apps. At times, the novelty value of both using technology and Show and Tell apps were factors in the students' high level of engagement. However, teachers with students already accustomed to using technology within their classroom also described a high level of student engagement because it was "hands on, interactive and fun" (Mary). When doing problem-solving with Show and Tell apps, the students had an "on-task busyness" where they got on with the task at hand with "more focus" (Olive). The students were vocal about their enjoyment of using the Show and Tell app, "asking every day … if they were able to go and use it" (Angela). Hannah described her students as "confident and gregarious" when using Show and Tell.

Students normally lacking in confidence with mathematics became more involved.

> Using the app helped students who didn't have confidence in maths brought in another element and increased their engagement because they wanted to give things a go. (Hannah)

In addition, Angela, Olive and Helen found that high-achieving students became more engaged because they enjoyed the mathematical challenge. Using the Show and Tell apps ensured the students explained their steps and built on others' thinking, allowing them to go "deeper into the problem-solving" (Angela). Sometimes students' mathematical thinking went beyond what the teacher expected.

> Some of their mathematical thinking went well beyond. Like I'm saying, how did you get that? (Helen)

The students were enthusiastic to share their problem-solving with others and disappointed if they did not get an opportunity to do this. Indeed, time became an issue for the teachers because so many students wanted to share back their recordings.

> The bell would ring and they'd still be wanting to share it back. (Helen)

## *Quality of Engagement*

The teachers noted the quality of the students' engagement when using Show and Tell apps for mathematical problem-solving. They believed the students concentrated, persevered, and remained engaged in the problems when they became difficult, because they were intimately involved with the mathematics of each problem. In other words, the students enjoyed exploring the possible solutions and the structures and patterns that emerged. They seemed to care about the process as well as the product. Their discussions and negotiations demonstrated that they had integrity; they cared if their answers were wrong or right, and they cared about their understanding that led to the answer.

> If they got stuck, they'd come and ask a question or they would go and talk to their classmates. (Mary)
>
> It's that confidence in themselves. Taking the time to slow down and make sure they are actually comprehending [the problem] when they are stuck. (Sara)
>
> Once she found a mistake she wanted to go back and do it again. (Cathy)
>
> They wanted to work on the problems and enjoyed exploring. (Ruth)

The beneficial impact of using Show and Tell apps had some limitations when used with younger students (aged between 5 and 6 years old). These students had difficulty in remaining engaged when the complexity of the task increased. In Cathy's multi-age class of 5–8 year olds, she found that pairing younger students with older classmates helped them to remain focused when working remotely.

Apart from these very young students, students worked well independently when using Show and Tell. They often demonstrated autonomy by making decisions about which problems to solve, who to work with, and how to report back on their problem-solving. They could "get their hands on it, take it away, and take charge" (Jennifer). Mary suggested that they enjoyed thinking aloud and recording away from the class because it was "non-invasive" and they had the time, space and independence to do it on their own terms, "without interference".

There were three ways that the use of Show and Tell apps particularly enhanced the quality of students' engagement in problem-solving activities. The processes used with the Show and Tell apps further enhanced the quality of student engagement through: (1) enabling the students to socially negotiate the mathematics; (2) making the students' learning and thinking visible; and (3) facilitating students' reflection on their mathematical learning and engagement skills.

### Socially Negotiating the Mathematics

Show and Tell apps were useful for scaffolding the social negotiation implicit in mathematical problem-solving, particularly for those students older than eight. When students experience mathematics, the meanings they get from those experiences either reinforce or alter their previous understandings (Hannula, Evans, Philippou, & Zan, 2004). Negotiating meanings socially is particularly powerful.

Angela invited students to work together by asking two students to separately record their thinking about the same problem on a Show and Tell app and then come together to share their recorded solutions. By doing this, each student had individual thinking time and therefore had ownership of the problem when it came to sharing the recording with the other person and negotiating a correct path or answer. The initial recordings were just the first step in negotiating meaning and the solution.

In general, however, teachers had students work on the tablets in groups. Some teachers only had enough tablets for one per group; others chose that way of working. Hannah's 14-year olds worked in the same groups throughout their Number Unit. When the students worked in a group, there was first work to be done on the group dynamics and there was some "initial wrangling" (Angela) about who would do the role of scribe. The students often took turns to be the scribe, with the others joining in at a particular point in the problem-solving process, reaching over to jot down their ideas as they found a connection or could move the problem forward. Working in this way meant "it got messy" (Hannah), as people talked over each other, worked on separate sections, and contributed to the written or spoken recording. Jennifer's students interacted because of the use of Show and Tell. They "showed their thinking to each other [which] provided good discussion and learning". The dialogue created in the process of recording and sharing these recordings meant that students were seamlessly justifying and negotiating their learning.

In general, the making of these recordings encouraged the students to clearly explain their thinking, which in turn helped them to discuss the solution and solve the problem. The process of thinking aloud for the recording, contributing to the cooperative problem-solving, and further justifying their thinking when sharing their recording with others, encouraged the students to explain their ideas.

> It makes you fully explain your ideas. (Roland, Year 10)
>
> Some ideas simply don't fall on paper and others just can't be said, but when you combine both, a combination of writing and speaking suddenly you can convey your ideas. (Mitchell, Year 10)

Nine of the thirteen teachers in the latter iterations gave the students the opportunity to share their recording with the class. Mary noted that it was a safe way for students to receive feedback because the recording started the initial conversation, rather than the student having to talk.

> It's a removed way of getting feedback in class [when they are sharing their recording]. Yes I'm on there, but I'm not standing up there giving you the answer. (Mary)

This sharing of the recording showed the range of strategies used and was a catalyst for dialogue as the class then discussed the recordings, made decisions on the most efficient strategy, or pointed out errors. Rather than the teacher solely providing student feedback, the students were co-constructing meaning through negotiation, as Helen found out when one group presented incorrect problem-solving.

> There was one group that got it all wrong. When they watched it back the other kids were able to see exactly the steps that they'd gone wrong and pointed it out to them, rather than me pointing it out, so that was, you know quite ... powerful, motivating for them. (Helen)

## Visible Thinking

According to the teachers, Show and Tell apps enabled the students' thinking to be visible to others. The recording, encompassing the dialogue, writing, and the drawing, captured the evolution of the problem-solving process in all its messiness. This enhanced the problem-solving for the students.

> They learnt from being able to see others' thinking visually. (Jennifer)

The app also made the students' learning processes explicit for the teachers. Two teachers found it difficult to assess the students when they worked in groups.

> When they were in groups, they worked together so I don't know who did actually what. (Ruth)

The remaining teachers found that they were able to closely monitor the thinking of individuals, even when they were working with others. By viewing the recording, teachers were able to critically analyse and assess students' understandings by differentiating the students' voices.

As they viewed the recordings, teachers of all ages of students were sometimes surprised about aspects of the students' understandings that may have been missed otherwise. Sometimes students' issues were more about misunderstanding the question, or for the younger children, their number formation, rather than their mathematical understanding per se. For example, Mike was surprised at how many students were writing the one's digit before the ten's digit in a two-digit number. At other times, the teachers found that deep understanding was not occurring, when on the surface the student appeared to understand. As Mary noted,

> Sometimes kids look like they understand it, but when you dig that bit deeper and look at them through the whole process, actually they're not. On a piece of paper sometimes you don't see all those things unfolding.

By playing back their recording, the students' problem-solving was visible and therefore they were able to reflect on their mathematics, find mistakes in their process, and self-correct. Mike described how one of his students, James, felt comfortable enough during playing the recording to scroll back through the pages and change an answer he realised was incorrect, before editing the rest of the pages in front of the audience. Mike identified the power of this:

> James had the understanding [and] he was really interested to go back and see what he did and unpack it … it's very powerful for their own learning to go back and see where they've missed a number or where they've misinterpreted something.

### Reflection

Usually scaffolded by the teacher, the students engaged in reflective dialogue about the affective aspects of their problem-solving, including the different ways that they and their classmates engaged. Using the Show and Tell apps with explicit teaching about engagement gave the students the opportunity to reflect on their engagement and learning during the problem-solving processes.

> The main thing I got out of this trial was the importance of reflecting on your learning. I think that's the best thing. Using [Show and Tell on] the tablets meant that we could be explicit about engagement. The boys thought way more about their engagement. Being stuck in maths. Satisfaction. Perseverance. What maths feels like. (Hannah)

> [Show and Tell] gives you the ability to … be able to review how you approach a problem. (Oscar, Year 10)

## Discussion and Conclusion

For these participants, Show and Tell apps were seen as a useful tool for problem-solving in mathematics. They worked well to record the 'messy' and iterative process of students' individual and cooperative problem-solving and, when shared, these recordings were beneficial for the co-construction of students' mathematical understandings. Teachers believed that students were both more engaged and engaged for longer in the problem-solving process when they were using Show and Tell apps. Furthermore, it appeared that using these apps enhanced the quality of student engagement. Students persevered with problems, cared about not just the answers but also the process of finding solutions, and were able to be autonomous as they worked to solve problems. The use of these apps supported the social negotiation process that is mathematical doing, made the thinking visible, and was a useful tool for student reflection regarding both the mathematical problem-solving process and the quality of their engagement in that process. The teachers were able to be more explicit about the importance, level, and quality of students' engagement when problem-solving.

Although Show and Tell apps show great promise for increasing student engagement in problem-solving in mathematics, it must be noted that there are a number of limitations to both the research done to date and the use of Show and Tell apps. The conclusions drawn are based on three small-scale qualitative studies, involving 15 primary and one secondary teacher from one city. In addition, there are a number of practical considerations that must be taken into account when deciding whether or not to use Show and Tell apps. Issues such as where data is stored, the availability of working technology, and the time required to make best use of the recorded material all need to be considered. It is clear, however, that Show and Tell apps show promise in their ability to enhance students' problem-solving.

The decision to use Show and Tell apps was made thoughtfully, based on a consideration of the affordances of the technology, and the desired outcome. In line with the TPACK framework, the technology, pedagogy, and content each had to be considered, with the choice then based on how best to integrate technology to facilitate engagement in problem-solving in mathematics. Rather than simply choosing a technology that could substitute or augment current approaches, a transformative tool was chosen. The Show and Tell apps appeared able to transform classroom practice, redefining how teachers could engage students in the problem-solving process.

## References


Attard, C. (2012). Engagement with mathematics: What does it mean and what does it look like? *Australian Primary Mathematics Classroom, 17*(1), 9–13.

Attard, C. (2014). "I don't like it, I don't love it, but I do it and I don't mind": Introducing a framework for engagement with mathematics. *Curriculum Perspectives, 34*(3), 1–14.

Attard, C., & Curry, C. (2012). Exploring the use of iPads to engage young students with mathematics. In J. Dindyal, L. P. Cheng, & S. F. Ng (Eds.), *Proceedings of the 35th Annual Conference of the Mathematics Education Research Group of Australasia* (pp. 75–82). Singapore, Singapore: MERGA.

Australian Curriculum, Assessment and Reporting Authority (2016). Retrieved from http://www.australiancurriculum.edu.au/mathematics/aims.

Blackwell, C. (2014). Teacher practices with mobile technology integrating tablet computers into the early childhood classroom. *Journal of Education Research, 7*(4), 1–25.

Bragg, L. A. (2012). The effect of mathematical games on on-task behaviours in the primary classroom. *Mathematics Education Research Journal, 24,* 385–401. https://doi.org/10.1007/s13394-012-0045-4.

Bransford, J., & Stein, B. (1984). *The IDEAL problem solver: A guide for improving thinking, learning, and creativity.* New York, NY: W.H. Freeman.

Bureau of Exceptional Education and Student Services. (2010). *Classroom cognitive and meta-cognitive strategies for teachers: Research-based strategies for problem-solving in mathematics K-12.* Tallahassee, FL: Florida Department of Education.

Calder, N. (2013). Mathematics in student-centred inquiry learning: Student engagement. *Teachers and Curriculum, 13,* 77–84.

Cavanagh, M. (2016). Introduction: The learning and teaching of mathematics. In G. Hine, R. Reaburn, J. Anderson, L. Galligan, C. Carmichael, M. Cavanagh, B. Ngu, & B. White (Eds.), *Teaching secondary mathematics* (pp. 2–27). Melbourne, Australia: Cambridge University Press.

Cavanaugh, C., Hargis, J., Kamali, T., & Soto, M. (2013). Substitution to augmentation: Faculty adoption of iPad mobile learning in higher education. *Interactive Technology and Smart Education, 10*(4), 270–284. https://doi.org/10.1108/ITSE-01-2013-0001.

Chen, Z.-H., Liao, C. C. Y., Cheng, H. N. H., Yeh, C. Y. C., & Chan, T.-W. (2012). Influence of game quests on pupils' enjoyment and goal-pursuing in math learning. *Educational Technology & Society, 15*(2), 317–327.

Christenson, S., Reschly, A. L., & Wylie, C. (2012). *Handbook of research on student engagement*. Singapore, Singapore: Springer.

Conole, G. (2012). *Designing for learning in an open world (Explorations in the learning sciences, instructional systems and performance technologies)*. London, United Kingdom: Springer.

Dotterer, A., & Lowe, K. (2011). Classroom context, school engagement, and academic achievement in early adolescence. *Journal of Youth and Adolescence, 40*(12), 1649–1660.

Hammond, M. (2010). What is an affordance and can it help us understand the use of ICT in education? *Education and Information Technologies, 15*(3), 205–217.

Hannula, M. S., Evans, J., Philippou, G., & Zan, R. (2004). Affect in mathematics education—Exploring theoretical frameworks. In *Proceedings of the Twenty-Eighth Conference of the International Group for the Psychology of Mathematics Education*. (Vol. 1, pp. 107–136). Bergen, Norway: PME.

Hiebert, J., Carpenter, T., Fennema, E., Fuson, K., Human, P., Murray, H., et al. (1996). Problem-solving as a basis for reform in curriculum and instruction: The case of mathematics. *Educational Researcher, 25*(12), 12–21.

Holton, D., Neyland, A., Neyland, J., & Thomas, B. (1999). *Teaching problem-solving: An introduction for primary and junior secondary teachers*. West Sussex, United Kingdom: Kingsham Press.

Ingram, N. (2011). *Affect and identity: The mathematical journeys of adolescents* (Doctoral dissertation). University of Otago, Dunedin, New Zealand. Retrieved from http://hdl.handle.net/10523/1919.

Ingram, N., Williamson-Leadley, S., Bedford, H., & Parker, K. (2015). Using Show and Tell tablet technology in mathematics. In R. Averill (Ed.), *Mathematics and statistics in the middle years: Evidence and practice* (pp. 18–34). Wellington, New Zealand: NZCER.

Ingram, N., Williamson-Leadley, S., & Pratt, K. (2016). Showing and telling: Using tablet technology to engage students in mathematics. *Mathematics Education Research Journal, 28* (1), 123–147.

Jonassen, D., Carr, C., & Yueh, H. (1998). Computers as Mindtools for engaging learners in critical thinking. *Techtrends, March*, 24–32.

Jones, M. (2013). *Mathematics problem-solving: What factors inhibit student achievement? What factors are effective in raising achievement?* Retrieved from http://www.educationalleaders.govt.nz/content/download/53112/441891/file/Mary%20Jones%20Sabbatical%20Report%202013.pdf.

Kim, M. C., & Hannafin, M. J. (2011). Scaffolding problem-solving in technology-enhanced learning environments (TELEs): Bridging research and theory with practice. *Computers and Education, 56*, 403–417.

Koehler, M. J., & Mishra, P. (2008). Introducing technological pedagogical content knowledge. In AACTE Committee on Innovation and Technology (Eds). *Handbook of Technological Pedagogical Content Knowledge (TPCK) for Educators* (pp. 3–29). New York, NY: Routledge.

Koehler, M., & Mishra, P. (2009). What is technological pedagogical content knowledge? *Contemporary Issues in Technology and Teacher Education, 9*(1), 60–70.

Kuiper, E., & de Pater-Sneep, M. (2014). Student perceptions of drill-and-practice mathematics software in primary education. *Mathematics Education Research Journal, 26*(2), 215–236. https://doi.org/10.1007/s13394-013-0088-1.

Ladel, S., & Kortenkamp, U. (2012). Early maths with multi-touch—an activity-theoretic approach. In *Proceedings of POEM 2012*. Retrieved from http://cermat.org/poem2012/main/proceedings_files/Ladel-Kortenkamp-POEM2012.pdf.

McLeod, D. (1992). Research on affect in mathematics education: A reconceptualization. In D. Grouws (Ed.), *Handbook of research on mathematics teaching and learning* (pp. 575–596). New York, NY: NCTM and Macmillan.

Ministry of Education (2012). *Secondary mathematics syllabus 1–4*. Singapore, Singapore: Ministry of Education. Retrieved from http://www.moe.edu.sg/education/syllabuses/sciences/.

Ng, W. (2015). *New digital technology in education: Conceptualizing professional learning for educators* (pp. 171–189). Cham, Switzerland: Springer.
O'Rourke, J., Main, S., & Ellis, M. (2013). So the kids are busy, what now? Teacher perceptions of the use of hand-held game consoles in West Australian primary classrooms. *Australasian Journal of Educational Technology, 29*(5), 735–747.
Pennant, J. (2013). *Developing a classroom culture that supports a problem-solving approach to mathematics*. Retrieved from https://nrich.maths.org/10341.
Puentedura, R. (2009). *Transformation, technology, and education*. Retrieved from http://hippasus.com/resources/tte/.
Reeve, J., Jang, H., Carrell, D., Soohyn, J., & Barch, J. (2004). Enhancing students' engagement by increasing teachers' autonomy support. *Motivation and Emotion, 28*(2), 147–169.
Reschly, A., Huebner, E., Appleton, J., & Antaramian, S. (2008). Engagement as flourishing: The contribution of positive emotions and coping to adolescents' engagement at school and with learning. *Psychology in the Schools, 45*(5), 419–431.
Schoenfeld, A. H. (1992). Learning to think mathematically: Problem-solving, metacognition, and sense making in mathematics. In D. Grouws (Ed.), *Handbook for research on mathematics teaching and learning*. New York, NY: MacMillan.
Selwyn, N., Potter, J., & Cranmer, S. (2009). Primary pupils' use of information and communication technologies at school and home. *British Journal of Educational Technology, 40*(5), 919–932. https://doi.org/10.1111/j.1467-8535.2008.00876.x.
Sinclair, N., Chorney, S., & Rodney, S. (2016). Rhythm in number: Exploring the affective, social and mathematical dimensions of using TouchCounts. *Mathematics Education Research Journal, 28*(1), 31–51. https://doi.org/10.1007/s13394-015-0154-y.
Skilling, K. (2014). Teacher practices: How they promote or hinder student engagement in mathematics. In J. Anderson, M. Cavanagh, & A. Prescott (Eds.), *Proceedings of the 37th Annual Conference of the Mathematics Education Research Group of Australasia* (pp. 95–102). Sydney, Australia: MERGA.
Sullivan, P., McDonough, A., & Harrison, R. T. (2004). Students' perceptions of factors contributing to successful participation in mathematics. In M. Johnsen Hoines & A. Berit Fugelstad (Eds.), *28th Conference of the International Group for the Psychology of Mathematics Education* (Vol. 3, pp. 289–296). Bergen, Norway: Bergen University College.
Williamson-Leadley, S., & Ingram, N. (2013). Show and tell: Using iPads for assessment in mathematics. *Computers in New Zealand Schools: Learning, Teaching, Technology, 25*(1–3), 117–137.
Williamson-Leadley, S. (2016). *New Zealand primary teachers' ICT professional development and classroom practices*. (Doctoral dissertation). Deakin University, Geelong, Victoria, Australia.

**Naomi Ingram** is a lecturer and researcher in primary and secondary mathematics education at the University of Otago, College of Education, New Zealand. Prior to this, she was a secondary school mathematics teacher in New Zealand and the Sultanate of Oman. In Oman she worked in an international school that has delivered a one-to-one iPad programme since 2010. Teacher registration and continuing contact with the teaching community through research, educating pre-service teachers and professional development are important aspects of Naomi's identity.

**Keryn Pratt** has a PhD in psychology and is a researcher and lecturer at the University of Otago, College of Education. She lectures in ICT in education, distance learning and quantitative research methodologies. Her research includes the experience of distance students; both distance postgraduate students and secondary school students who are involved in distance learning using videoconferencing and other technologies.

**Sandra Williamson-Leadley** Originally from Canada, Sandra Williamson-Leadley has lived in New Zealand for over 30 years. She is an experienced primary teacher who has worked as a researcher and lecturer at the University of Otago, College of Education since 2012. She teaches curriculum, e-learning, ICT and technology and her research interests include professional development, the use of ICT in schools and e-learning.

# Index




N. Calder et al. (eds.), *Using Mobile Technologies in the Teaching and Learning of Mathematics*, Mathematics Education in the Digital Era 12,
https://doi.org/10.1007/978-3-319-90179-4

# Index of Apps





N. Calder et al. (eds.), *Using Mobile Technologies in the Teaching and Learning of Mathematics*, Mathematics Education in the Digital Era 12,
https://doi.org/10.1007/978-3-319-90179-4

The manufacturer's authorised representative in the EU is Springer Nature Customer Service Centre GmbH, Europaplatz 3, 69115 Heidelberg, Germany. If you have any concerns regarding our products, please contact ProductSafety@springernature.com

Printed and bound by CPI Group (UK) Ltd, Croydon, CR0 4YY

30/09/2026

02225662-0009